METHODS IN MOLECULAR BIOLOGY

For further volumes:
http://www.springer.com/series/7651

Parasite Genomics Protocols

Second Edition

Edited by

Christopher Peacock

University of Western Australia, Nedlands, WA, Australia

Humana Press

Editor
Christopher Peacock
University of Western Australia
Nedlands, WA, Australia

ISSN 1064-3745 ISSN 1940-6029 (electronic)
ISBN 978-1-4939-1437-1 ISBN 978-1-4939-1438-8 (eBook)
DOI 10.1007/978-1-4939-1438-8
Springer New York Heidelberg Dordrecht London

Library of Congress Control Number: 2014950239

Printed on acid-free paper

Humana Press is a brand of Springer
Springer is part of Springer Science+Business Media (www.springer.com)

Preface

Parasitic diseases caused by both protozoal and helminth pathogens afflict hundreds of millions of people across the globe and are a significant cause of morbidity and mortality. While the main burden of disease occurs in the tropical and subtropical regions of the world, they have a significant impact in all societies. Unlike other infectious agents, there is a total lack of vaccines and chemotherapy options are often limited or have to be carefully managed to minimize toxicity to the patient. Many of these diseases, in particular those caused by vector borne parasites restricted to tropical regions of the world, lack investment in drug development such that frontline treatments utilize medicines developed many decades ago. The advent of whole genome sequencing of larger organisms such as parasites has led to renewed impetus in identifying druggable targets for future development. This together with increasing funding from philanthropic and not for profit organizations has led to new investment in targeting these neglected diseases.

The advent of affordable next generation sequencing has had a dramatic effect on the methodologies employed to research almost all areas of human health. The declining cost and increasing availability of sequencing has led to an era of massive data generation and provision of publically available annotated genomes. This has provided the platform for the generation of new large-scale methodologies that go beyond the static genome allowing for renewed insight into the structure, function, and adaptability of these pathogens. Although the size and complexity of parasite genomes has resulted in this field not benefiting from this technology as rapidly as other fields in infectious disease research, we are now in an era where large-scale genomic, transcriptomic, proteomic, and metabolomic data are being made readily available via public data repositories. Utilizing this data from these "omics" studies is the challenge faced by researchers with access to such huge data resources. *Parasite Genomics Protocols, Second Edition* is designed to detail the methodologies that have been adapted to research these unique organisms in the postgenomic era. While the parasites investigated in these studies represent the most significant of the protozoan and helminth pathogens involved in human disease, these methods are equally relevant to other related organisms of either veterinary or ecological importance.

One of the most important resources available to the researcher is the public databases and repositories that house genomic and postgenomic data. These have evolved from static data repositories used predominantly by a select few to complex relational databases accessible by anyone with an Internet browser that can be comprehensively interrogated by users with minimal training or bioinformatics skills. As well as the ability to construct complex queries, users can also download selected data to their computer for separate analysis and publication. Cooperation between research centers has also led to the linking up of databases providing a one-stop solution to users accessing a platform they are familiar with. The first chapter of *Parasite Genomics Protocols, Second Edition* details the methods used to use and query EuPathDB the largest of the parasite-specific data resources that is made up of a family of genus-specific datasets all linked together and utilize the same format such that a single set of instructions can be applied to all databases either singularly or collectively.

Sequencing of parasite genomes has until relatively recently been the domain of large well-funded institutes with extensive bioinformatics resources available for assembly, annotation, and analyses. The advent of affordable benchtop sequencers together with the availability of cheap outsourcing has led to research groups with modest budgets being able to complete genome sequence projects. However, while generating the raw data is relatively straightforward, the assembly, annotation, and meaningful interpretation of the data still require both significant computational resources and bioinformatics expertise. While some user-friendly commercial software is available, the research community is very active in providing free open source bioinformatics software. The second chapter in this book provides the detailed framework for developing a sequence assembly and annotation pipeline that is designed to generate a high-quality finished genome that can be analyzed and published using freely available software. Other chapters early on in *Parasite Genomics Protocol, Second Edition* describe the use of sequence data to examine genetic variation within these parasites to inform on evolution, genetic diversity, and determinants of virulence. A majority of the chapters in *Parasite Genomics Protocols, Second Edition* describe protocols for undertaking other large-scale "omics" methodologies such as those to determine the epigenome, transcriptome, proteome, and metabolome. Some parasites such as the trypanosomatids that include the causative agents of leishmaniasis, African trypanosomiasis and Chagas disease, have unusual mechanism of gene transcription and expression that require modifications to methods utilized for other eukaryotic organisms. Regulatory mechanisms involved in controlling gene expression are discussed in several chapters including transcriptional control, mRNA decay, RNA interference, and posttranslational modifications to proteins. Access to whole genome data also lends itself to developing tools and methods to help answer more targeted questions such as those associated with genetic manipulation, examining the effect of genetic variation, identifying virulence factors, or selecting potential vaccine candidates. The remaining chapters provide some examples of studies in these more application-based approaches.

The first edition of this book was published just after the completion of the sequencing projects for the first protozoan pathogens. The editor of that edition speculated on the rapid progress that the postgenomic era would precipitate in this field. Less than a decade later it is indeed incredible to see how rapidly this arena has evolved to generate volumes of data that could hardly have been imagined when the first parasite genome projects were initiated. Sequencing projects for most if not all of the major human pathogenic protozoan and helminth pathogens are either complete or in progress. More recently the projects for the most significant vectors have also been published. While we have yet to see significant impact of this new era of science on the really important area of patient care and disease prevention, one can draw comfort from the fact that other areas of infectious disease research have seen tangible benefits in terms of new diagnostic tools, disease management and control, and drug and vaccine development. While it is hard to imagine that personalized medicine will make any impact in this field, development of affordable therapies and more importantly protective vaccines do appear to be a step closer.

Finally, I would like to thank all of the many authors who have taken the time to contribute to this second edition of *Parasite Genomics Protocols, Second Edition* and to all the many researchers who have been involved in the development of these methods.

Nedlands, WA, Australia ***Christopher Peacock***

Contents

Contributors

LOREN BAUGH • *Seattle BioMed, Seattle, WA, USA*
KATJA BECKER • *Biochemistry and Molecular Biology, Justus Liebig University Giessen, Giessen, Germany*
STEVEN M. BEVERLEY • *Department of Molecular Microbiology, Washington University School of Medicine, St. Louis, MO, USA*
FRÉDÉRIC BRINGAUD • *Centre de Résonance Magnétique des Systèmes Biologiques (RMSB), UMR 5536 CNRS, Université de Bordeaux, Bordeaux, France*
MARIE-CHRISTINE BROTHERTON • *Centre de Recherche en Infectiologie; CHU-Québec research centre, Québec, QC, Canada*
PAUL CAPEWELL • *Wellcome Trust Centre for Molecular Parasitology, Institute of Biodiversity, Animal Health and Comparative Medicine, College of Medical, Veterinary and Life Sciences, University of Glasgow, Glasgow, UK*
JENNIFER M. CHAMBERS • *Department of Biochemistry and Molecular Biology, Bio21 Institute of Molecular Science and Biotechnology, University of Melbourne, Parkville, VIC, Australia*
CAROLINE CLUCAS • *Wellcome Trust Centre for Molecular Parasitology, Institute of Biodiversity, Animal Health and Comparative Medicine, College of Medical, Veterinary and Life Sciences, University of Glasgow, Glasgow, UK*
ANNELI COOPER • *Wellcome Trust Centre for Molecular Parasitology, Institute of Biodiversity, Animal Health and Comparative Medicine, College of Medical, Veterinary and Life Sciences, University of Glasgow, Glasgow, UK*
AKRAM A. DA'DARA • *Department of Infectious Disease and Global Health, Molecular Helminthology Laboratory, Cummings School of Veterinary Medicine, Tufts University, North Grafton, MA, USA*
JEZIEL D. DAMASCENO • *Departamento de Biologia Celular e Molecular e Bioagentes Patogênicos, Faculdade de Medicina de Ribeirão Preto, Universidade de São Paulo, Ribeirão Preto, SP, Brazil*
DENISE L. DOOLAN • *Molecular Vaccinology Laboratory, QIMR Berghofer Medical Research Institute, Herston, QLD, Australia*
PATRICK DRIGUEZ • *Molecular Parasitology Laboratory, QIMR Berghofer Medical Research Institute,, Herston, QLD, Australia*
PHIL L. FELGNER • *Department of Medicine, University of California Irvine, Irvine, CA, USA*
ELISABETH FISCHER • *Biochemistry and Molecular Biology, Justus Liebig University Giessen, Giessen, Germany*
ROBIN GASSER • *Faculty of Veterinary Science, The University of Melbourne, Parkville, VIC, Australia*
ELODIE GHEDIN • *Center for Genomics & Systems Biology, Department of Biology, New York University, New York, NY, USA; Centre for Vaccine Research, University of Pittsburgh, Pittsburgh, PA, USA*
OMAR S. HARB • *University of Pennsylvania, Philadelphia, PA, USA*
ANDREW HAYDOCK • *Seattle BioMed, Seattle, WA, USA*

DANIEL C. JEFFARES • *Research Department of Genetics, Evolution and Environment, University College London, London, UK*
AARON R. JEX • *Faculty of Veterinary Science, The University of Melbourne, Parkville, VIC, Australia*
KAMI KIM • *Albert Einstein College of Medicine, Bronx, NY, USA*
NIKOLAY G. KOLEV • *Department of Epidemiology of Microbial Diseases, School of Public Health, Yale University, New Haven, CT, USA*
SASA KONCAREVIC • *Proteome Sciences R&D GmbH & Co. KG, Frankfurt am Main, Germany*
MICHAEL D. LEWIS • *London School of Hygiene and Tropical Medicine, London, UK*
D. TIMOTHY LITTLEWOOD • *Department of Life Sciences, Natural History Museum, London, UK*
MARTIN S. LLEWELLYN • *London School of Hygiene and Tropical Medicine, London, UK*
ALEX LOUKAS • *Centre for Biodiscovery and Molecular Development of Therapeutics, James Cook University, Cairns, QLD, Australia*
ANNETTE MACLEOD • *Wellcome Trust Centre for Molecular Parasitology, Institute of Biodiversity, Animal Health and Comparative Medicine, College of Medical, Veterinary and Life Sciences, University of Glasgow, Glasgow, UK*
GREG MATLASHEWSKI • *Microbiology and Immunology, McGill University, Montreal, QC, Canada*
MALCOLM J. MCCONVILLE • *Department of Biochemistry and Molecular Biology, Bio21 Institute of Molecular Science and Biotechnology, University of Melbourne, Parkville, VIC, Australia*
DONALD P. MCMANUS • *Molecular Parasitology Laboratory, QIMR Berghofer Medical Research Institute, Herston, QLD, Australia*
LOUISA A. MESSENGER • *London School of Hygiene and Tropical Medicine, London, UK*
MICHAEL A. MILES • *London School of Hygiene and Tropical Medicine, London, UK*
DOUGLAS M. MOLINA • *Antigen Discovery Incorporated, Irvine, CA, USA*
LIAM MORRISON • *Roslin Institute, University of Edinburgh, Easter Bush, Midlothian, UK*
MICHAELA MÜLLER-MCNICOLL • *Institute for Cell Biology and Neuroscience, Goethe University Frankfurt, Frankfurt, Germany*
PETER J. MYLER • *Seattle BioMed, Seattle, WA, USA*
SHEILA C. NARDELLI • *Instituto Carlos Chagas-Fiocruz, Fiocruz-Paraná. Rua Prof. Algacyr Munhoz Mader, Curitiba/PR, Brazil*
MILICA NG • *Department of Biochemistry and Molecular Biology, Bio21 Institute of Molecular Science and Biotechnology, University of Melbourne, Parkville, VIC, Australia*
THOMAS D. OTTO • *Wellcome Trust Sanger Institute, Hinxton, UK*
MARC OUELLTE • *Centre de Recherche en Infectiologie; CHU-Québec research centre, Québec, QC, Canada*
PRASAD K. PADMANABHAN • *Department of Immunology and Infectious Diseases, Harvard School of Public Health, Boston, MA, USA*
BARBARA PAPADOPOULOU • *Research Centre in Infectious Diseases, CHU de Quebec Research Center, Quebec, QC, Canada; Department of Microbiology Infectious Disease and Immunology, Faculty of Medicine, Laval University, Quebec, QC, Canada*
JUDITH HELENA PRIETO • *Chemistry Department, Western Connecticut State University, Danbury, CT, USA*
JAMES PYKE • *Metabolomics Australia, Bio21 Institute of Molecular Science and Biotechnology, The University of Melbourne, Parkville, VIC, Australia*

GINA RACINE • *Centre de Recherche en Infectiologie; CHU-Québec research centre, Québec, QC, Canada*
GOWTHAMAN RAMASAMY • *Seattle BioMed, Seattle, WA, USA*
MARIO DOS REIS • *Research Department of Genetics, Evolution and Environment, University College London, London, UK*
MATTHEW ROGERS • *Centre for Vaccine Research, University of Pittsburgh, Pittsburgh, PA, USA*
DAVID S. ROOS • *University of Pennsylvania, Philadelphia, PA, USA*
ELEANOR C. SAUNDERS • *Department of Biochemistry and Molecular Biology, Bio21 Institute of Molecular Science and Biotechnology, University of Melbourne, Parkville, VIC, Australia*
AARTHI SEKAR • *Seattle BioMed, Seattle, WA, USA*
PATRICK J. SKELLY • *Helminthology Laboratory, Department of Infectious Disease and Global Health, Cummings School of Veterinary Medicine, Tufts University, North Grafton, MA, USA*
VICTOR SOJO • *Centre for Mathematics and Physics in the Life Sciences and Experimental Biology (CoMPLEX), University College London, London, UK; Research Department of Genetics, Evolution and Environment, University College London, London, UK*
DAVID P. DE SOUZA • *Metabolomics Australia, Bio21 Institute of Molecular Science and Biotechnology, The University of Melbourne, Parkville, VIC, Australia*
YUTAKA SUZUKI • *Laboratory of Functional Genomics, Department of Medical Genome Sciences, Graduate School of Frontier Sciences, University of Tokyo, Shirokanedai, Minato-ku, Chiba, Japan*
ANDY TAIT • *College of Medical, Veterinary and Life Sciences, University of Glasgow, Glasgow, UK*
MONICA TERRAO • *Departamento de Biologia Celular e Molecular e Bioagentes Patogênicos, Faculdade de Medicina de Ribeirão Preto, Universidade de São Paulo, Av. Bandeirantes, Ribeirão Preto - SP, Brazil*
LI-MIN TING • *Albert Einstein College of Medicine, Bronx, NY, USA*
BARTŁOMIEJ TOMICZEK • *Centre for Mathematics and Physics in the Life Sciences and Experimental Biology (CoMPLEX), University College London, London, UK; Research Department of Genetics, Evolution and Environment, University College London, London, UK*
LUIS R.O. TOSI • *Departamento de Biologia Celular e Molecular e Bioagentes Patogênicos, Faculdade de Medicina de Ribeirão Preto, Universidade de São Paulo, Ribeirão Preto, SP, Brazil*
ANGELA TRIEU • *Molecular Vaccinology Laboratory, QIMR Berghofer Medical Research Institute, Herston, QLD, Australia*
CHRISTIAN TSCHUDI • *Department of Epidemiology of Microbial Diseases, Yale School of Public Health, New Haven, CT, USA*
ELISABETTA ULLU • *Department of Internal Medicine, School of Medicine, Yale University, New Haven, CT, USA; Department of Cell Biology, School of Medicine, Yale University, New Haven, CT, USA*
HELI VAIKKINEN • *Wellcome Trust Centre for Molecular Parasitology, Institute of Biodiversity, Animal Health and Comparative Medicine, College of Medical, Veterinary and Life Sciences, University of Glasgow, Glasgow, UK*

WILLIE WEIR • *Wellcome Trust Centre for Molecular Parasitology, Institute of Biodiversity, Animal Health and Comparative Medicine, College of Medical, Veterinary and Life Sciences, University of Glasgow, Glasgow, UK*
JUNYA YAMAGISH • *Tohoky Medical Megabank Organization, Tohoku University, Aoba-ku, Sendai, Japan*
JOHN YATES • *Department of Chemical Physiology, The Scripps Resesarch Institute, La Jolla, CA, USA*
MATTHEW YEO • *London School of Hygiene and Tropical Medicine, London, UK*
WEN-WEI ZHANG • *Microbiology and Immunology, McGill University, Montreal, QC, Canada*
DAN ZILBERSTEIN • *Faculty of Biology, Technion-Israel Institute of Technology, Haifa, Israel*

Chapter 1

The Eukaryotic Pathogen Databases: A Functional Genomic Resource Integrating Data from Human and Veterinary Parasites

Omar S. Harb and David S. Roos

Abstract

Over the past 20 years, advances in high-throughput biological techniques and the availability of computational resources including fast Internet access have resulted in an explosion of large genome-scale data sets "big data." While such data are readily available for download and personal use and analysis from a variety of repositories, often such analysis requires access to seldom-available computational skills. As a result a number of databases have emerged to provide scientists with online tools enabling the interrogation of data without the need for sophisticated computational skills beyond basic knowledge of Internet browser utility. This chapter focuses on the Eukaryotic Pathogen Databases (EuPathDB: http://eupathdb.org) Bioinformatic Resource Center (BRC) and illustrates some of the available tools and methods.

Key words Eukaryotic, Pathogen, Parasite, EuPathDB, Genomic, Database, Search strategy, Bioinformatics

1 Introduction

The EuPathDB BRC [1] is mainly funded by the National Institutes of Health with additional funding for the kinetoplastid component (TriTrypDB) [2] coming from the Bill and Melinda Gates Foundation and the Wellcome Trust and in collaboration with GeneDB [3]. The overarching goal of EuPathDB is to incorporate genomic and postgenomic data from the global research community and making it possible to interrogate the data in an integrative manner.

While EuPathDB includes a family of databases supporting various eukaryotic pathogens (Table 1), the look and feel of these databases have been streamlined to facilitate mobility between databases without the need for reeducation. Hence, protocols described herein can be used universally on any EuPathDB website. In addition, a number of collaborative efforts with groups using the EuPathDB infrastructure [4] extend this usability to

Christopher Peacock (ed.), *Parasite Genomics Protocols*, Methods in Molecular Biology, vol. 1201,
DOI 10.1007/978-1-4939-1438-8_1, © Springer Science+Business Media New York 2015

Table 1
This table lists EuPathDB resources, their web addresses, and the included organisms

Database	Web address	Supported organisms
EuPathDB	http://eupathdb.org	All EuPathDB organisms listed below
AmoebaDB	http://amoebadb.org	*Acanthamoeba castellanii, Entamoeba histolytica, E. dispar, E. invadens, E. moshkovskii, E. nuttalli*
CryptoDB	http://cryptodb.org	*Cryptosporidium parvum,C. parvum, C. muris*
GiardiaDB	http://giardiadb.org	*Giardia lamblia* assemblages A, B, and E
MicrosporidiaDB	http://microsporidiadb.org	*Anncaliia algerae, Edhazardia aedis, Encephalitozoon cuniculi, E. hellem, E. intestinalis, E. romaleae, Enterocytozoon bieneusi, Hamiltosporidium tvaerminnensis, Nematocida parisii, Nosema ceranae, Vavraia culicis, Vittaforma corneae*
PiroplasmaDB	http://piroplasmadb.org	*Babesia bovis, B. microti, Theileria annulata, T. parva*
PlasmoDB	http://plasmodb.org	*Plasmodium berghei, P. chabaudi, P. cynomolgi, P. falciparum, P. gallinaceum, P. knowlesi, P. reichenowi, P. vivax, P. yoelii*
ToxoDB	http://toxodb.org	*Toxoplasma gondii, Eimeria tenella, Gregarina niphandrodes, Neospora caninum*
TrichDB	http://trichdb.org	*Trichomonas vaginalis*
TriTrypDB	http://tritrypdb.org	*Crithidia fasciculata, Trypanosoma brucei, T. congolense, T. cruzi, T. vivax, Leishmania major, L. infantum, L. braziliensis, L. donovani, L. Mexicana, L. panamensis, L. tarentolae, Endotrypanum monterogeii*
OrthoMCL	http://orthomcl.org	Includes proteins from over 150 organisms across bacteria, archaea, and eukarya

other genomic resources including FungiDB (http://fungidb.org) [5], SchistoDB (http://schistodb.net) [6], TBDB (http://www.tbdb.org/wdk/) [7], and BetaCell (http://www.betacell.org) [8].

Searches in EuPathDB are categorized based on the type of returned results. Data in EuPathDB is obtained from publications (or directly from researchers), and from sequence and data repositories such as GenBank, sequencing centers (i.e., the Sanger Institute, the Broad Institute, the J. Craig Venter Institute). Information regarding the source of the data is available on multiple pages within EuPathDB resources and in the extensive data set section (*see* Subheading 4, below).

2 Materials

1. Computer (desktop, laptop tablet, or smartphone).
2. Internet browser such as Firefox, Safari, Internet Explorer, or Chrome.
3. Internet access with sufficient bandwidth for web surfing.

3 Methods

3.1 Building a Search Strategy (In Silico Experiment)

Searches in EuPathDB resources start by executing an initial query from any of over 80 different available searches. Searches can be used to define sets of genes, isolates, SNPs, genomic segments (i.e., DNA motifs), expressed sequence tags (ESTs), open reading frames (ORFs), or SAGE tags (Fig. 1a). Searches are organized in expandable categories (click on the plus symbol to expand a category) (Fig. 1b). Results of a query are placed into a search strategy, which may be expanded by combining these results with those of additional searches. Results can be combined with each other using intersect, union, or minus operations. To build a search strategy, follow these steps:

1. Define the question you are interested in asking. For example, one may be interested in finding all genes in all apicomplexan parasites available in EuPathDB that are secreted, contain at least four transmembrane domains, have evidence of expression in any parasitic stage based on RNA-sequence evidence, and do not have orthologs in mammals. An answer to such a question is attainable using the integrated search strategy developed by EuPathDB.
2. Identify the searches that will allow you to answer your question. Searches in EuPathDB are triggered against the underlying data such as finding genes with defined characteristics (i.e., genes that have a predicted signal peptide or a specified number of transmembrane domains). An initial search starts by selecting the appropriate link on the home page, clicking on the plus symbol next to a search category, and then selecting the search of interest (Fig. 1b).
3. Define the search parameters and run your first search (Fig. 1c). Species of interest may be selected from the taxonomically organized checklist. Once you are satisfied with your parameters, click on the "Get Answer" button. This will initiate a search strategy that includes a step with the results of the signal peptide search (Fig. 1d).
4. Grow your search strategy by adding additional steps (Fig. 2a). Adding steps is done by clicking on the add step button, selecting a search from the pop-up window, and choosing how to

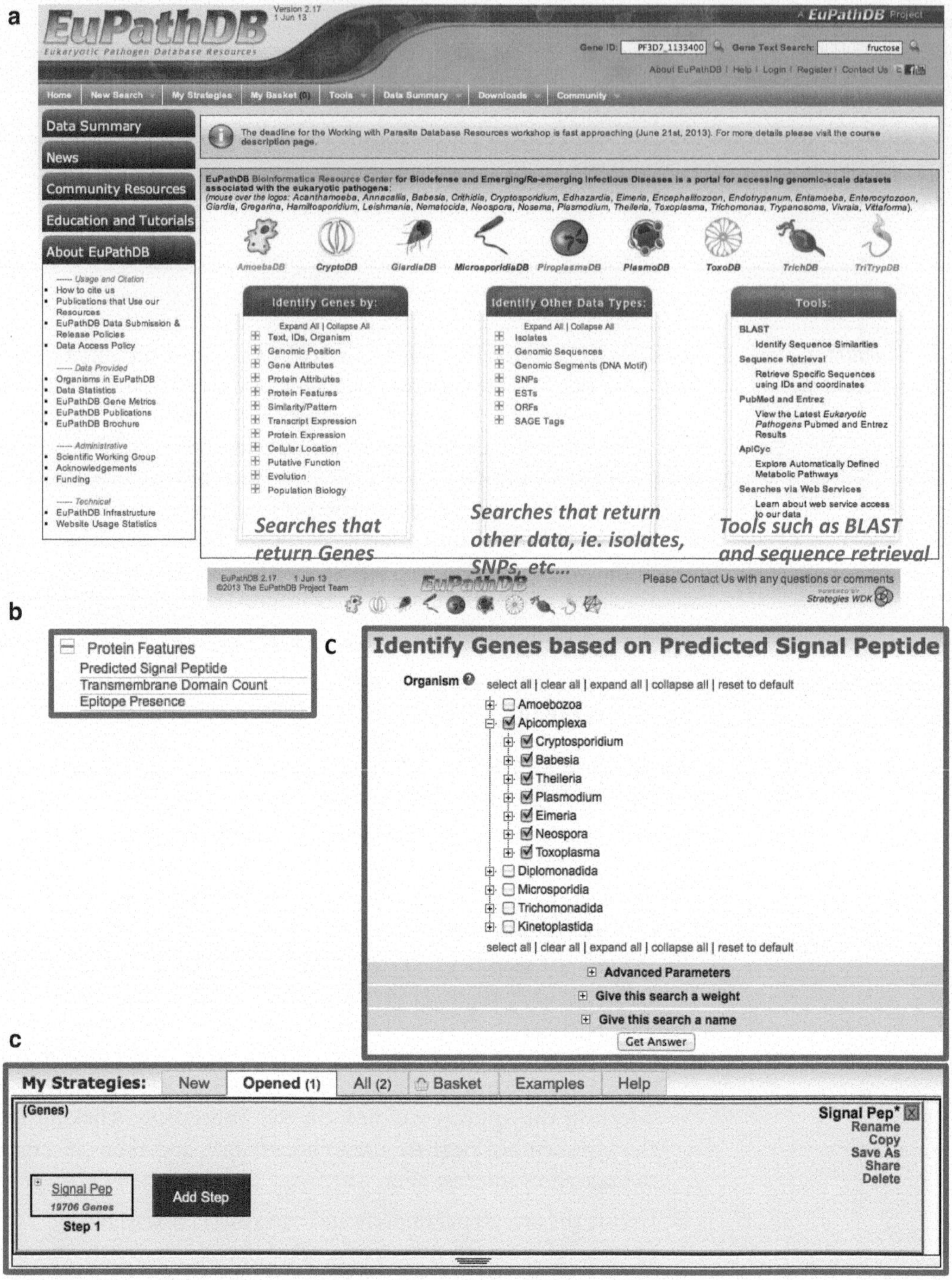

Fig. 1 Screenshots from EuPathDB depicting the home page and example first search. (**a**) The EuPathDB home page, searches are organized based on the data type they return. (**b**) Categories can be expanded by clicking on the plus symbol to reveal specific searches. (**c**) Once a search is selected the next web page provides search options. In this example, the search page for genes with predicted signal peptides is displayed. Organisms are taxonomically organized and species of interest may be selected. (**c**) Once a search is engaged a search strategy is revealed. This example shows the results of running a signal peptide search on all Apicomplexan organisms in EuPathDB

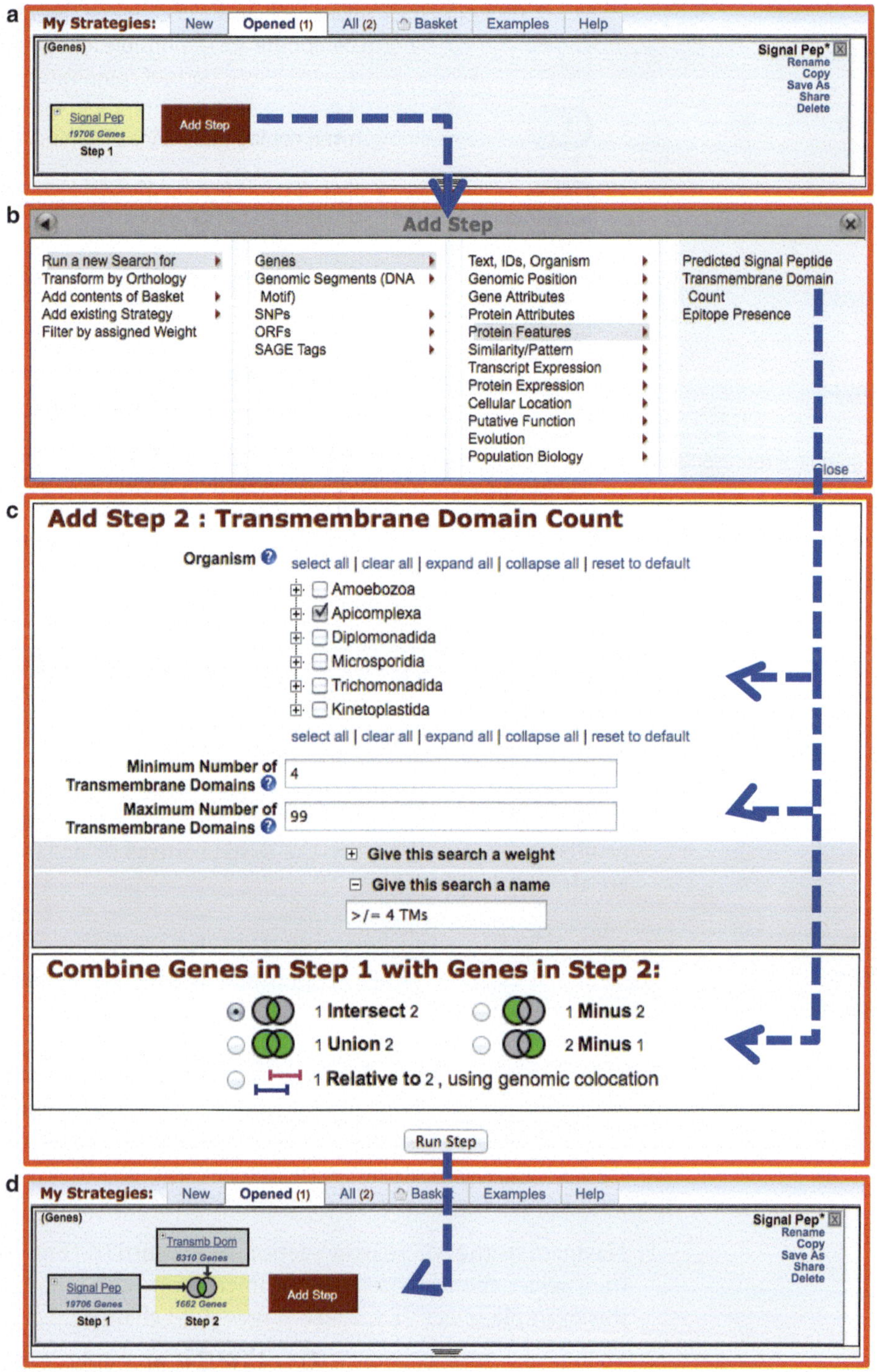

Fig. 2 Screenshots from EuPathDB depicting the process of adding a step to a search strategy. (**a**) Click on the "Add Step" button to reveal a pop-up window with all available searches in (**b**). Navigate and select the search of interest. (**c**) Once a search is selected a second pop-up window with search parameters becomes available. In addition to selecting search parameters, the method of combining step results needs to be selected (*see* Fig. 3). (**d**) Clicking on the "Run Step" button adds the results of the new search to those of the first search resulting in a two-step strategy

Name of operation	Symbol of operation in EuPathDB	Definition
Intersect		If A = {1,2,3} and B = {1,2,4,5} then A intersect B = {1,2}
Union		If A = {1,2,3} and B = {1,2,4,5} then A union B = {1,2,3,4,5}
Difference	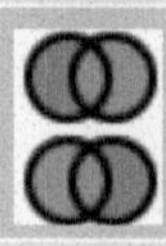	If A = {1,2,3} and B = {1,2,4,5} then A -B = {3}
Colocation	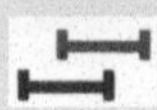	A is defined by its genomic location relative to the genomic location B.

Fig. 3 A graphical representation of the available operations for combining results in a search strategy. Note that the colocation option requires additional parameter selections (described elsewhere in this chapter)

combine the results of this search with those of the previous one. Figure 3 illustrates the type of available operations and their definitions. These include union, intersect, difference, and colocation.

5. Results from all searches are displayed below a search strategy and are dynamically updated as additional steps are added, revised, or deleted. As with any experiment determine if the results are sound: What are the false positives or negatives and are the results plausible?

3.2 Using the Orthology Transform Tool

Genes may be identified based on their characteristics defined by experimental data. Typically, experimental data (i.e., microarray, mass spectrometry, RNA-seq) are collected from a single species of a parasite due to the interest of a lab or experimental accessibility. Orthology may be used to leverage data collected from other species to define genes in your organism of interest. For example, the orthology transform tool enables you to define *Plasmodium falciparum* and *P. vivax* orthologs of genes expressed in liver stages from a microarray experiment performed on *P. yoelii* [9]:

1. Navigate to the Microarray section of PlasmoDB (Fig. 4a) and then select the microarray experiment you wish to query (for this example select "P.y. Liver Stages (fold change))" (Fig. 4b).
2. Define the search parameters. For this example, select up-regulated genes by at least twofold in the blood stage (BS) vs. liver stage 40-h (LS40) comparison (Fig. 4c).
3. Click on the "Get Answer" button. This will start a search strategy with a result of 70 *P. yoelii* genes that are up-regulated in liver stages compared to blood stages (Fig. 4d).

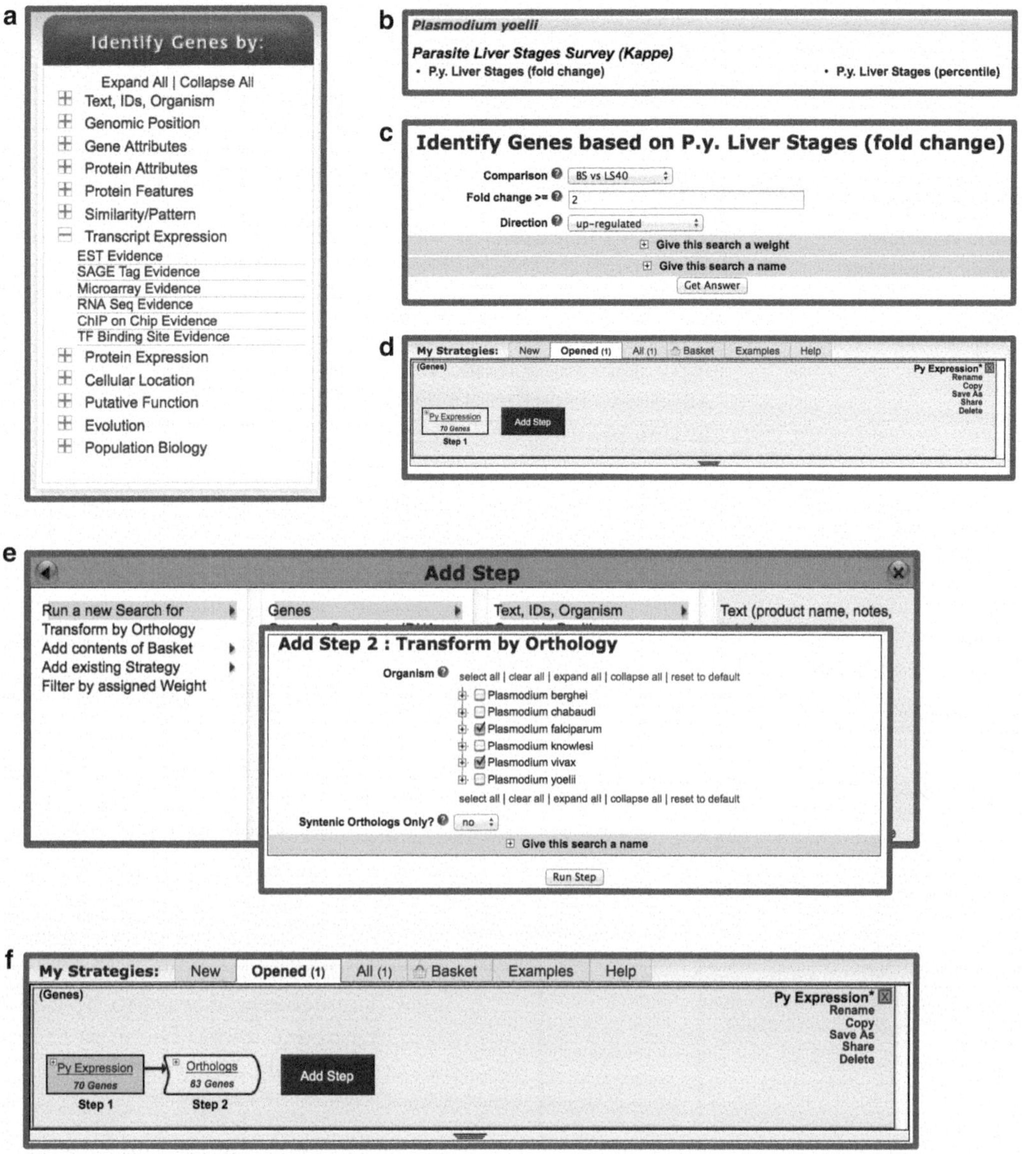

Fig. 4 Screenshots depicting the utility of the orthology transform tool. (**a**) The "Identify Genes by" portion of the PlasmoDB home page with the "Transcript Expression" category expanded. (**b**) A portion of the microarray expression page depicting the experiment chosen for the search. (**c**) Once an experiment and analysis are selected, search parameters are revealed. (**d**) A search strategy depicting results from a microarray experiment specific to *P. yoelii*. (**e**) Transforming results from one species to another requires adding a step and then selecting the "Transform by orthology" option. The "Transform by Orthology" pop-up window allows the selection of species to transform to. (**f**) A search strategy with the *P. yoelii* results transformed to orthologs in *P. vivax* and *P. falciparum*

4. To define the orthologs of these genes in *P. falciparum* and *P. vivax*, click on add step, in the pop-up select the "Transform by Orthology" option (Fig. 4d), then select the species you wish to transform your results to (Fig. 4e), and click on

"Get Answer." The results are any *P. vivax* and *P. falciparum* genes that are orthologs of the *P. yoelii* genes (Fig. 4f).

Note that orthology in EuPathDB databases is determined using OrthoMCL [10–12].

3.3 Building a Search Strategy to Define Secreted Kinases

1. Finding genes using keywords:

   There are a variety of ways to reach a specific gene record in EuPathDB databases. The most straightforward approach is to use the text search option using a specific keyword to identify a gene of interest. This type of approach relies on text available form the annotation, community user comments, genome ontology, InterPro domains, BLAST similarity, etc. The following protocol describes how to identify kinases in PlasmoDB (http://plasmodb.org):

   1. Enter the keyword "kinase" (without quotations) in the search box using either option (a) or (b). Click on the search icon if using option (a), or on the get answer button if using option (b).

      (a) In the "Gene Text Search" box at the top right of any webpage (Fig. 5a).

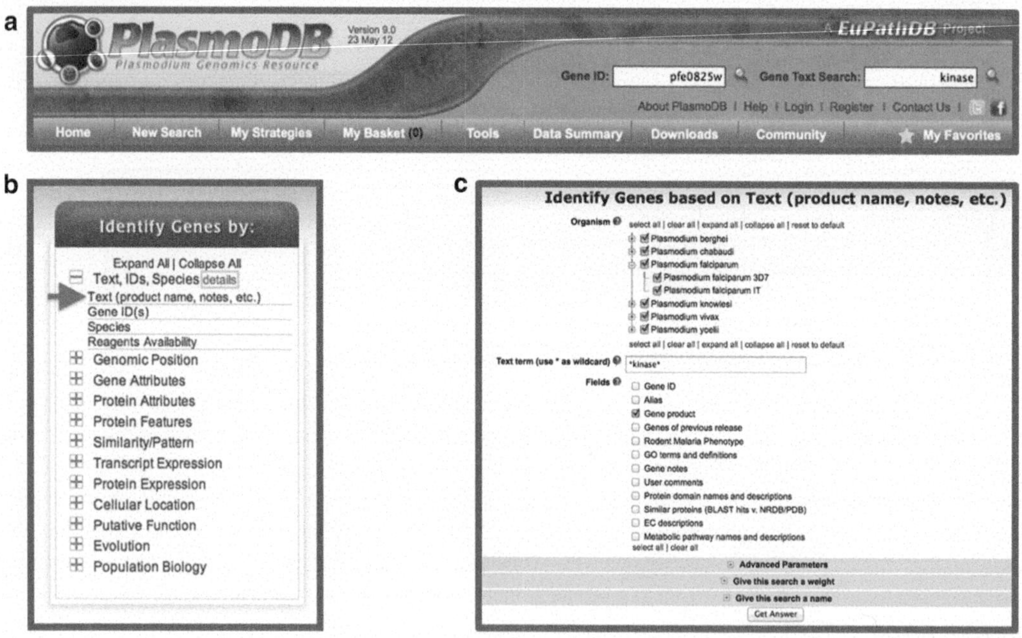

Fig. 5 Screenshots from PlasmoDB depicting text search options. (**a**) The banner section of PlasmoDB, which includes a text search window in the upper right-hand corner. (**b**) The "Identify Genes by" section of the PlasmoDB home page with the "Text, IDs, Species" category expanded. (**c**) Selecting "Text (product names, notes etc.)" opens a pop-up window that enables specifying organisms to search, the text term, and the fields to search

(b) Via the text search query page which can be accessed by clicking on the Text query link under "Text, IDs, Species" section located in the "Identify Genes by:" column on the home page (Fig. 5b, c).

2. The search above will miss words like "6-phosphofructokinase" or "kinases." To retrieve genes containing such words you may use a wild card in your search—try "kinase*," "*kinase," and/or "*kinase*" (without quotations).

3. There are two places where a keyword may be entered to search for genes:

2. Finding genes that contain a predicted secretory signal peptide. Add a step to the kinase results that searches for genes with predicted secretory signal peptides. The search for signal peptides can be found under the "Cellular Location" search (Fig. 6a, b).

3. Adding genes with predicted transmembrane domains. Grow this search strategy to also include genes that have predicted

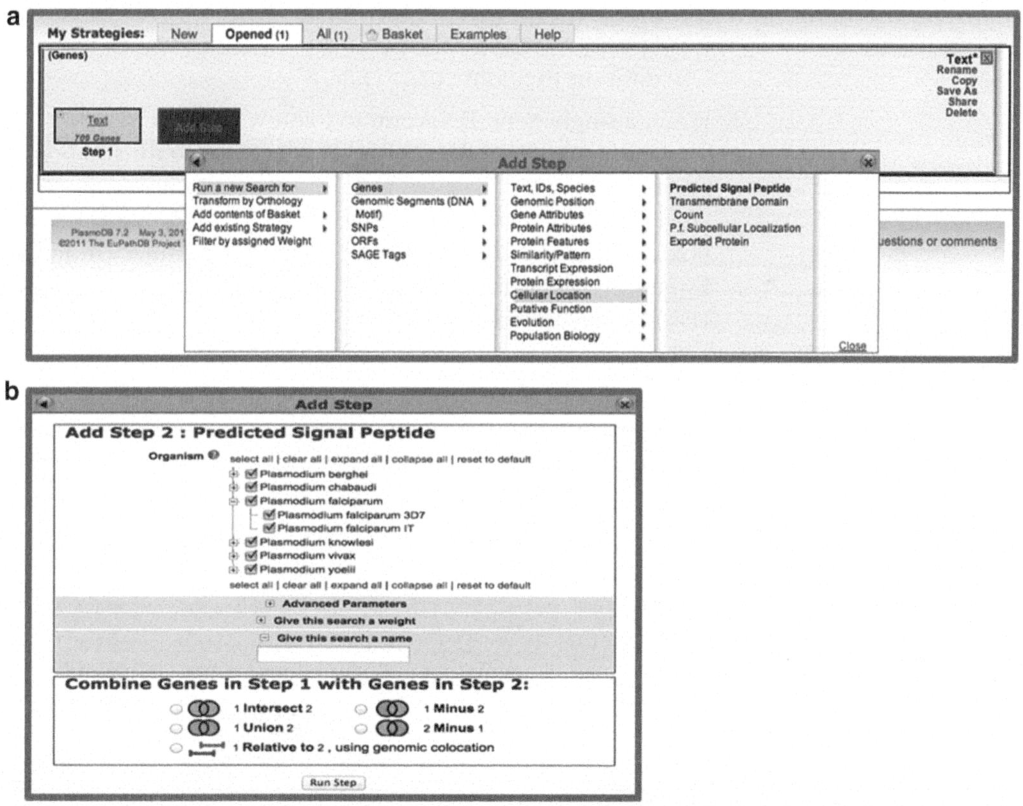

Fig. 6 Screenshots from PlasmoDB depicting adding a step. (**a**) Clicking on the "Add Step" button reveals a pop-up window with all searches in PlasmoDB. Selecting "Predicted Signal Peptide" under the "Cellular Location" category reveals pop-up that enables the customization of this search (**b**)

transmembrane domains. In this case the goal is to define kinases that have a predicted signal peptide, at least one transmembrane domain or both. Hence, it is critical to expand the signal peptide step into a nested strategy, the results of which will be combined with the list of kinases (Fig. 7). Note that without the option of creating a nested strategy, results would be combined sequentially resulting in very different consequences (Fig. 7c, d).

3.4 Identifying Genes Based on Their IDs

Genes may be identified based on their unique identifiers (IDs). EuPathDB maps old IDs to new ones enabling searching with old archival IDs in updated versions of the databases. IDs may be entered one at a time or in bulk—the following protocol employs the ID search in http://PlasmoDB.org.

1. You can find genes based on their IDs, one at a time or in bulk. There are two places where you can enter a gene ID(s):
   (a) The "Gene ID" search box at the top of the home page (Fig. 5a).
   (b) Using the Gene ID query, which can be accessed by clicking on the Gene ID(s) query link under "Text, IDs, Species" section located in the "Identify Genes by:" column on the home page (Fig. 8).
2. When a single gene ID is entered you will be taken directly to the gene page. For example, enter the gene ID for the bifunctional dihydrofolate reductase-thymidylate synthase (DHFR-TS) gene (PF3D7_0417200) in the Gene ID search box and click on the search icon next to the box (note that EuPathDB databases provide ID mapping of old or alternative IDs to current official gene IDs).
3. Multiple gene IDs may be used as the input in the Gene ID query. This is useful if you have a list of gene IDs from your own experiments or a publication that you would like to further investigate in EuPathDB. In this example a list of gene IDs were obtained from a publication [13]:

   PFF0615c, Pf13_0338, PFE0395c, PF14_0201, PFF0995c, PF10_0346, PF10_0347, PF10_0348, PF10_0352, PF13_0197, PF13_0196, MAL13P1.174, PF13_0193, MAL13P1.173, Pf13_0191, PF13_0192, PF13_0194, PFL1385c, PFB0340c, MAL7P1.208, PF13_0348, PF10_0144, PF14_0102, PFE0080c, PFE0075c, PFD0955w

Fig. 7 (continued) (**b**) Selecting "Make Nested Strategy" expands the step into a substrategy that can be expended as an independent branch of the search strategy. (**c**) Results of the nested strategy are combined with a step in the main search strategy. (**d**) An illustration of what the results would look like if a nested strategy is not used

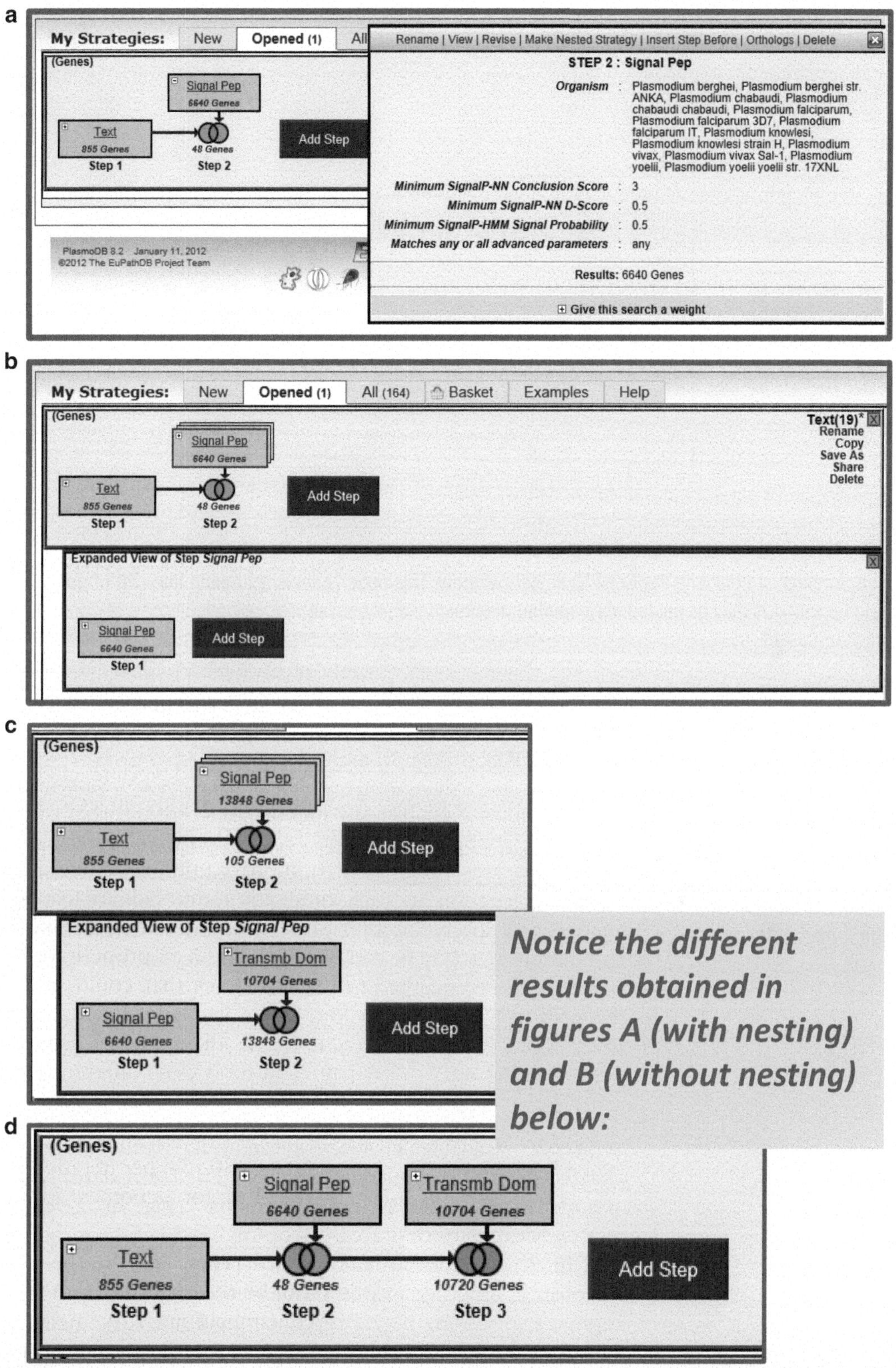

Fig. 7 Screenshots from PlasmoDB depicting the conversion of a step into a nested strategy. (**a**) Click on the name of the step to be made into a nested strategy. In this image, the signal peptide step was selected. A pop-up window enables the selection of several options including revise, delete, insert step before, and make nested strategy.

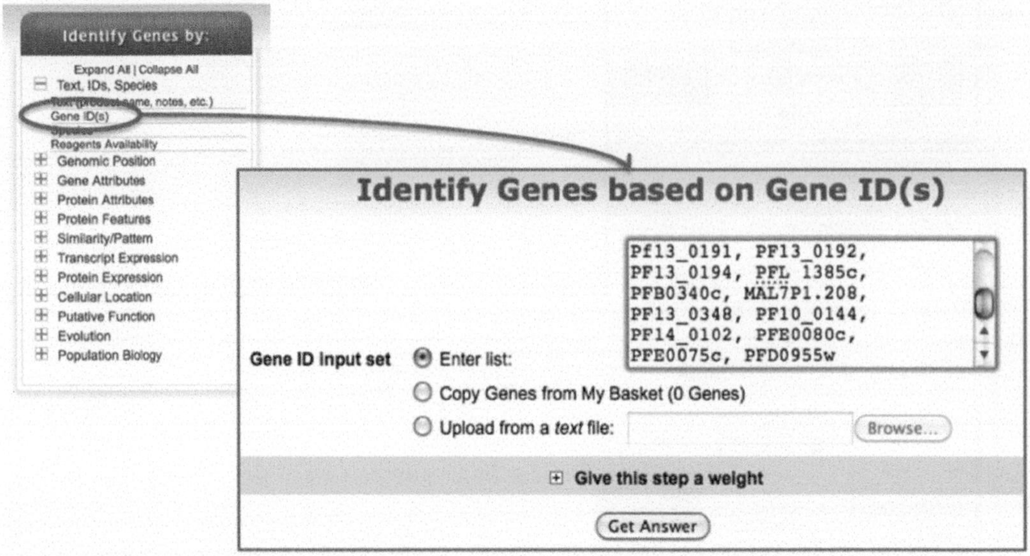

Fig. 8 Screenshot depicting the Gene ID search window. This option allows searching for a list of gene IDs in one lump sum. IDs may be pasted from another document (i.e., a publication), uploaded from a file, or imported from your basket

Paste these IDs into the ID query and click on the get answer button to retrieve this list of genes in PlasmoDB (Fig. 8).

3.5 Using the Colocation Tool to Find Genes Within a Defined Distance from a DNA Motif

The colocation tool enables the identification entities that can be mapped on a genome (i.e., genes, restriction sites, transcription factor-binding sites, single-nucleotide polymorphisms) based on their relative location to each other (genomic colocation). For example, genes located within 500 nucleotides of transcription factor-binding sites may be defined since both genes and transcriptions factors can be mapped to specific coordinates on a genome. In the protocol presented below, all genes located within 500 nucleotides of a BamHI restriction site are identified using http://MicrosporidiaDB.org.

1. Find all BamHI restriction sites in all microsporidia genomic sequences available in MicrosporidiaDB. BamHI sites are defined by the DNA motif GGATCC. The DNA motif search is under the heading "Genomic Segments" (Fig. 9a). Selecting "DNA Motif Pattern" reveals a pop-up window where the specific nucleotide motif may be defined (Fig. 9b). Note that in addition to entering a DNA motif as a string of IUPAC code, regular expressions may be utilized to defined less stringent motifs. Take a look at your results; notice the Genomic location and the Motif columns (Fig. 9c).
2. Find genes that are 500 nucleotides downstream of the BamHI sites: Add a "Genes by Organism" step to the motif search,

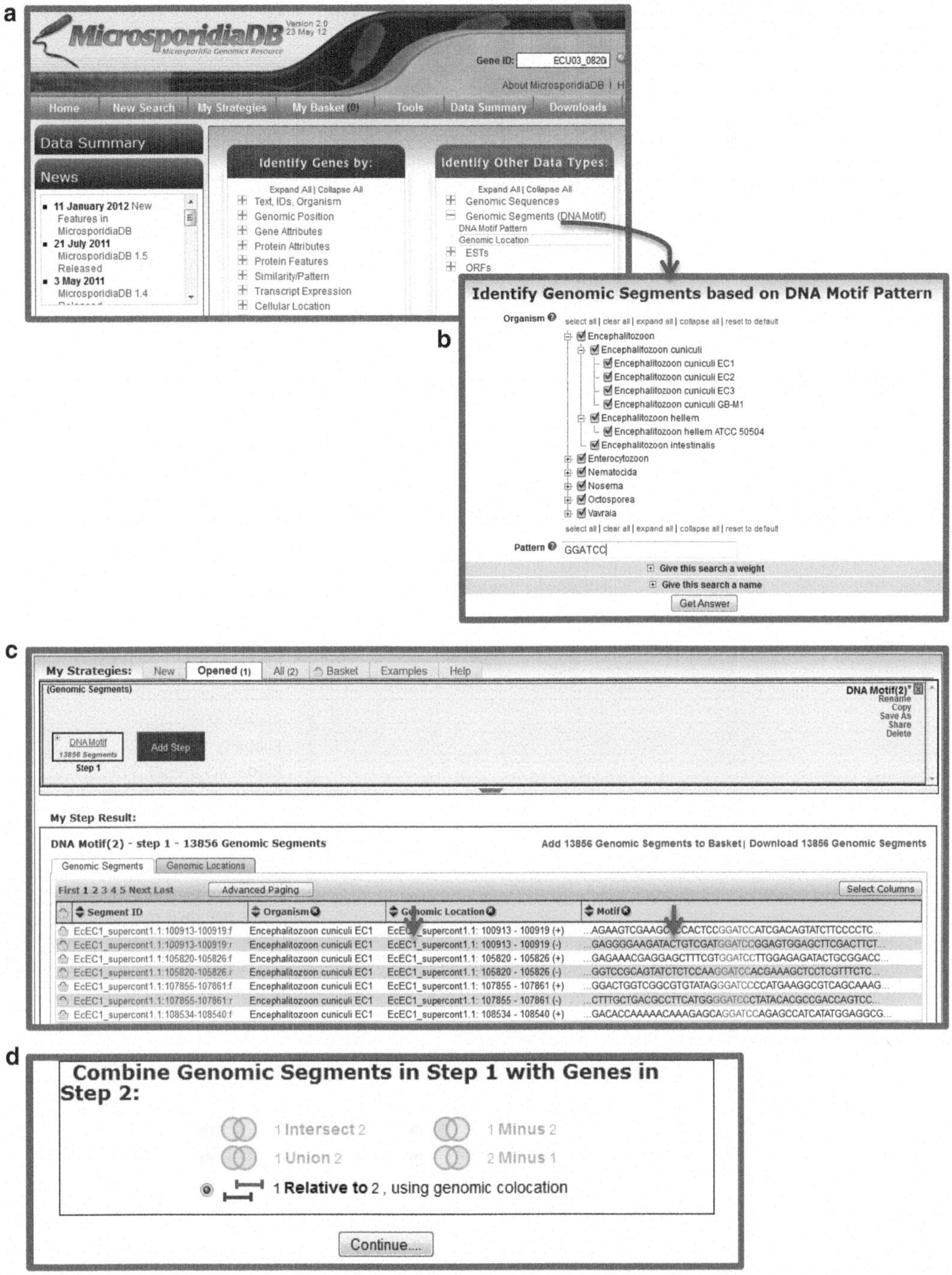

Fig. 9 Screenshots from MicrosporidiaDB depicting a search for a DNA motif. (**a**) A portion of the MicrosporidiaDB home page with the "Genomic Segments" category expanded. (**b**) Selecting "DNA Motif Pattern" reveals a pop-up window where the specific nucleotide motif may be defined. (**c**) Results of a DNA motif query are displayed as a search strategy. DNA motif records are dynamically generated and displayed as a list of results under the search strategy. (**d**) Results from a DNA motif query may be combined with other types of results (i.e., genes) using the genomic colocation option

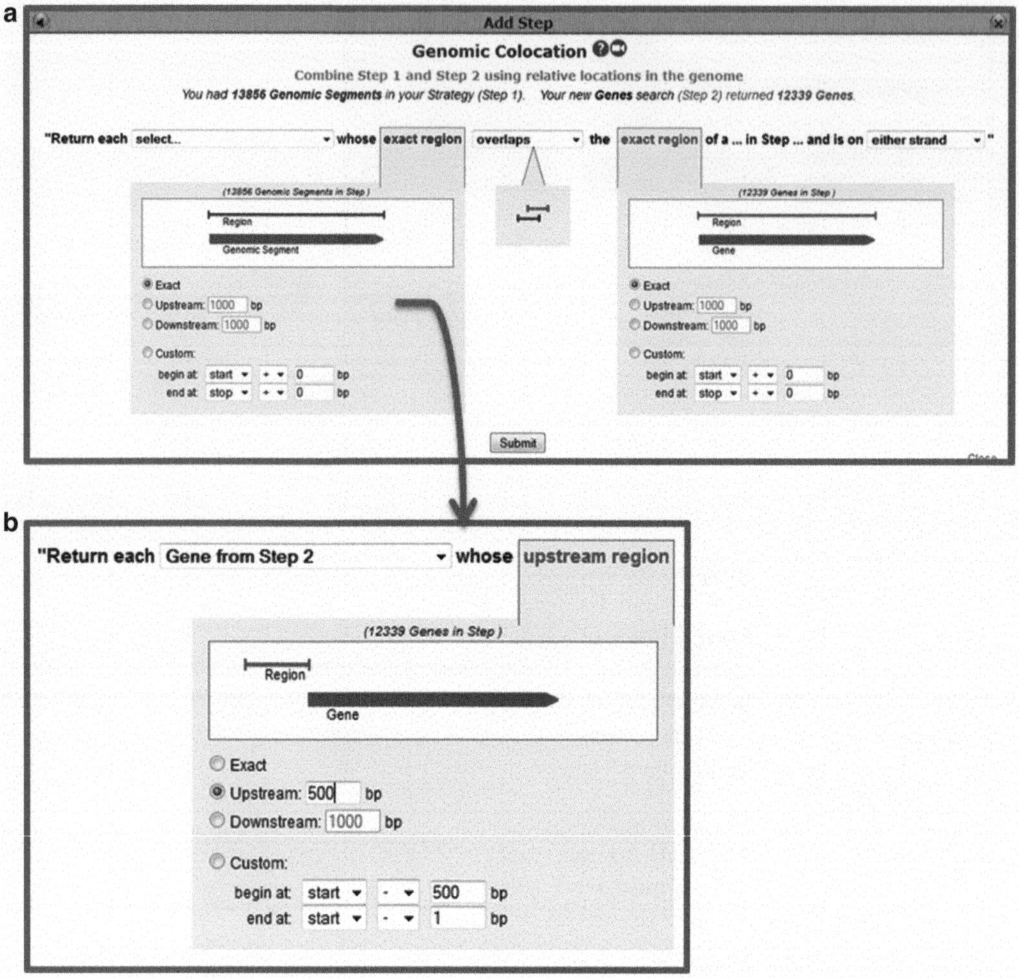

Fig. 10 A screenshot of the genomic colocation pop-up window. (**a**) The genomic colocation pop-up includes a logic statement that allows selecting the results to return and to define the relationship between the results based on their relative genomic locations. (**b**) An enlarged section of the genomic colocation pop-up window showing a dynamic graphical interface that illustrates the selected relationship

select the "1 relative to 2, using genomic locations" option (Fig. 9d), and click on continue.

3. Use the logic statement at the top of the pop-up window (Fig. 10a) to define which results to return (genes or motifs) and the desired relationship between the genes and the motifs. For this example select genes from step two whose upstream 500 nucleotides contain the motif (BamHI) (Fig. 10b)

3.6 Defining Proteins with Specific Amino Acid Motifs

Genes which translated products contain a specific amino acid motif may be identified using the protein motif pattern search. This query allows defining a motif based on an exact string of amino acids or using a regular expression. The protein motif pattern search can be found under the heading "Similarity/Pattern" in the "Identify gene by" section of EuPathDB home pages.

Regular expressions are straightforward to compose as illustrated in the example below that finds all proteins in *Trypanosoma cruzi* that contain a signature motif for trans-sialidases using http://TriTrypDB.org. The search strategy described in this protocol may be accessed here:http://tritrypdb.org/tritrypdb/im.do?s=a905e36f634f7b42

1. *T. cruzi* has an expanded family of trans-sialidases. In fact, if you run a text search for any gene with the word "trans-sialidase," you return over 3,500 genes among the strains in the database!!! Try this and see what you get (Fig. 11a).

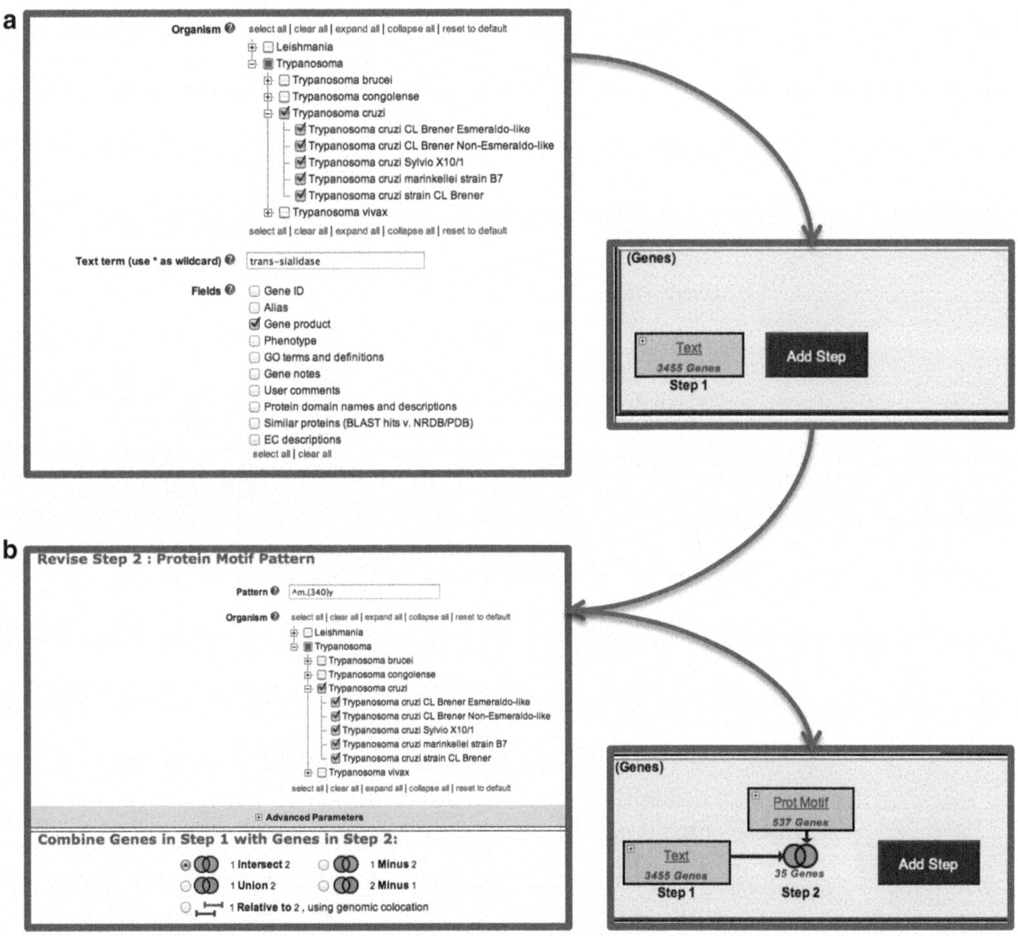

Fig. 11 Screenshots representing a text search in (**a**) combined with the protein motif search in (**b**). (**a**) A text search for all gene products containing the keyword "trans-sialidase" in *Trypanosoma cruzi* returns 3,455 genes. (**b**) A protein motif pattern search for all *T. cruzi* proteins that start with a methionine, followed by 340 amino acids of any kind and a tyrosine (Y) at position 342—represented by the regular expression ^m.{340}y—returns 537 genes. The intersection of both searches is 35 genes

2. However, not all of these are predicted to be active. It is known that active trans-sialidases have a signature tyrosine (Y) at position 342 in their amino acid sequence. Add a motif search step to the text search in "a" to identify only the active trans-sialidases. Note that for this regular expression the first amino acid should be a methionine (start of the protein), followed by 340 of any amino acid, followed by a tyrosine "Y" (Fig. 11b).

3.7 Identifying Isolates Based on Associated Metadata

EuPathDB sites integrate isolate data from multiple sources including GenBank. The genetic background of isolates may be defined by single-locus sequencing, single-nucleotide profiling (SNP-Chip), or high-throughput genomic sequencing. Isolate searches are available under the "Isolates" heading in the "Identify Other Data Types" section of EuPathDB home pages (Fig. 12a). The following protocol uses http://CryptoDB.org to identify Cryptosporidium isolates from Europe that were isolated from feces.

1. Find all Cryptosporidium isolates identified from Europe. This is achieved by running an isolate by geographic location search (Fig. 12a, b) and defining Europe as the geographic location.
2. Add a search for isolates based on isolation source (Fig. 12a), select "feces" (Fig. 12c), and combine the results of this search with those from **step 1** (Fig. 12d).
3. Isolate data is displayed in tabular format and can also be viewed graphically on a dynamic world map by clicking on the "Isolate Geographic Location" tab.

4 Notes

- Additional exercises used in EuPathDB workshops: http://workshop.eupathdb.org/current/
- Online tutorials: http://tinyurl.com/eupathdbTutorials
- Updated EuPathDB data content summary: http://tinyurl.com/eupathdbSummary
- EuPathDB data set information: http://tinyurl.com/eupathdbdatasource
- EuPathDB news: http://eupathdb.org/eupathdb/aggregateNews.jsp
- EuPathDB data submission standard operating procedure: http://eupathdb.org/EuPathDB_datasubm_SOP.pdf
- Request a workshop or webinar: help@eupathdb.org

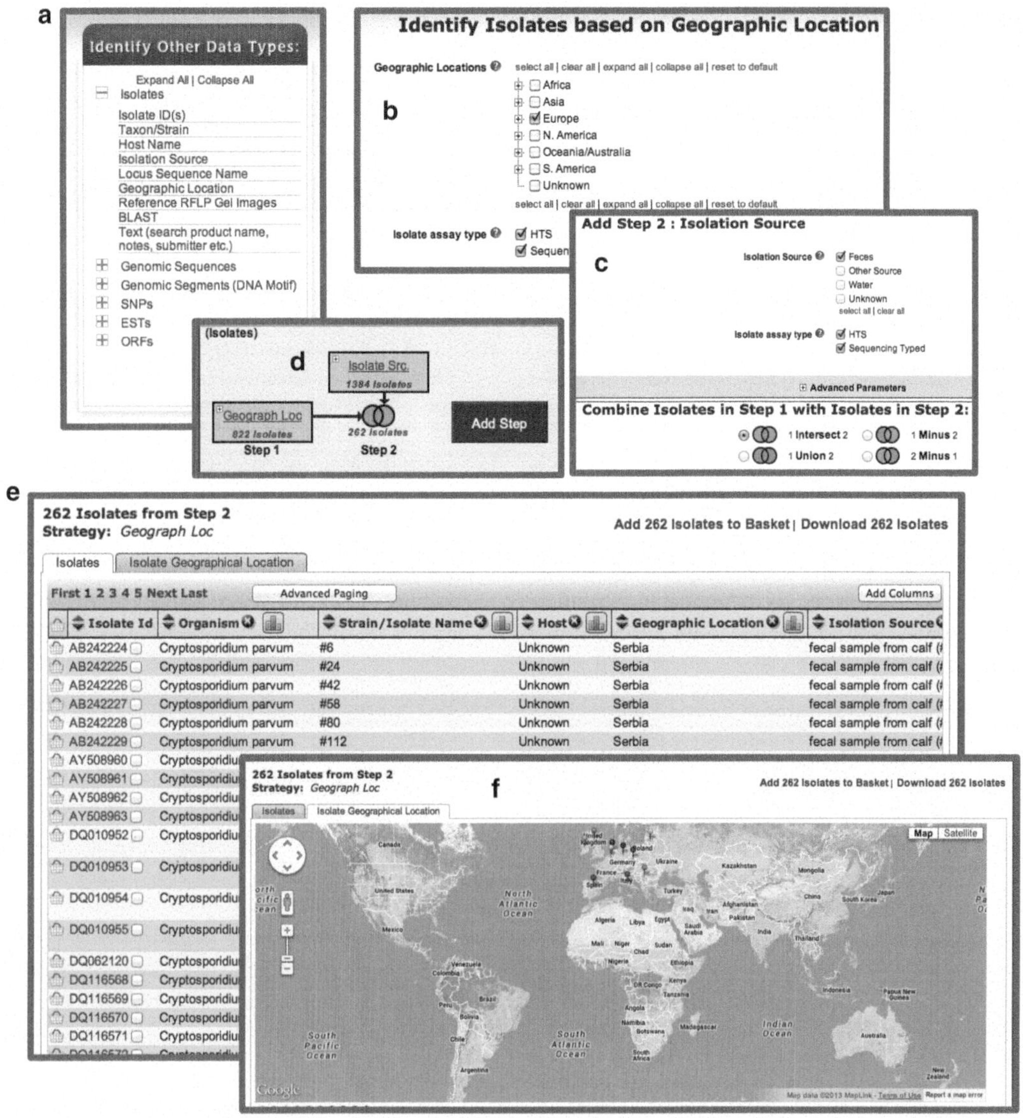

Fig. 12 Screenshots of an isolate query in CryptoDB. (**a**) A number of searches for isolates are available under the "Isolates" heading in the "Identify Other Data types" section of EuPathDB home pages. (**b**) The search for isolates based on geographic location allows the selection of entire continents or specific countries. (**c**) Isolates may be identified based on their isolation source. (**d**) A combination of geographic location and an isolation source defining 262 Cryptosporidium isolates identified in Europe from feces. (**e**) Isolate search results are listed in a dynamic table that can be sorted and expanded. (**f**) Isolate results may also be visualized graphically on a world map by clicking on the "Isolate Geographic Map" tab above the result list

References

1. Aurrecoechea C, Brestelli J, Brunk BP et al (2010) EuPathDB: a portal to eukaryotic pathogen databases. Nucleic Acids Res 38:D415–9. doi:10.1093/nar/gkp941
2. Aslett M, Aurrecoechea C, Berriman M et al (2009) TriTrypDB: a functional genomic resource for the Trypanosomatidae. Nucleic Acids Res 38:D457–D462. doi:10.1093/nar/gkp851
3. Logan-Klumpler FJ, De Silva N, Boehme U et al (2012) GeneDB—an annotation database for pathogens. Nucleic Acids Res 40:D98–108. doi:10.1093/nar/gkr1032
4. Fischer S, Aurrecoechea C, Brunk BP, et al. (2011) The strategies WDK: a graphical search interface and web development kit for functional genomics databases. Database (Oxford) 2011:bar027. doi: 10.1093/database/bar027
5. Stajich JE, Harris T, Brunk BP et al (2012) FungiDB: an integrated functional genomics database for fungi. Nucleic Acids Res 40:D675–81. doi:10.1093/nar/gkr918
6. Zerlotini A, Heiges M, Wang H et al (2009) SchistoDB: a Schistosoma mansoni genome resource. Nucleic Acids Res 37:D579–82. doi:10.1093/nar/gkn681
7. Galagan JE, Sisk P, Stolte C et al (2010) TB database 2010: overview and update. Tuberculosis (Edinb) 90:225–235. doi:10.1016/j.tube.2010.03.010
8. Mazzarelli JM, Brestelli J, Gorski RK et al (2007) EPConDB: a web resource for gene expression related to pancreatic development, beta-cell function and diabetes. Nucleic Acids Res 35:D751–5. doi:10.1093/nar/gkl748
9. Tarun AS, Peng X, Dumpit RF et al (2008) A combined transcriptome and proteome survey of malaria parasite liver stages. Proc Natl Acad Sci 105:305–310. doi:10.1073/pnas.0710780104
10. Li L, Stoeckert CJ, Roos DS (2003) OrthoMCL: identification of ortholog groups for eukaryotic genomes. Genome Res 13:2178–2189. doi:10.1101/gr.1224503
11. Chen F, Mackey AJ, Stoeckert CJ, Roos DS (2006) OrthoMCL-DB: querying a comprehensive multi-species collection of ortholog groups. Nucleic Acids Res 34: D363–8. doi:10.1093/nar/gkj123
12. Fischer S, Brunk BP, Chen F, et al. (2011) Using OrthoMCL to assign proteins to OrthoMCL-DB groups or to cluster proteomes into new ortholog groups. Curr Protoc Bioinformatics 1–19 Chapter 6:Unit 6.12. 1–19
13. Tetteh KKA, Stewart LB, Ochola LI et al (2009) Prospective identification of malaria parasite genes under balancing selection. PLoS One 4:e5568. doi:10.1371/journal.pone.0005568

Chapter 2

From Sequence Mapping to Genome Assemblies

Thomas D. Otto

Abstract

The development of "next-generation" high-throughput sequencing technologies has made it possible for many labs to undertake sequencing-based research projects that were unthinkable just a few years ago. Although the scientific applications are diverse, e.g., new genome projects, gene expression analysis, genome-wide functional screens, or epigenetics—the sequence data are usually processed in one of two ways: sequence reads are either mapped to an existing reference sequence, or they are built into a new sequence ("de novo assembly"). In this chapter, we first discuss some limitations of the mapping process and how these may be overcome through local sequence assembly. We then introduce the concept of de novo assembly and describe essential assembly improvement procedures such as scaffolding, contig ordering, gap closure, error evaluation, gene annotation transfer and ab initio gene annotation. The results are high-quality draft assemblies that will facilitate informative downstream analyses.

Key words Mapping, De novo assembly, Assembly improvement, Local assemblies, Bin assemblies, Annotation

1 Introduction

The aim of sequence assembly is to join short sequences of nucleotides (sequence reads 35–1,000 bp in length) into contiguous sequences (contigs) that represent the sequenced DNA. Sequence assembly is needed when no reference genome is available, or when the sequenced DNA is too different from a potential reference genome. In contrast, when strains or isolates are similar enough to a reference sequence, reads can be mapped against this reference by finding the unambiguous place where an alignment generates the highest score for a given read, similar to a BLAST search. Figure 1 shows an example of reads from the *Plasmodium falciparum* IT clone mapped to the *MSP3* (Merozoite surface protein) gene of the reference genome. Most of the regions are covered by mapped reads and genetic variation is represented by red lines in the alignments. But some regions are too polymorphic for reads to map. In this case, only the comparison of the reference with the de novo assembly reveals an insertion in the *MSP3* gene in the IT strain.

Christopher Peacock (ed.), *Parasite Genomics Protocols*, Methods in Molecular Biology, vol. 1201,
DOI 10.1007/978-1-4939-1438-8_2, © Springer Science+Business Media New York 2015

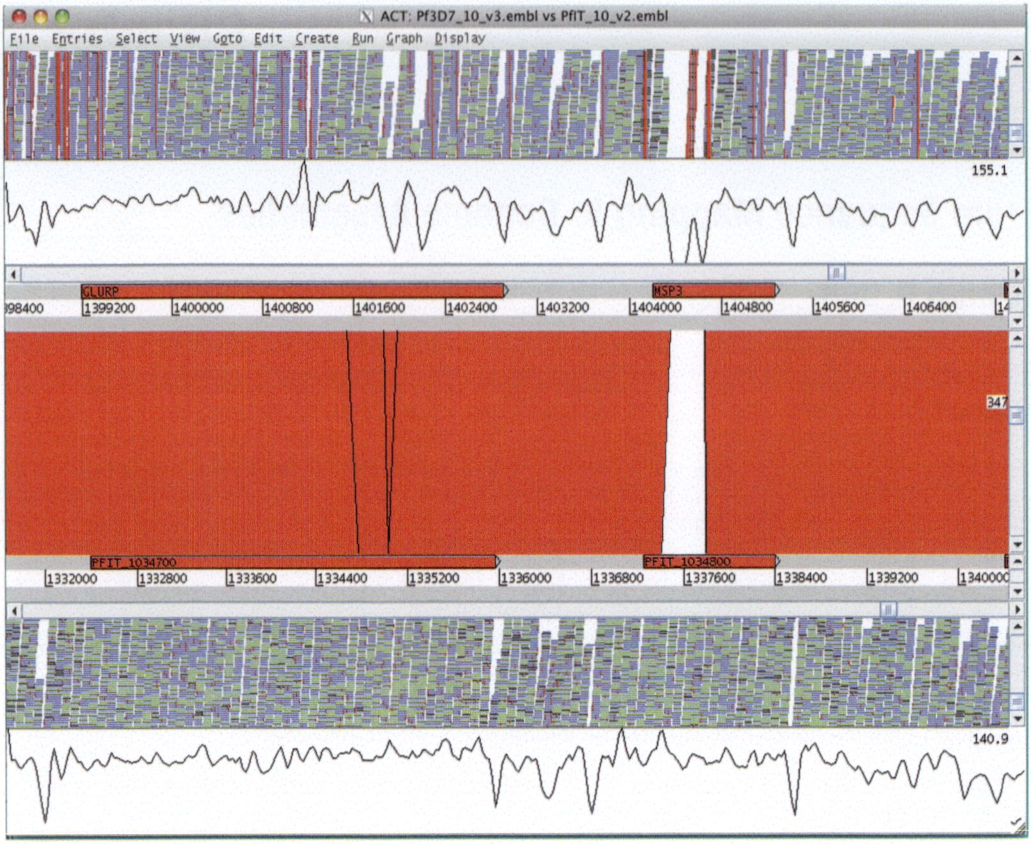

Fig. 1 Mapping versus assembly. Two genes of *P. falciparum* 3D7 (*red boxes*) can be seen at the top. The *horizontal green* and *blue lines* are mapped sequencing reads from the IT clone. *Red points* in the reads are differences between the IT reads and the 3D7 reference. The lower part shows the *de novo* assembly of IT. The *vertical bars* are blast hits. The graphs are the coverage plots. Some regions of *MSP3* in 3D7 are not covered by mapped IT reads. The de novo assembly has an insertion, indicated by the shape of the blast hit. Reads map even over this new assembled region

The basic idea of sequence assembly can be summarized as follows. First, all the reads are compared against each other to find shared identical sequences, as is done by the programs like CAP3 [1] and Celera [2]. Next, through joining reads by their overlaps (identical sequence) the consensus sequence, usually in discrete sequences called contigs, is generated (Fig. 2a). But due to the high number of reads generated through recent sequencing technologies, the step of comparing reads to each other takes too much time to be practical. One way around this is to use a more efficient representation of read similarity. Instead of looking for overlaps, it is more efficient to index all words of a specific length (k-mers) in all reads. Then an algorithm can generate contigs by traversing a graphical representation (de Bruijn graph) of the k-mers. Many high-throughput read assemblers use this approach, like ABYSS [3], Velvet [4] or see CITATION 1 in work document. A good introduction to the de Bruijn graph is

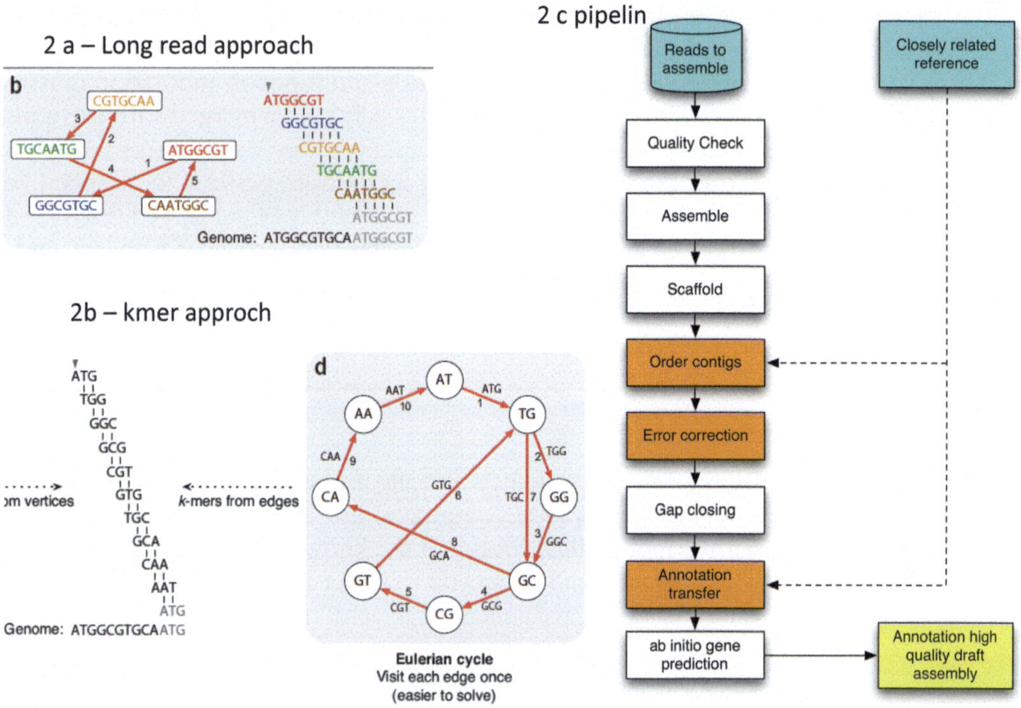

Fig. 2 (**a**) Assembly with longer reads: Nearly identical overlap between reads enable the generation of the consensus. (**b**) Assembly with short reads, using de Bruijn graph: First the reads are index and the k-mer are stored in a hash table, including the k-mer and the frequency. With a k-mer length of 3 the k-mer TCG is non unique. Due to this non unique k-mer, the graph quite complicated. (**c**) Overview of typical pipeline for de novo assembly and annotation

[5]. Unfortunately, assemblers rarely if ever generate one contig per chromosome using short reads. The causes are usually repetitive sequences, uneven or the complete lack of reads for particular genome regions [6]. For example, reads from different copies of a repeat will collapse into one contig, rather than into separate copies. To improve the contiguity of assemblies, large insert size libraries can be used to bridge the difficult regions and join contigs into scaffolds (also called supercontigs) [2, 7, 8]. The limitation is the insert size, i.e., the distance between the paired reads, which determines the size of a problematic region that can be bridged.

If a sequence is reasonably similar to a reference, scaffolds can be further joined by ordering them against the reference [9, 10]. Comparing against a reference helps to reveal where the genomes are different, such as synteny breakpoints, insertions/deletions, or differences in gene content. Other post-assembly improvements are to close sequencing gaps (in the scaffolds) [11, 12] and to correct single-base errors [13, 14]. Methods for fixing the latter are based on mapping the reads against the assembly. The distance between the two mates of a mapped read pair can also be used to identify assembly errors [3]. If gene annotation (i.e., the positions of exons and introns) is available for the reference this annotation can be transferred to the new assembly in regions where the two genomes are syntenic [15]. For regions of the new assembly without synteny, ab initio gene prediction and function annotation must be done. The resulting genome models can be merged with the transferred gene models.

In this chapter we present methods to perform local assemblies (for example of single genes), an assembly of unmapped reads (a so-called bin assembly, for example of very diverse gene families), and a complete assembly of genomes. Further, we describe methods to improve the quality of an assembly and do a first pass annotation.

2 Materials

2.1 Installation and Resources

Bioinformatics analysis, especially in the assembly process, requires not only appropriate computers, but also the right environment with many installed tools and the knowledge of how to run them. This chapter should help you to understand and apply the different tools. To do so, we generated a tarball that contains all the needed software packages (Table 1) of the work described here. This protocol is designed to work with the Linux operating system. To facilitate the application of the protocol, we generated a test data set that can be used to go through the different steps. Finally, you will need to bear in mind how much memory your computer will need to process the data. For genomes up to 5 Mb we would recommend up to 6 GB of memory. Genomes of 20, 100, and 200 Mb require up to 20, 200, and 500 GB of memory. Those numbers will vary depending on the structure of the genome, the quality of the reads, the software used, and preprocessing of the reads. For the presented example the computer would require around 2 GB of memory.

2.1.1 How to Install the Programs

The best way is to download the latest version of the programs from their web sites (Table 1) and install them. Nevertheless need to download the tar ball, see blow, as it contains some custom scripts used in this chapter. The custom scripts from the tarball, which are described in Table 2.

Alternatively it is possible to use a preinstallation, where all tools are already installed and the necessary dependencies are set. The requirement is a 64 bit Linux operating system. If this is the case, do following steps to download and install it. Switch to the bash shell and create a directory in which to install the software:

```
$ bash
$ mkdir -p ~/bin/Assembly
$ cd ~/bin/Assembly
```

Next download the file contains the software from the ftp server. This file has to be extracted in a directory and the system-wide variables have to be set:

Table 1
Description of the tools used in this chapter

Name	Description	
Read quality		
SGA [19]	String graph assembler that has functions to quality trim and correct reads	https://github.com/jts/sga
Trimmomatic [8]	Trims adapter from sequences	www.usadellab.org/cms/?page=trimmomatic
Mappers		
SMALT	Maps reads to a reference	ftp://ftp.sanger.ac.uk/pub4/resources/software/smalt/
SAMTOOLS [20]	Processes alignment files (SAM/BAM)	http://samtools.sourceforge.net/
Assemblers		
Velvet[4]	Assembler based on de Bruijn graphs	http://www.ebi.ac.uk/~zerbino/velvet/
Post-assembly genome improvement		
REAPR [3]	Assesses quality of sequences and can break assemblies	ftp://ftp.sanger.ac.uk/pub4/resources/software/reapr/
SSPACE [7]	Scaffolder	http://www.baseclear.com/landingpages/basetools-a-wide-range-of-bioinformatics-solutions/sspacev12/
ABACAS [9]	Tools to order contigs against a reference sequence	http://abacas.sourceforge.net/
IMAGE [11]	Closes sequencing gaps and extends contigs by local assembly	http://sourceforge.net/projects/image2/files/
ICORN [13]	Corrects 1–3 bp errors in sequences	http://icorn.sourceforge.net/
PAGIT [24]	Toolkit that joins ABACAS, IMAGE, ICORN, and RATT	http://www.sanger.ac.uk/resources/software/pagit/
Annotation		
RATT [15]	Transfers annotation from a reference to a query, based on synteny	http://ratt.sourceforge.net/
AUGUSTUS [23]	Gene prediction software for Eukaryotic organisms	http://augustus.gobics.de/binaries/
Glimmer [22]	Gene prediction software for bacteria	http://www.cbcb.umd.edu/software/glimmer/
Prokka [7]	Software to annotate bacterial genomes	http://vicbioinformatics.com/

```
$ wget
ftp://ftp.sanger.ac.uk/pub/resources/software/
pagit/ParasiteProtocols.tgz
$ tar xzf ParasiteProtocols.tgz
$ ./installme.sh
```

Table 2
Custom scripts from the tarball

Scripts from the tarball	
stats	Returns assembly statistics
map.smalt.sh	Wrapper script for smalt
revcompFastq.pl	Reverse complements fastq files
sga2readpair.pl	Generates two paired fastq files from a merged fastq file (SGA correction output)
deNovoPlus.sh	Script to run the assembly and the correction step in one call
gff2gb.sh	Transforms an Artemis gff to a genbank file
augustusAnnotate.sh	Takes an Augustus gtf and annotates it with the first blast hit as embl file
annotation.MergeAnnotationSecondAway.pl	Joins gene models as embl, and excludes a model from the second set, if it overlaps with a model from the first set
excludeGeneEMBL.pl	Deletes gene models in an EMBL file from a given list.
annotation.giveIDCDS.pl	Generates automatically geneIDs
AllCommands.sh	All the commands used in this chapter, adapted to the latest version of the software and correct for possible errors
RemoveSequencesSmaller.pl	Removes fasta entries that are smaller than a given parameter
BAM2consensus_reads.pl	Script that takes mapped reads and returns reads with consensus sequence

Each time you want to run one of the programs, do following step:

```
$ source ~/bin/Assembly/sourceme.sh
```

Alternatively, include the last command at the end of the ~/.bashrc system file.

2.1.2 Software

There are several software components installed in the package, which are summarized in Table 1. They are ordered by groups: read processing tools, assemblers, scaffolders, post-assembly improvement tools, annotation tools, and custom Perl scripts that will be needed in this protocol, Table 2. All the tools will be discussed in detail through this chapter.

2.1.3 Test Dataset

To help the user to understand the protocol, we included a test dataset for each section. The data are from a re-sequencing project, concerning the Malaria parasite *P. falciparum*. Here we only consider chromosome 10 of the IT clone. The complete genome can be found on Gene DB [16]. As reference we will be using the 3D7 clone.

To work with the data change to the directory:

```
$ cd $ASSEMBLY_HOME/testdata
$ ls
```

The test reads are called reads_1.fastq, reads_2.fastq, reads_3k_1.fastq, and reads_3k_2.fastq. The reference chromosome 10 from 3D7 is called ref.fa. There are four scripts: Mapping.sh, LocalAssembly.sh, BinAssembly.sh, and deNovoAssembly.sh. Type:

```
$ cat *.sh
```

to see all the commands

2.2 Sequencing Technology

Several sequencing technologies are available to date but it is not our aim to discuss them here [4]. For a successful assembly the reads should ideally have the following properties:

1. Be fairly uniformly distributed across the genome sequence.
2. Have enough coverage of the genome, i.e., 80× coverage with 100 bp reads. Longer reads will reduce the coverage need.
3. Read pair information seems to be vital to make a good assembly. The fragment size should be around 500 bp.
4. Large insert size libraries will help to scaffold more complex and larger genomes. Those libraries are called from time to time also mate pairs.

This protocol should be applicable to sequences of the length of 76–250 bp in sufficient depth, as provided by SOLID, Illumina, or Ion torrent. For scaffolding the large insert size libraries (8/20 kb) of the 454 technology are also helpful.

The importance of uniformity of the read distribution is often underestimated. This means that the amount obtained from each region of the genome should be similar. But due to PCR amplification steps, extreme GC content is amplified differently. This can result in uneven coverage that hinders the performance of assemblers. The following publication might be useful for further details [17]. In our experience, good DNA quality and good library preparation are the most crucial steps for a good de novo assembly.

3 Methods

Here we describe the methods of the sections: read preprocessing, mapping, local assembly, de novo assembly, and annotation. For most of the step we provide further information in Subheading 4.

3.1 Read Preprocessing

Before the reads can be used for assembly, sequencing adapters have to be trimmed. Bad quality regions of reads could also be trimmed. When not enough coverage is available (<80×), it is advisable to correct the reads. However, we would recommend trying the assembly without read correction first. If the assembly fails due to runtime and memory requirements, you should use the read correction.

1. To clip the reads for adapter you can use a program like trimmomatic [8]. Assume that your reads are called reads_1.fastq

   ```
   $ java -Xmx1000m -jar $PAGIT_HOME/trimmomatic-0.32.jar PE reads_1.fastq reads_2.fastq trimmed_1.fastq trimmo_unpaired_1.fq trimmed_2.fastq trimmo_unpaired_2.fq ILLUMINACLIP:adapters.fasta:2:10:7:1:MINLEN:50
   ```

   Though the call looks a bit long, the reads with the trimmed adapters are in the files trimmed_1.fastq trimmed_2.fastq. The file "adapters.fasta" contains the adapters used in the sequencing process.

2. To cut low-quality ends of reads, it is possible to use the program "preprocess" from the assembler SGA [19].

   ```
   $ sga preprocess -m 51--ermute-ambiguous -f 3 -q 3 -p 1 reads_1.fastq reads_2.fastq>reads_trimmed.fastq
   ```

3. To correct reads from sequencing errors, SGA also has a function. But first the reads have to be indexed.

   ```
   $ sga index reads_trimmed.fastq
   $ sga correct -k 51 -x 5 -o
   reads_corrected.fastq reads_trimmed.fastq
   ```

   Here, a k-mer of 41 bp in the reads (-k) that occurs less than five times (-x), will be corrected to a k-mer that occurs with the expected frequency.

4. The output of SGA will be one merged file, where reads might have been discarded due to general bad quality. To generate again forward and reverse reads (or read one and two), use following command:

   ```
   $ sga2readpair.pl reads_corrected.fastq reads_corr
   ```

3.2 Mapping the Reads

A good first analysis step is to map the reads against a closely related reference (i.e., 90 % nucleotide identity), if one exists. Here we are going to use the mapper SMALT. The final output of this mapping process will be a BAM file, which contains all the reads, including their sequence and quality, as well as mapping information (*see* Fig. 3). Assuming your reference is called ref.fa and your reads reads_1.fastq and reads_2.fastq, the following steps need to be done:

1. For pragmatic reasons, it might be good to link the name of your files to the names used here:

   ```
   $ ln -s YOUR_REFERENCE ref.fa
   $ ln -s YOUR_READS_1.fastq reads_1.fastq
   $ ln -s YOUR_READS_2.fastq reads_2.fastq
   ```

2. To see the differences between the fasta and the fastq format (Fig. 3) use the Linux command "head":

   ```
   $ head ref.fa reads_1.fastq
   ```

3. First your reference has to be indexed:

   ```
   $ smalt index -k 17 -s 3 ref.fa ref.fa
   ```

   Here, all words (so-called k-mer) of 13 bp in the genome are counted and stored in an efficient way through hashing. A short summary of the number of k-mers and the runtime is given. The output will be two files: ref.fa.sma and ref.fa.smi.

4. Comparing those k-mers between the reference and the reads, the mapper is able to decide where in the genome each read could be placed.

   ```
   $ smalt map -i 1000 -m 50 -r 0 -f samsoft -o
   Mapped.sam ref.fa reads_1.fastq reads_2.fastq
   ```

a

```
>Pf3D7_10_v3
taaaccctgaaccctaaaccctgaaccctaaaccactaaccctaaaccctgaaccctgaa
ccctgaaccctaaaccctaaacctaaaccctgaaccctaaaacctaaaccctgaaccctg
aaccctaaaccctaaaccctaaaccctaaaccctgaaccctaaaccctaaaccctgaacc
ctaaaccctgaaccctgaaccctaaaccctgaaccctgaaccctaaaccctgaaccacta
```

b

```
@IL39_6014:8:110:3699:4595#3/1
TATTTGAACTGACAATTTTATAAGATCCATATATATGAAGATCTCAAAAAAATATATGTTTTTTTGAAAATTTTCA
+
CCB@@*CCC<GCHHEHGGGGGEEGDDGGGGGEGGHEGCEGGFFCHGDBCB?BH?GGBH=BEBB@@E=B>=EBECGA
@IL39_6014:8:88:4857:8768#5/1
TATTTAAACTGACAATTTTATAAGATCCATATATATGAAGATCTCAAAAAAATATATGTTTTTTTGAAAATTTTCA
```

c

```
IL39_6014:8:110:3699:4595#3    83          PfIT_10_v2             1404405    60          76M
          =            1404336    -145
          TGAAAATTTTCAAAAAAACATATATTTTTTTGAGATCTTCATATATATGGATCTTATAAAATTGTCAGTTCAAATA
          AGCEBE=>B=E@@BBEB=HBGG?HB?BCBDGHCFFGGECGEHGGEGGGGGDDGEEGGGGGHEHHCG<CCC*@@BCC
          AS:i:73
```

Fig. 3 Examples of different file formats. (**a**) fasta: Each sequence starts with a ">" and a name. Then the sequence is followed. (**b**) fastq: Similar to fasta, but with the quality coded in ASCII. (**c**) SAM format: First column is the name of the read. Next column is the mapping flag that can be used for querying a BAM file. Third and fourth, seven and eight columns are mapped to the reads and its mate, respectively. Column nine is the fragment size. The information how well the reads map is in column five and six, mapping quality and cigar string, respectively. The sequence and the quality of the reads are stored in column ten and eleven. The last column can have many different information, like an alignment score, other possible position to map repetitively. This depends on the mapper

As parameters you can set the maximum expected fragment size for a read pair to be properly paired (-i), place reads repetitively (-r), and exclude reads that map with a lower Smith-Waterman alignment score than 50 (-s). The reads are stored in the file Mapped.sam in SAM format [20] (-o -f samsoft). An example of the SAM format can be seen in Fig. 3. It is a well-defined format, including for each read how and where it is mapped (column 2–6), where its mate is mapped, the sequence of the read and its quality and finally some tags.

5. If you want to map a large insert library (more than 1 kb), first the reads have to be reverse complemented.

```
$ revcompFastq.pl reads_3k_1.fastq rev.reads_3k_1.fastq
```

```
$ revcompFastq.pl reads_3k_2.fastq rev.reads_3k_2.fastq
```

For the mapping the settings for the fragment size have to be adapted. With a library of 3 kbp, a limit of 5 kbp should be set.

```
$ smalt map -i 5000 -j 1000 -m 50 -r 0 -f samsoft -o Mapped_3K.sam ref.fa rev.reads_3k_1.fastq rev.reads_3k_2.fastq
```

The newly introduced -j parameter limits the minimal distance for mates. If you have more libraries, repeats this step.

6. Next, we will transform the SAM file into a binary version, called BAM. This will enable us to do more analysis, and save disc space, as long as you delete the SAM file after the transformation:

```
$ samtools view -b Mapped.sam -t ref.fa.fai | samtools sort - Mapped
```

```
$ samtools index Mapped.bam
```

```
$ rm Mapped.sam
```

For mapping of the large insert, just adapt the commands by changing the file name Mapped to Mapped_3K.

7. If you would like to visualize the mapping you could use Artemis BAMview [21],

```
$ art -Dbam=Mapped.bam ref.fa
```

3.3 Local Assemblies

Although mapping is a powerful method, there are limitations: Some regions in the reference might be too polymorphic for reads to be mapped. Nor can larger insertions be detected. In this section we will first present steps showing you how to analyze those polymorphic regions by reassembling reads that map around it. Then we show how to assess larger insertions or new DNA elements through the assembly of non-mapped reads, the so-called *bin assembly*.

1. To reassemble a specific region we will need to gather the reads of this region (or at the border of it) and save them in the SAM format. We use samtools for this:

```
$ samtools view Mapped.bam Chr:From-To | sort>Region1.sam
```

   "Chr" is the name of the replicon and "From"-"To" the position of the target region. For this example use the Pf3D7_10_v3:1404400-1405500.

2. It is always good to have a look at the extracted reads to check if they come from the correct region, see column three and four:

```
$ head Region1.sam
```

3. Those reads can now be assembled.

```
$ velveth Assembly.55 55 -sam -short Region1.sam
$ velvetg Assembly.55 -exp_cov auto
```

   The first step generates the so-called de Bruijn graph. The next step is to generate the contigs from it. The parameters specify the input format (-sam), short reads (-short), and the expected median k-mer coverage (-exp_cov auto) here determined automatically.

4. Both programs generate a lot of output: # of reads, # k-mers, average coverage etc. To obtain statistics of the assembly, look at the last line: The number of nodes indicates the number of contigs, so the pieces obtained from the assembly. The *n50* is a continuity metric, *max* is the length of the largest contig, total the size of the assembly, and the last two numbers are the amount of reads used in the graph versus the total amount. Another way to look at the same statistics is the program stats:

```
$ stats Assembly.55/contigs.fa
```

5. As explained in Subheading 4.3, **step 2** the k-mer has the strongest impact on the assembly. It is good to iterate through different k-mer values in an automated fashion to optimize the assembly:

```
$ for ((kmer=31;$kmer<=73;kmer+=6)) ; do
velveth Assembly.$kmer $kmer -sam -short Region1.sam>out.velh.$kmer.txt;
velvetg Assembly.$kmer -exp_cov auto> out.velg.$kmer.txt
done
```

   This time each assembly output is written to a different file, through the ">" command.

6. To analyze the different assemblies, we "grep" the line that starts with "Final" in all the output file of velvetg and different k-mers:

```
$ grep "^Final" out.velg.*.txt
```

Which assembly is the best? For local assemblies you would expect one contig that represents the targeted region.

7. The result of the assembler is the fasta file Assembly.55/contigs.fa (or another k-mer depending on your genome). One way to analyze it would be to load it into Artemis or blast it against a public database. But in some cases the local assembly didn't return one contigs, but several. Our approach has two caveats: Some reads are too divergent to map, or an insertion occurred and we are not using the mate pairs. The following command will pull in the read's mate, even if it doesn't map:

```
$ samtools view Mapped.bam | awk '($3=="Chr" && $4>=From && $4<=To) || ($7=="Chr" && $8>=From && $8<=To)' | sort>Region2.sam
```

If you are following the example use Pf3D7_10_v3, 1394400, and 1400000 for the parameters "Chr," "From," and "To," respectively.

8. To assemble paird reads, just adapt the Velvet call as follow:

```
$ velveth AssemblyRP.55 55 -sam -shortPaired Region2.sam
```

```
$ velvetg AssemblyRP.55 -exp_cov auto -ins_length 400 -ins_length_sd 30 -min_pair_count 15
```

The changes tell Velvet that the input file contains mate pairs and that their fragment size is 400 with a standard deviation of 30 % of the library. "-min_pair_count" is the number of read pairs needed to join two contigs into a scaffold. n.b. Here the fragment size is the median, rather than the maximal fragment size, as is the case with SMALT.

9. In case of large insert libraries, repeat **step 8** to gather those reads:

```
$ samtools view Mapped_3K.bam | awk '($3=="Chr" && $4>=From && $4<=To) || ($7=="Chr" && $8>=From && $8<=To)' | sort>Region2_3K.sam
```

The following parameters are added to the Velvet commands to include a second library:

```
$ velveth AssemblyRP_3K.55 55 -sam -shortPaired Region2.sam -sam -shortPaired2 Region2_3K.sam
```

```
$ velvetg AssemblyRP_3K.55 -exp_cov auto -ins_length 400 -ins_length_sd 30 -ins_length2 3000 -ins_length2_sd 30
```

Again, have a look at the statistics. The number of bases in the assembly should have increased significantly.

10. The next step would be to iterate again through the k-mers as shown in **step 6**. In this part we showed you how to assemble a specific region of the genome. We would encourage the

reader to apply those commands to the example of the *MSP3* and S-antigen gene from the exercise, *see* Subheading 2.1.1. This procedure is appropriate if reads map to the reference but have many differences. Next we show how to get hold of the sequences that are completely different to the reference, like plasmids or very divergent multigene families.

11. The following command returns all the reads that don't map as proper pairs:

```
$ samtools view -F 2 Mapped.bam | head
```

But as discussed before, we would like to get just the mate pairs that don't map. The flag 4 is set if the read is not mapping and the flag 8 is set if the mate is not mapping. Adding the flags will return when both don't map:

```
$ samtools view -f 12 Mapped.bam | sort>
NotMapped.sam
```

12. Now we can assemble the reads as before. You might want to run the last call also for the large insert library. Here is the assembler call for one library:

```
$ velveth Bin.55 55 -sam -shortPaired
NotMapped.sam

$ velvetg Bin.55 -exp_cov auto -ins_length 400
-min_contig_lgth 300 -cov_cutoff 5
```

Two new parameters are introduced. First we want to ignore contigs smaller than 300 bp (min_contig_lgth). Next, regions (or nodes in the de Bruijn graph) that have a coverage of less than 5 k-mer are ignored from the assembly (cov_cutoff). This will minimize the possibility of false joins. If the read coverage is even you can set this option to (auto). The value will be set to half of the "exp_cov" parameter. We chose the name "bin" as many users tend to forget about the non-mapped reads.

13. Depending on the size of the organism whose genome you are assembling, the number of contigs might be significantly higher than in our little example. Now it is even more important to use different k-mers. The call would look like:

```
$ for ((kmer=31;$kmer<=73;kmer+=6)) ; do

velveth Bin.$kmer $kmer -sam -shortPaired
NotMapped.sam>out.velh.$kmer.txt;

$ velvetg Bin.$kmer -exp_cov auto -ins_length 400
-min_contig_lgth 300 -cov_cutoff 5>out.velg.
$kmer.txt

done
```

14. To look at the results you could again use the grep call, or use the little stats script:

```
$ stats Bin.*/contigs.fa
```

It is important to keep in mind, that those statistics do not tell you, how good (in terms of errors) the assembly really is. In the next section we are going to introduce a tool called REAPR that can evaluate the quality of the assembly, and return corrected assembly statistics.

15. Now include the mate pair library. Redo the **step 11** with the Mapped_3K file and run the assembly like in **step 9**. Use the stats command to see the impact of the library, especially the N_count.
16. At this step you might have generated larger sequences (several kbp) of the target region. In the example those will be subtelomeric regions with the genes of different gene families. To look at it, you could for example load it into Artemis (file Bin.55/contigs.fa) and detect open reading frames (ORF), *see* Subheading 4. Alternatively you run ab initio gene finding tools like Glimmer [22] or Augustus [23], *see* Subheading 3.5.

3.4 Whole Genome Assembly

Although the bin and local assemblies are powerful ways to get results quickly, in many cases a complete de novo assembly is necessary. Reasons are that no reference is available or is too divergent, or the mapping and SNP calls aren't accurate enough. Also, it is more difficult to combine a local assembly with the reference into a contiguous sequence than to do a de novo assembly. We assume that the reader has understood the earlier steps, as this section builds on them.

In the sections before, we performed de novo assemblies on a limited read set. Now we are going to use all the reads. The assembly call won't be very different, but we are going to do improvement of the assembly. Let's again assume that your short insert reads are called reads_1.fastq and reads_2.fastq and the reads of the large insert library are reverse complemented and called rev.reads_3k_1.fastq rev.reads_3k_2.fastq, *see* Subheading 3.2, **step 5**.

1. To run the assembly is straight forward:

```
$ for ((kmer=31;$kmer<=73;kmer+=6)) ; do

velveth deNovo.$kmer $kmer -fastq -shortPaired -separate reads_1.fastq reads_2.fastq -fastq -shortPaired2 -separate rev.reads_3k_1.fastq rev.reads_3k_2.fastq>out.velh.$kmer.txt;

velvetg deNovo.$kmer -exp_cov auto -cov_cutoff 5 -ins_length 400 -min_contig_lgth 300 -ins_length2 3000 -ins_length2_sd 30>out.velg $kmer.txt;

done
```

2. As before you can now look into the assembly with the stats script, or grep the line in the output files, *see* Subheading 3.3, **step 4**.
3. You might not necessarily want to optimize the assembly based on the n50. One example would be to increase the number of

genes of a specific gene family, which is very repetitive. So instead of looking at a large n50 you want to increase the numbers of genes. In general, a higher k-mer will better separate the different copies. Also the modification of the "-max_divergence" parameter might help.

4. The next step is to check the quality of the assembly. Just because the assembly has good statistics, doesn't mean it is a good one with no error.

   REAPR [3] is a tool that can find errors in assembled sequences by remapping the reads. First, the reads, ideally from a large insert library, have to be mapped against the assembly. This can be done with the commands of Subheading 3.2, or with this little program from the tarball:

   ```
   $ map.smalt.sh deNovo.55/contigs.fa rev.reads_3k_1.fastq rev.reads_3k_2.fastq MappedNovo55 5000
   ```

   This will generate the BAM "MappedNovo55.bam" of the mapped mate pairs on the chosen assembly. As parameter the script uses "-x," "-r 0," and "-y 0.8."

5. Now we can run REAPR:

   ```
   $ reapr pipeline deNovo.55/contigs.fa MappedNovo55.bam Reapr.55
   ```

   Different metrics are going to be applied to decide which bases are correct and which are wrong; scaffolds will be broken where there are errors. The important outputs are in the report file with the new statistics of the assembly (05.summary.report.txt) and the new assembly file 04.break.broken_assembly.fa.

6. In choosing the best assembly it is better to compare the corrected n50s rather than those given by the assembler. For each assembly the mapping and REAPR would need to be run.

7. Once we choose the best assembly, we are going to do another round of scaffolding using the program SSPACE. Though assemblers themselves have a scaffolding step, other scaffolding might improve the assembly. First we are going to iterate through different settings of the short library, and then the mate pair library. To prepare the call type:

   ```
   $ echo "LIB1 reads_1.fastq reads_2.fastq 400 0.3 FR">lib1
   ```

8. The resulting file will provide SSPACE with the fragment size (400 bp), the standard deviation (30 %), and the read orientation FR (Forward/Reverse). If your library has a different fragment size, adapt the command in **step 7**. Now run SSPACE:

   ```
   $ SSPACE_Basic_v2.0.pl -l lib1 -s Reapr.55/04.break.broken_assembly.fa -k 200 -n 31 -b out.200
   ```

The parameters indicate the nature of the reads (-l lib1), the input file, the result from REAPR (-s), the number of mates needed to join two contigs/supercontigs (-t) to a supercontig, how many bases must overlap to merge two contigs rather than scaffolding them and -b, the output. The file out.200.summaryfile.txt gives a summary of the mapping and scaffolding and out.200.final.scaffolds.fasta holds the current assembly.

9. In the step before, we used 200 mates to join two contigs. This might sound a lot, but we are looking at fragment coverage, rather than read coverage. Also, the way SSPACE works, the best results are obtained by first making the most high scoring joins and then running SSPACE again with a decreasing k parameter.

```
$ SSPACE_Basic_v2.0.pl -l lib1 -s out.200.
final.scaffolds.fasta -k 100 -n 31 -b out.100
```

```
$ SSPACE_Basic_v2.0.pl -l lib1 -s out.100.
final.scaffolds.fasta -k 50 -n 31 -b out.50
```

```
$ SSPACE_Basic_v2.0.pl -l lib1 -s out.50.
final.scaffolds.fasta -k 10 -n 31 -b out.10
```

Looking at the statistics of the scaffolding results you should see a clear decrease in the number of contigs/scaffolds.

```
$ stats out*.final.scaffolds.fasta
```

We would encourage the reader to try to scaffold the output directly with 10 read pairs for the -k parameter to compare the effect on their assembly.

10. Now we are going to scaffold with the mate pair library. One important point must be made. Small contigs of less than 500 bp, which belong between two larger contigs, might not be included in the scaffold, as the number of large-insert reads between the large contigs is higher than between them and the smaller contig. To our knowledge no scaffolder solves this problem in a satisfying manner. Therefore we normally exclude contigs smaller than 500 bp. The hope is that in the later stages we can regenerate the sequence by doing gapclosing, **step 18**. Leaving the contigs in would make this more difficult. The size limitation can be added in the velvetg step or with the following PERL script:

```
$ RemoveSequencesSmaller.pl out.10.final.
scaffolds.fasta 500>SSPACE.1.fasta
```

11. Here the command to prepare and run SSPACE on the mate pair library:

```
$ echo "LIB2 reads_3k_1.fastq reads_3k_2.
fastq 3000 0.3 RF">lib2
```

Note, you don't have to use the reverse complemented reads; you can set the direction to RF rather than FR.

```
$ SSPACE_Basic_v2.0.pl -l lib2 -s SSPACE.1.fasta -k 500 -n 31 -b out2.500
```

Next rerun the command, decrease k, as shown before for the short insert library.

12. Compare the number of scaffolded contigs between the use of short and large insert size libraries. Generally, large insert libraries have a strong impact on the contiguity of the sequence. They enable bridging of repetitive regions. This is very valuable for parasites which may have repetitive subtelomeric sequences and to improve comparison between different isolates for structural variation.
13. Remember that the assembler already did some scaffolding. It might be advantageous to tell velveth not to use the large insert size library for scaffolding. Scaffolder can use the complete length of the reads to place reads (rather than just the k-mer) and they can deal with PCR duplicates. The call would look as follows for a k-mer of 55:

```
$ velveth deNovoSE.55 55 -fastq -shortPaired -separate reads_1.fastq reads_2.fastq -fastq -short -separate reads_3k_1.fastq rev.reads_3k_2.fastq
```

```
$ velvetg deNovoSE.55 -exp_cov auto -cov_cutoff 5 -ins_length 400 -min_contig_lgth 300
```

14. In some projects a mix of different sequencing technologies are used. The scaffolding step might be the best step at which to combine the different technologies. Assuming you have a BAM file of the mapped reads, do:

```
$ samtools view -F12 Mapped454.bam | awk '$7!="="' | sort | BAM2consensus_reads.pl Assembly.fa Reads_Scaff
```

15. We are again using awk, sort, PERL, and pipes. As parameter for the PERL program BAM2consensus.pl you have to provide the assembly sequence (Assembly.fa) and the result prefix for the new read files (Reads_Scaff).
16. As mentioned before most of the errors are introduced to the assembly in the scaffolding step. Therefore we recommend that you rerun REAPR. Caution must be taken for the fact that SSPACE renames the scaffolds, including a pipe symbol. The following function of REAPR can be used to rename them:

```
$ reapr facheck out2.10.final.scaffolds.fasta ForReapr.fa
```

Now we have long scaffolds with sequencing gaps and some base errors. In the next steps we are going to try to close sequencing gaps with IMAGE ([11]) and correct base errors using ICORN ([13]). If you have a closely related reference you can order your scaffolds against it and transfer the annotation, using the tools ABACAS and RATT. These programs are part of the PAGIT pipeline [24]. PAGIT has an automated way to invoke the tools; however here we are presenting the specific program calls. For more in-depth information we recommend to read the PAGIT protocol paper.

17. If you don't have a reference sequence available, do

```
$ PAGIT.noRef.sh ResultReapr.fa read_1.fastq
reads_2.fastq 500 Final.fa
```

    To call the script successfully give it your current assembly, the reads, the insert size, and the final result name.

18. The PAGIT script will first run nine iterations of gapclosing and contig extension, decreasing the k-mer length every three iterations from 71, to 55 and then 41. The calls look like:

```
$ image.pl -scaffolds ResultReapr.fa -prefix
reads -iteration 1 -all_iteration 3 -dir_
prefix ite -kmer 71 -smalt_minScore 60
$ restartIMAGE.pl ite3 55 3 partitioned
$ restartIMAGE.pl ite6 41 3 partitioned
```

    Local assemblies are done for each sequencing gap and at the ends of contigs, by including the mate pairs that don't map, as done in Subheading 3.3, **step 7**. The k-mer for the assembly can be changed as can the minimal score for a read to be placed with SMALT (smalt_minScore - the "-s" parameter in Subheading 3.2, **step 4**). Again, we encourage the reader to change the parameters and analyze the impact. In the end, the contigs are joined and placed in the file Res.image.fasta by

```
$ contigs2scaffolds.pl ite9/new.fa
ite9/new.read.placed 300 10 Res.image
```

19. After the IMAGE step, 50–80 % of the sequencing gaps should be closed and many scaffold ends extended. Next, we are going to apply the tool ICORN to correct base errors. Compared to REAPR it looks for 1–3 bp errors and corrects them. REAPR scores single bases rather than correcting them. Reads are mapped with SMALT and differences between reads and the reference are found and corrected.

```
$ icorn2.start.sh Reads 500 Res.image.fa 1 3
```

    As before, the reads, fragment size, and the input reference from IMAGE are passed to the script. The last two parameters are the iteration start and stop. The output will be a summary file (Res.image.fasta.summary.txt) and the corrected sequence file (Res.image.fasta.4). If run through the PAGIT pipeline the final result will be called Final.fa. The next step for the analysis would be to start with the ab initio annotation of genes, *see* Subheading 3.5.

20. In cases where a reference sequence is available, the PAGIT pipeline has a tool to order the contigs against the reference, and to transfer the annotation. This following call will also invoke the gap closing (IMAGE) and correction (iCORN) steps:

```
$ PAGIT.sh ResultReapr.fa read_1.fastq reads_
2.fastq 500 Final ref.fa AnnotationDIR
```

    The call is very similar to the above one, with the exception that the reference and a directory with the annotation of the reference are given.

21. In the first step, the scaffolds will be ordered against a reference. A certain caution must be taken however. If it is known that the species under study has many synteny breaks relative to the reference, the resulting order might not be correct. Furthermore, we recommend deleting regions (or replace them with n's—the symbol for an ambiguous base) where there is evolutionary pressure, as in virulence factors or the subtelomeres of species such as *Plasmodium* or *Trypanosomes*. To put it another way, you would like the scaffolds to be ordered only against the well-conserved parts of the reference.

22. As discussed, the PAGIT pipeline will order the contigs with the following command:

```
$ abacas.pl -r ref.fa -q ResultReapr.fa -p
nucmer -d
```

    If you work with more divergent species, you might want to change *nucmer* to *promer* in the PAGIT.sh script. Instead of using nucleotide similarity, an amino acid comparison will be done. The result will be a multifasta file. Scaffolds ordered against a reference chromosome will be joined with n's and are now named after the reference replicon.

23. After this step, IMAGE and ICORN will be run again (**steps 16** and **17**).

24. Now it is possible to transfer the annotation onto the improved assembly with RATT. The call used by the PAGIT script is:

```
$ start.ratt.sh AnnotationDIR ForRatt.fa
Transfer Species
```

    Similar to ABACAS, the last parameter determines the similarity. Due to the nature of the program it won't work with amino acid comparisons. The parameter *Species* is the most robust. If your reference is very similar, and the assembly is contiguous, in pieces larger than 10 kb, we would recommend the value "Assembly" or "Strain" for this parameter. The value "Transfer" is a prefix for the result files. "AnnotationDIR" is the position of the reference annotation in embl format. Note that you might need to adapt the configuration file of RATT, with a simple editor. Here is the position of the file:

```
$ echo $RATT_CONFIG
```

This configuration file enables you to set the start codons, splice sites and if pseudo genes should also be corrected.

25. The result of RATT is one annotation file for each replicon, starting with Transfer.*.final.embl (ordered and unordered scaffolds). You can open them in Artemis. To compare these with the reference, use ACT. ACT can be seen as two Artemis view joined by a similarity comparison, *see* Fig 1. First we will generate this comparison file for a single chromosome by extracting the chromosome sequence from the multifasta file, preparing it and blasting it.

```
$ samtools faidx ref.fa RefChr>RefChr.fa
$ formatdb -p F -i RefChr.fa
$ blastall -p blastn -m 8 -e 1e-6 -d RefChr.fa
-i Sequences/deNovoSuper -o comp.RefChr.blast
```

where RefChr.fa is the name of the reference chromosome. Without going into too much detail, the reference chromosome is being compared to scaffolds or ordered scaffolds, from the PAGIT pipeline. These sequences are in the folder "Sequences." To start ACT, use the reference file, which should be in the folder "embl," the comparison file you just generated. The result file from RATT is Transfer.deNovoSuper.final.embl.

```
$ act AnnotationDIR/RefChr.embl comp.RefChr.
blast Transfer.deNovoSuper.final.embl
```

26. In ACT, it is possible to see insertions, deletions, and rearrangements in the comparison window. To evaluate the RATT transfer, load onto the reference chromosome the file Transfer.RefChr.NOTTransfered.embl (click on File->2nd option ->Read an Entry). This file will show the gene models that weren't transferred. Lastly, load the GFF file from the "Query" folder, and look for the synteny tag. These regions have no synteny to the reference, and are probably insertions. Furthermore, those will need to be annotated separately. Figure 1 is an example of an ACT view.
27. Using ACT, it would be possible to look for Open Reading Frames (ORFs), not overlapping RATT-transferred annotation and follow the description of Subheading 4.3, **step 10**.
28. Although at this stage we have an improved, annotated genome, one quality check remains to be done. We recommend doing a "bin" assembly as described in Subheading 3.3. During the process, we deleted small contigs, which might contain important sequences. Also, some assemblers exclude reads with an extreme k-mer coverage, for example those derived from mitochondrial or plasmid DNA. Furthermore,

maybe something went wrong in the process—there could have been a truncated file. If the bin assembly contains interesting sequences, join it with the current, improved assembly:

```
$ cat ForRatt.fasta bin.55/contigs.fa>
Joined.Assembly.fasta
```

29. Each program generates temporary files, which use a lot of space. For example, IMAGE generates ite*/ directories, SSPACE creates several mapping directory or ICORN does ICORN2_*/ directories. Those should be deleted on a regular base with the rm command, i.e:

```
$ rm -rf ICORN2_*/ ite*/ reads/ bowtieouput/
```

30. Some of you may have noticed that it would be possible to first run IMAGE to extend contigs, then run the scaffolder. Or, perhaps it would be good to scaffold the bin contigs into the improved assembly. Both these points are true and we hope to have given the reader the impression that each process can be seen as a module. The order can be changed, and processes iterated. The aim of this protocol is to show the reader the possibilities of generating assemblies and improving them.

3.5 Annotation

In the previous section we showed how to improve the quality of a genome sequence and how to transfer annotation from a reference onto the assembly. But the genome is far from being well annotated: Where its sequence is not similar to the reference, like in multigene families or insertions, the annotation would be transferred poorly or not at all. To annotate those regions, an ab initio gene prediction must be done. (The same would need to be done, when no closely related reference exists.) Here we present a method of predicting gene models and a first pass functional gene annotation. In the end, we show how to merge the new annotation with the RATT transferred annotation.

Although the method presented here for gene prediction and functional annotation is valid when a closely related reference exists, we highly encourage the reader to further study the subject of gene prediction and functional annotation, as each program and method has its strength and weakness [5, 6].

1. First, the gene predictor has to be trained. For simplicity we assume that we can use the genes from the reference genome.
2. Load the annotation from the reference genome (you can use the one of the test dataset) into Artemis and save it in the GFF format (File ->Save An Entry As ->GFF Format). For simplicity, save it as ChrX.gff. Accept all the warnings that some classifiers won't be saved.
3. Next we require to know the names of each chromosome in the reference fasta file ref.fa. The command

```
$ grep '>' ref.fa
```

will do the job. Remember the name of the chromosome of which you generated the GFF file. In this example her for simplicity we assume it is called ChrX.

4. This command will transform the gff into the gb format needed for Augustus.

```
$ gff2gb.sh ChrX.gff ChrX ChrX.gb
```

In case that you have more than one chromosome, redo the steps from number 2 and concatenate the files with:

```
$ cat ChrX.gb ChrX2.gb … ChrXn>All.gb
```

5. Next, we initiate Augustus

```
$ new_species.pl -- pecies=NEW
$ etraining -- Species=NEW --
stopCodonExcludedFromCDS=false ChrX.gb
```

The first command will generate a training instance for your reference, called "NEW." This instance is than trained in with the gene models saved in Artemis second command. Read carefully the output of the programs. Gene models that seem to be wrong for Augustus will be excluded.

6. Now we can apply the trained model to the new assembly:

```
$ augustus -- Species=NEW ImprovedAssembly.
fasta>abintio.gtf
```

The output is a gtf file that contains all the predicted models.

7. Those models obviously don't have any functional annotation. The next command will attribute to each model the first BLAST hit with an E-value of at least 1e-40 of the new model against the proteome of the reference genome "ref.aa.fa" in the next command. The output will be EMBL files in the "Augustus" directory.

```
$ augustusAnnotate.sh ImprovedAssembly.fasta
abintio.gtf ref.aa.fa
```

At this point we have the annotation of the ab initio gene models. Now we present here how to merge them with the gene models of the RATT transfer.

8. The first step is to delete gene models that contain still errors in the RATT transfer. Although RATT tries to correct gene models, this step can fail. The next command will use the statistic of each gene stored in the "report" files:

```
$ cat Transfer.*txt | perl -nle '@ar=split(/\t/);
if    (($ar[8]+$ar[9]+$ar[10]+$ar[11]+$ar[12]+
$ar[13])>0){print $ar[0]}'>exclude.txt
```

This command counts the columns 8–13 that represent specific errors, such as wrong start codon, frame shifts, or incorrect splice sites.

9. The next call will exclude all the models that are flagged to be wrong from the EMBL files. (The new files will be stored in the directory RATT_excluded):

```
$ mkdir RATT_excluded
$ for x in ` ls Transfer.*final.embl ` ; do
cat $x | excludeGeneEMBL.pl exlude.txt>
RATT_excluded/$x;
done
```

10. At this stage it is possible to join the models from the RATT model with the models from the gene prediction. The rule is that if a predicted gene overlaps with a RATT transferred model, it will be deleted.

    First we generate a directory to store the files:

```
$ mkdir Joined
$ for x in `grep '>' assembly.fa | sed
's/>//g'`; do annotation.MergeAnnotation
SecondAway.pl
RATT_excluded/Transfer.$x.final.embl
Augustus/$x.embl>Joined/$x.embl;
done
```

11. Now it is possible to examine the annotation in Artemis.

```
$art Joined/ChrX.embl
```

    Further you can also add the models from the ab initio gene prediction as a separate track (Menu: File -> Read Entry, and select the gff from the "Augustus" directory. If genes are wrong, it is advisable to delete those from the RATT transfer and redo **step 10**.

12. The last step would be to give the genes systematic ids. This can be done in Artemis or with following script:

```
$ mkdir final
$ cat Joined/chrX.embl | annotation.giveIDCDS.
pl NameID_01 T>final/chrX.embl
```

    This command has to be run for each new chromosome/supercontigs, in this case "chrX.embl." The first parameter (NameID_01) would be the ID plus the chromosome number, here 01. The second parameter is optional: RATT transfers also the locus tag from the reference. If this has the prefix of the second parameter (in this case "T") the reference locus_tag will be stored in the ratt_orthologs tag. If not, it gets deleted.

4 Notes

This protocol uses a wide range of tools and commands. We assume that a novice user will probably need a week to work through the protocol when assembling a bacterial genome. It is important to have a computer with all the tools installed and enough memory, around 6 GB. When using the preinstalled tarball (Subheading 2.1, **step 1**) be sure to run it on a computer with at least 6 GB of memory. To run through the provided example will take around 1 day.

4.1 Preprocessing

It is important to keep in mind that reads can be of poor quality and that sequencing might generate insufficient depth. Although assembly will still be possible, the representation of the genome as a whole will be poor, and the range of meaningful downstream analyses will be quite small.

Another important point is the possibility of contamination. Depending on the source of material, host contamination is common. If host contamination is present, and there is a reference sequence for the host, map the reads against the host genome (Subheading 3.2, **step 3** ff) and extract those from the results file with the command in Subheading 3.2, **step 5**. This will filter most of the contamination. For the rest, once the assembly is done, blast the contigs against the reference. You can also compare the GC content of the contigs. The contaminants normally have a different GC content from that of the target genome.

1. There are many programs to trim adapters. The key thing is to have a file of adapter sequences. In our experience, it is most often mate pair libraries and bad quality runs that have many adapters.
2. Trimming should be done for very bad quality reads. But the best cutoff is tricky to set. A base with a quality value of ten still has a 90 % chance of being correct. For the assembly, one out of ten reads will be excluded and this generates noise. With enough coverage, and without doing read correction, choose a quality cut off of 20.
3. The SGA correction will first index the reads and then do the correction by looking for k-mers in reads. A read is corrected where it contains a k-mer with too low abundance. Depending on the complexity of the genome and the read coverage, this step takes between 4 h and 4 days. The biggest impact of the correction will be on the memory and runtime requirement for the assembly.

4.2 Mapping Reads

There are many tools that can be used to map reads against a reference [25]. SMALT is a tool that gives us more control when a read is mapped, using the parameters for minimum score or fragment size.

1. The indexing step can be optimized in terms of speed and sensitivity. The k-mer is the length of the identical words and the parameter *s* is the step size. With a step size of one every k-mer is taken, with a step size of two only every second k-mer is used. Low k and *s* parameters will result in many small pieces that represent the genome. More divergent reads can be mapped, as if you have an SNP every 14 bases a k-mer of 13 can be found in the reads, but not a higher k-mer. The price will be the runtime. Higher parameters like $k=20$ and $s=11$ will result in a more sparse representation. Reads with many differences to the reference might not get mapped. But the mapping time is significantly shorter. A k-mer smaller than 13 should not be used, due to the runtime and the fact that k-mers which occur very often will be ignored.
2. For the mapping step many parameters can be set ($ smalt map -H). Interesting parameters include "-y" to limit the placement of the reads by identity, "-n" to use more than one thread, or "-d" to allow multiple hits for each read. In terms of runtime, for a 5 MB genome with 100× read coverage, we estimate a mapping time of 1 h. The runtime increases linearly with the number of reads or the genome size. Normally not more than 2 GB of memory is needed.
3. To map large insert libraries, reverse complement the Illumina reads. For 454 reads, you just need to reverse complement the second read. Although the mappers have functions to reverse complement, some just recalculate the flag, rather than really reverse complementing the reads. This is generally okay, but as our analysis builds on the mapped reads, we have to have them in the correct orientation.
4. The reason for sorting and indexing the reads is a faster access to reads at specific positions. Later this will become more obvious. This step takes around 10 % of the total mapping time. To get a statistic of the mapping, do $ samtools flagstat Mapped.bam.

4.3 Local Assemblies

1. The samtools command "view" prints all the reads mapped to the specified chromosome "Chr" between the positions "From" to "To." Select a region slightly larger than the one you are interested in. For example if you chose a specific gene, extend the border by 100 bp.
2. Those two commands will do the assembly. The first command builds the graphical representation of the reads. As for the mapping, we select a specific k-mer, which will represent the nodes of the graphs. Two reads are basically joined into a contig if they share an identical k-mer. The k-mer setting will have the strongest impact on the quality of the assembly. If the

k-mer is too long, two reads that should be joined might not get joined due to sequencing errors. On the other hand, shorter k-mers are more likely to be repetitive, which is a big issue for the assembler. This could lead to many small contigs, or in some rare cases also to mis-assemblies. The second command cleans the graph and finds an Euler path through the graph. A very important parameter is the expected coverage. This can be determined automatically by Velvet. The value can be obtained manually through the "stats.txt" file, see velvet manual. Obviously, there are many assemblers that could be used at this point. Although the results will vary slightly, we think that it will be more important here to explain the general procedure than to list all the different assembler calls. In the end, they have very similar parameters and ways to be called. Both Velvet programs have many parameters. The most important will be explained below. To see them all, just type ($ velveth or velvetg).

3. The overview values of the Velvet and the stats program have the following meaning. The sum is the total number of bases in the assembly. "n" represents the number of contigs in the assembly; "mean" and "largest" relate to the contig size. The N50 is the length L such that 50 % of the assembly lies in contigs of at least length *L*. N60 is defined analogously, but for 60 % of the genome, etc. The N_count is the amount of n's contained in the assembly.

4. The for loop is a feature of bash, the Linux environment you are working in. It basically iterates over several values of the variable $kmer from 31 to 73, with a step size of 6. This command will take roughly eight times as long as a single Velvet call. For larger read sets you might want to run different Velvet calls on different computers, but this kind of optimization is not part of this chapter.

   Depending on the quality of the reads, the amount of coverage and the base composition of the genome, a certain k-mer will generate a more contiguous and larger assembly. If you have coverage over 80×, it is likely to be a larger k-mer. It is possible to iterate the k-mer with a lower step size, around the optimum. Please note that a k-mer must always be an odd integer value.

5. This command will take some minutes to run. All reads are handed (or piped) to the awk command. This one looks to see whether a read or its mate map on the specified chromosome and position (columns 3, 4 and 7, 8). These two conditions are connected with a logical OR (||), so if at least the mate or the read fulfill the condition, the mapping information of the read is piped to a sort command. The sorting is necessary, as the assembly step will require a SAM file, sorted by read name, so all reads in a pair are together. To write the output into a file,

use the ">" command. As we ensure to collect both mates, a local assembly should be able to reconstruct an insertion of nearly twice the fragment size. As a look ahead, this method to pull in non-mapping reads through mate pairs is the basic idea of gap closing software like IMAGE, part of PAGIT that will be discussed later.

6. Velvet will finally try to scaffold the contigs using the mate pairs. Basically the reads are mapped back to the graph (using the k-mer as seeds), and if a certain number (value of -min_pair_count) maps to two contigs, these will be joined into a scaffold. Ns are inserted between the contigs, the number of which depends on the expected gap size. This step is the most likely source of mis-assemblies. Therefore higher values will tend to generate more conservative assemblies, with fewer errors and more contigs. A good setting is generally to set the value to the median coverage, which is returned in the second last line of the velvetg output. To set this value, the velvetg call would therefore need to be run twice. Finally, although the output of Velvet now has supercontigs, the file will still be called "contigs.fa" The quickest way to check if the results are contigs or scaffolds is to use the stats program—if the N_count is not zero then the results are scaffolds.

7. Although including the large insert size library reduced the number of scaffolds, the effect might not be seen in local assemblies.

8. We chose to select only read pairs where neither map, to avoid chimeras entering into the read set. Although these could be filtered out later (parameter -cov_cutoff), they would slow down the process and introduce noise. To obtain these reads we query their flags (the second column of the BAM file), in this case 12, read and mate don't map. Next, each line of the SAM file is passed via the pipe command to the sort program.

9. The mate pair library should make a huge difference to the statistics of the supercontigs. Later we are going to discuss further problems with scaffolding using large insert libraries. But depending on the organism, the final result should be one scaffold per amplicon, which is rarely achieved.

10. To detect and annotate ORFs do the following in Artemis: click on Menu Create ->mark open reading frame. Choose a minimum length of 200 and enable the option to break at contig boundaries. Next, you can blast the obtained ORFs against a uniprot database: Select ->Select all CDS; Run ->Run blastp on selected feature ->Uniprot_eukaryota. Choose Uniprot_bacteria, if you work with bacteria. The blastp gets run. Once done you can see the results by selecting a CDS and pressing the keys crtl+back quote.

4.4 De Novo Assembly

1. As all the reads are going to be used, the runtime will be increased. Genomes below 5 MB will run in around 20 min, and need just 2 GB of memory. Larger genomes, up to 30 MB, will take around 1 h and the memory can increase up to 30 GB. To lower the memory requirements, work with a higher k-mer (>49) and correct the reads before running the assembler (Subheading 3.1, **step 3**). For very large parasites, Velvet might need more than 200 GB memory. In this case, the SGA assembler is a useful alternative, which normally doesn't use more than 60 GB of memory. When using large insert size libraries, a fraction of reads can point in the wrong direction, and may result in incorrect scaffolding. To avoid this, set the shortMatePaired parameter to yes.

   Some users might want to try a tool called velvetoptimser.pl. It automatically tries different settings in velvet to optimize a specific value, such as a large N50 or assemblies with many bases. Interestingly, in our experience, manually iterating through the parameters generates still better results for larger genomes.
2. To test the different assemblies for the best representation, one could blast conserved motifs against the different assemblies and assume that the one with the highest number might best represent the gene family. Obviously, one must find a balance with the other statistics, such as the N50.
3. It is a very common procedure to join several steps of processing into one script. This reduces not only the amount of time, but also the likelihood of typing errors.
4. REAPR generates many statistics. A very useful feature is the generation of the per-base quality. Every base will be scored for correctness, try $ reapr perfectmap. REAPR with only break scaffolds if the error is over a gap (n's). If an error is within a contig, the bases are replaced with Ns and the deleted sequence is written into the "bin" file (Contig errors can also be broken with reapr break -a). Plots for the different errors can be loaded into Artemis. The command $ reapr plot<chromosomeName> will generate all the plots and a script to start Artemis automatically. For genomes under 4 MB REAPR takes less than 10 min. For larger genome, the runtime and memory requirement is similar to the mapping step.
5. The application of automatically mapping reads and correcting assemblies needs a bit of scripting. Supplied in the tarball is a script called "deNovoPlus.sh." It is a very trivial script, which has as parameters the k-mer and the insert size. The following commands will be run:

```
$kmer=$1

$insersize=$2
```

```
velveth deNovo.$kmer $kmer -fastq -shortPaired
-separate reads_1.fastq reads_2.fastq velvetg
deNovo.$kmer -exp_cov auto -cov_cutoff 5 -ins_
length $insersize -min_contig_lgth 300

map.smalt.sh deNovo.$kmer/contigs.fa rev.
reads_3k_1.fastq rev.reads_3k_2.fastq Mapped
Novo$kmer 5000

reapr pipeline deNovo.$kmer/contigs.fa Mapped
Novo$kmer Reapr.$kmer
```

The script basically joins all the commands explained before, and assumes a very stringent naming convention. This is a simple example how powerful and easy scripts can be. One would call it through a for loop and use different k-mers, or start the job on different computers to save time.

6. Compared to the assemblers, a scaffolder will use the complete length of the reads to determine its position in the assembly. Also it should ignore duplicate reads, where mate pairs that came from the same DNA fragment are overrepresented due to the PCR amplification step. We use SSPACE as it is straightforward to use and was one of the first using Illumina data. As mentioned later in Subheading 3.4, **step 12**, the scaffolding of SSPACE is better than that of velvet. Therefore one could disable the scaffolding with the large insert size library.
7. At first it might sound weird to iterate through different settings, decreasing the evidence. But again, this optimizes the results, without generating more errors.
8. As for most of the tools, there are limitations. To further scaffold you would need combinatorial PCR. Alternatively you can order contigs of at least 40 kb using optical maps.
9. This step will have more impact on larger genomes, where more k-mers are repetitive, more duplicate reads are expected, and the order of contigs is not so obvious. There are also limitations of the existing scaffolders. If three copies of a repeat are joined into one contig, the scaffolder should not join the contigs. It would be better to exclude those reads from the assembly, and hope that local assemblies (*see* Subheading 3.4, **step 17**) will regenerate them, or split the collapsed repeat into three copies. This is not a problem if the repeat is smaller than the largest mate pair library.
10. SSPACE is designed to work with Illumina reads. But you may have 454 8 kb/20 kb libraries available for scaffolding. Here we present a script that returns fake reads from a BAM file, with the consensus sequence taken from the reference using the correct read length. This preprocessing step will also reduce the reads that have to be analyzed by SSPACE, by excluding reads not holding scaffolding information (as mapping for

example onto the same contig). First you have to reverse complement your reads so that they point towards each other (*see* Subheading 3.2, **step 5**). Next map them with SMALT (Subheading 3.4, **step 4**). Then you can do **step 14**. The reads can now be given to SSPACE. If you have a large genome, you can also do this to speed up the runtime of SSPACE significantly.

11. There is always a trade-off between automated pipelines and running the tools one by one. Here we provide an automated approach, while also explaining the main parameters for each tool.
12. IMAGE is a powerful tool to improve your assemblies. The price is a long runtime. In each iteration, the reads are mapped, gathered for each gap or scaffold end, and a local assemblies are done. Subsequent iterations are faster, as properly paired reads will not be remapped and regions that could not be improved in the previous iteration will not be touched. But this method is still much faster than filling gaps by PCR. Interestingly, other gapfilling tools close different types of gaps than IMAGE. For the remaining sequencing gaps, you would need to do PCR, if considered necessary.
13. ICORN is an iterative tool that takes 30 min per iteration for genomes around 4 Mb and 8–24 h for genomes around 200 GB.
14. Some scaffolds will not get ordered. These may be the most interesting sequences, as they will be the most different from the reference (if they are not contamination). It is important not to forget them!

    After running ABACAS, you should rename your ordered scaffolds. Naming is always important and mostly needs to be done in a manual matter, through an editor. For those who would like a more automated way, look into the Linux "sed" program.
15. As mentioned RATT can only transfer annotation where synteny exists. But those regions without synteny are the ones to be examined in more detail. Furthermore, it will be necessary to annotate them separately, Subheading 3.5.
16. It might be surprising, but indeed errors occur that no one expected. To do a "bin" assembly is an efficient way to double-check the assembly for errors.

4.5 Gene Prediction

Although stated before, we must iterate that the ab initio gene prediction and functional annotation we present is not the most sophisticated, but valid as a first pass annotation if a closely related reference genome exists. The reference genome will be used to train the gene finder and as a database for the functional annotation. General errors of the gene prediction are overprediction, missing exons, and wrong splice sites. In the functional annotation, it can be wrong to assume homology due to similarity to homology, especially when paralogous exists. Nevertheless, as a starting point (and merged with the RATT transfer) the method presented here will be extremely helpful.

For bacterial gene prediction we would recommend Glimmer or Prokka [7], which are easy to use.

1. In case that you don't have a closely related reference, you will need to generate 200–400 high-quality gene models. This training set should cover as many different type of genes, not just the core genes, like predicted from CEGMA [26].
2. A lot of time in bioinformatics is spent on transforming files to different formats. To ensure that this won't be a problem for this protocol, we generated the gff file through Artemis. Users that are more experienced with scripting will have their own methods, especially, when the genome has many chromosomes.
3. This script hides some ugly code. From the gff a gtf is generated, with the name of the sequence. This is then transformed to a genbank file, using an augustus script. For more advanced users, it might worth to look into the script and modify the parameters to obtain better results.
4. If the second command returns a lot of errors, like wrong models, then in the transformation step something went wrong. One solution might be to name the gene models with locus_tag in Artemis, without using any special characters in the name.
5. All the above steps run within minutes. This step might need several hours if the genome is over 30 Mb.
6. To generate a full EMBL file the program "`ratt.main.pl doEMBL`" can be used. It combines the sequence (fasta file) with the annotation (EMBL format).
7. In some case, it is desirable to exclude also specific gene families, as it is likely that the transfer will be wrong, for example missing exon, and the *ab initio* gene prediction might be better. `$ cat Transfer.*txt | grep "variant erythrocyte" | cut -f 1 >>exclude.txt` will add all genes ids that have the annotation "variant erythrocyte" as product to the list to exclude genes.
8. Depending on the expected quality of the annotation, here a lot of time can be spent. Further, several information from the reference transferred onto the new assembly might not be relevant. This information should be deleted.
9. Actually, the geneID should be obtained from a database like EBI, to agree with their submission format.

Acknowledgements

I would like to thank Adam Reid, Martin Hunt, and Bernardo Foth for proofreading the chapter.

References

1. Huang X, Madan A (1999) CAP3: a DNA sequence assembly program. Genome Res 9(9):868–877
2. Myers EW et al (2000) A whole-genome assembly of Drosophila. Science 287:2196–2204
3. Simpson JT et al (2009) ABySS: a parallel assembler for short read sequence data. Genome Res 19(6):1117–1123
4. Zerbino DR, Birney E (2008) Velvet: algorithms for de novo short read assembly using de Bruijn graphs. Genome Res 18:821–829
5. Compeau PE, Pevzner PA, Tesler G (2011) How to apply de Bruijn graphs to genome assembly. Nat Biotechnol 29(11):987–991
6. Alkan C, Sajjadian S, Eichler EE (2011) Limitations of next-generation genome sequence assembly. Nat Methods 8(1):61–65
7. Boetzer M et al (2011) Scaffolding pre-assembled contigs using SSPACE. Bioinformatics 27(4):578–579
8. Pop M, Kosack D, Salzberg S (2004) Hierarchical scaffolding with bambus. Genome Res 14:149–159
9. Assefa S et al (2009) ABACAS: algorithm-based automatic contiguation of assembled sequences. Bioinformatics 25(15): 1968–1969
10. van Hijum S et al (2005) Projector 2: contig mapping for effecient gap-closure of prokaryotic genome sequence assemblies. Nucleic Acid Res 33:560–566
11. Tsai IJ, Otto TD, Berriman M (2010) Improving draft assemblies by iterative mapping and assembly of short reads to eliminate gaps. Genome Biol 11:R41
12. Boetzer M, Pirovano W (2012) Toward almost closed genomes with GapFiller. Genome Biol 13(6):R56
13. Otto TD et al (2010) Iterative correction of reference nucleotides (iCORN) using second generation sequencing technology. Bioinformatics 26(14):1704–1707
14. Ronen R et al (2012) SEQuel: improving the accuracy of genome assemblies. Bioinformatics 28:i188–i196
15. Otto TD et al (2011) RATT: rapid annotation transfer tool. Nucleic Acids Res 39:e57
16. Logan-Klumpler FJ et al (2012) GeneDB—an annotation database for pathogens. Nucleic Acids Res 40(Database issue):D98–D108
17. Quail MA et al (2012) Optimal enzymes for amplifying sequencing libraries. Nat Methods 9:10–11
18. Simpson JT, Durbin R (2012) Efficient de novo assembly of large genomes using compressed data structures. Genome Res 22(3):549–556
19. Li H et al (2009) The sequence alignment/map format and SAMtools. Bioinformatics 25(16):2078–2079
20. Carver T et al (2012) BamView: visualizing and interpretation of next-generation sequencing read. Brief Bioinform 14:203–212
21. Delcher AL et al (1999) Improved microbial gene identification with GLIMMER. Nucleic Acids Res 27(23):4636–4641
22. Stanke M, Morgenstern B (2005) AUGUSTUS: a web server for gene prediction in eukaryotes that allows user-defined constraints. Nucleic Acids Res 22:W465–W467
23. Swain MT et al (2012) A post-assembly genome-improvement toolkit (PAGIT) to obtain annotated genomes. Nat Protoc 7(7): 1260–1284
24. Fonseca NA et al (2012) Tools for mapping high-throughput sequencing data. Bioinformatics 28:3169–3177
25. Parra G, Bradnam K, Korf I (2007) CEGMA: a pipeline to accurately annotate core genes in eukaryotic genomes. Bioinformatics 23(9): 1061–1067

Chapter 3

Sequencing and Annotation of Mitochondrial Genomes from Individual Parasitic Helminths

Aaron R. Jex, D. Timothy Littlewood, and Robin B. Gasser

Abstract

Mitochondrial (mt) genomics has significant implications in a range of fundamental areas of parasitology, including evolution, systematics, and population genetics as well as explorations of mt biochemistry, physiology, and function. Mt genomes also provide a rich source of markers to aid molecular epidemiological and ecological studies of key parasites. However, there is still a paucity of information on mt genomes for many metazoan organisms, particularly parasitic helminths, which has often related to challenges linked to sequencing from tiny amounts of material. The advent of next-generation sequencing (NGS) technologies has paved the way for low cost, high-throughput mt genomic research, but there have been obstacles, particularly in relation to post-sequencing assembly and analyses of large datasets. In this chapter, we describe protocols for the efficient amplification and sequencing of mt genomes from small portions of individual helminths, and highlight the utility of NGS platforms to expedite mt genomics. In addition, we recommend approaches for manual or semi-automated bioinformatic annotation and analyses to overcome the bioinformatic "bottleneck" to research in this area. Taken together, these approaches have demonstrated applicability to a range of parasites and provide prospects for using complete mt genomic sequence datasets for large-scale molecular systematic and epidemiological studies. In addition, these methods have broader utility and might be readily adapted to a range of other medium-sized molecular regions (i.e., 10–100 kb), including large genomic operons, and other organellar (e.g., plastid) and viral genomes.

1 Introduction

Mitochondrial (mt) genomes have long been used as markers for molecular population genetic and systematic studies [1, 2]. Despite their broad utility, the availability of information on mt genomes is severely limited for many metazoan groups, particularly small invertebrates, such as parasitic helminths. Indeed, of the ~1,200 complete mitochondrial genomes publicly accessible via GenBank (www.ncbi.nih.nlm.gov), ~70 % represent species of vertebrates. This bias of information most likely relates to the need for a relatively simple and cost-effective technique for mt genomic sequencing, particularly from minute quantities of DNA. Traditionally, research of mt genomes has relied on the isolation of mtDNA from

Christopher Peacock (ed.), *Parasite Genomics Protocols*, Methods in Molecular Biology, vol. 1201,
DOI 10.1007/978-1-4939-1438-8_3, © Springer Science+Business Media New York 2015

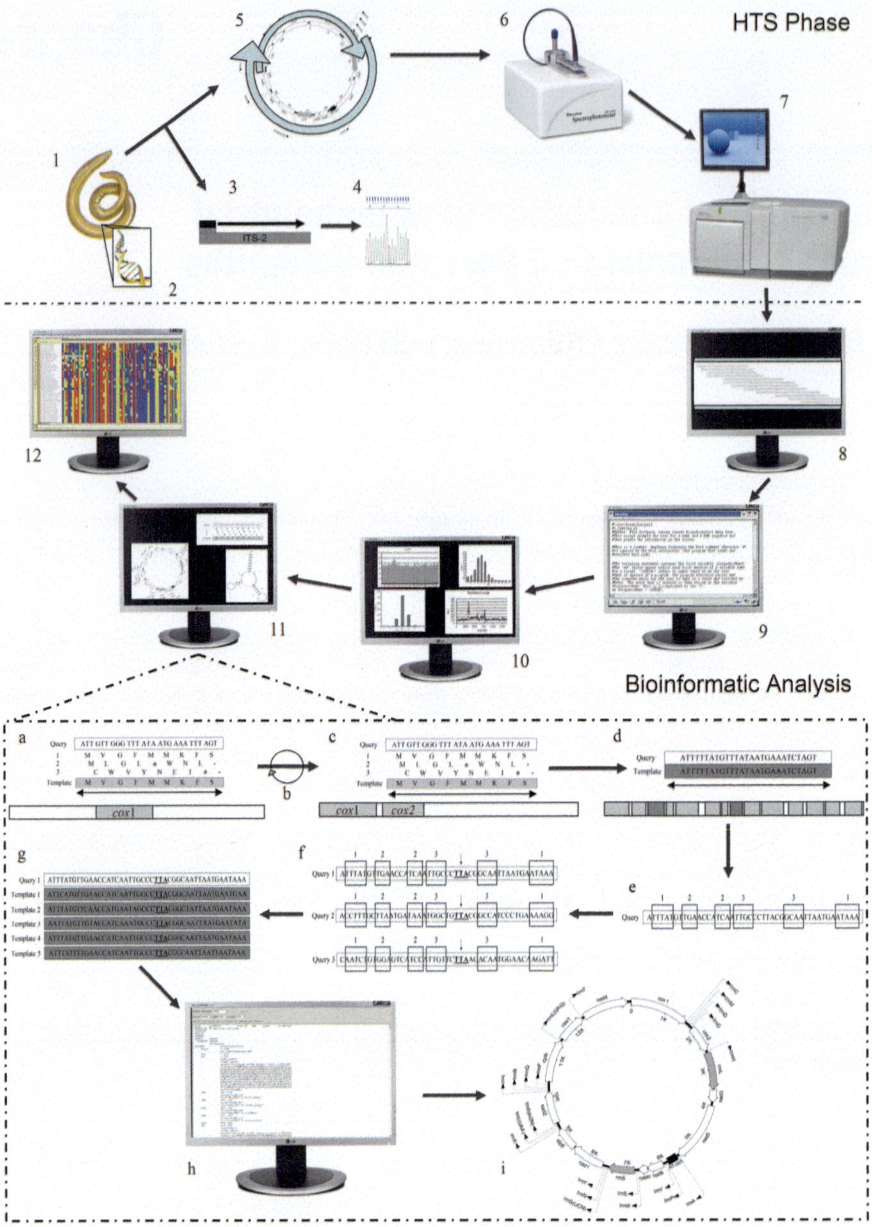

Fig. 1 Flow-diagram for high-throughput mt genomic sequencing, annotation, and analysis. Individual stages of the high-throughput pipeline are numbered *1–12* as follows: *1* = morphological identification of the parasite, *2* = total genomic extraction, *3* = independent species identification by PCR-based sequencing of a diagnostic marker (i.e., second internal transcribed spacer (ITS-2) of the nuclear ribosomal DNA), *4* (direct sequencing), *5* = long-PCR amplification of the complete mt genome as two overlapping fragments, *6* = quantification of each long-PCR amplicon by spectrophotometry (NanoDrop), *7* = simultaneous sequencing of the complete mt genome of each specimen by HTS technology, *8* = read assembly, *9* and *10* = bioinformatic analysis of raw read data, *11* = automated annotation (*see* Subheading 3.4.2), *12* = estimation of sequence error rate, base-calling, and selective re-sequencing of small regions by conventional means, and comparative analysis against published reference sequence data (from the GenBank database). Figure adapted from Jex et al. [14]

an organism, followed by sequencing (directly or via amplification), assembly, and annotation to then determine the genome structure and gene order [3–9]. For vertebrates and large invertebrates, micrograms or milligrams of mtDNA can be isolated from individuals for subsequent sequencing (either with or without PCR and/or cloning). However, for microscopic organisms, such as many parasitic helminths, direct isolation and sequencing of mtDNA is not possible from individual worms, and the substantial sequence heterogeneity that exists among individuals [5, 10, 11] generally precludes the pooling of multiple individuals. Thus, the development of a sensitive protocol for long-range PCR amplification of complete mt genomes from total DNA extracts represented a significant advance, allowing complete mt genome sequencing by PCR-mediated primer walking [11]. The long-range PCR approach (Fig. 1) has a number of advantageous features: (1) Time-consuming and laborious steps required for mt DNA isolation and purification are circumvented; (2) DNA can be readily isolated from individual nematodes for effective amplification and sequencing of mt genomes; (3) at least five PCRs can be performed from an individual nematode, using as little as ~20 ng total genomic DNA, thus allowing multiple analyses, or cloning of amplicons; and (4) direct sequencing of large amplicons prevents the potential to generate artifacts through cloning. However, the substantial AT-richness associated with some helminths, particularly nematodes, makes sequencing by primer walking a costly and time-consuming process, particularly across the AT-rich control region [12], and, most notably, when the sequencing of large numbers of mt genomes is required. The advent of next-generation sequencing platforms [13] provided a unique opportunity to overcome this major bottleneck, allowing the rapid and highly parallelized sequencing of complete mt genomes from pooled amplicons produced by long-range PCR. This high-throughput approach proved more reliable, cost-effective and time-efficient than conventional (Sanger) sequencing [14–16]. Coupled to indexing technologies [17], which allow amplicons of up to 384 individuals to be uniquely labeled with an oligonucleotide bar code, and the recent release of low-cost, desktop NGS platforms (e.g., 454 GS Junior, Ion Torrent, and/or Illumina MiSeq) [18], the major limitations to high-throughput mt genomic sequencing have been largely overcome.

However, the application of NGS approaches to high-throughput sequencing presents significant bioinformatic challenges. By the nature of their technology, NGS platforms generate large, highly fragmented datasets, which need to be quality-controlled, processed and assembled prior to analysis, which, given the low complexity associated with highly AT-rich templates, can be a challenging prospect when short-read sequence data are used. In addition, the ability to rapidly sequence large numbers of mt genomes reveals the limitations linked to the manual annotation and curation of these data. Recently, we described a prototypical pipeline to allow the rapid and automated annotation of tens or hundreds of complete mitochondrial genomes [14]. In the present chapter, we provide a protocol that we have established for the amplification (using conventional primer sets) and subsequent sequencing and annotation of entire mt genomes from individual parasitic nematodes (*see* Fig. 1), which overcomes most previous technical challenges. The present protocol provides a useful platform to determine and annotate the mt genomes for a range of helminths, may to be applied to any metazoan (invertebrates and vertebrates) with a mt genomes, and can be readily adapted to the sequencing of other organelle genomes (e.g., plastids) or small viral genomes as well.

2 Materials

Prepare all solutions using reverse osmosis deionized water and analytical grade reagents. Prepare and store reagents as recommended by the manufacturer. Strictly follow regulations for waste disposal. We do not add sodium azide to reagents.

2.1 Reagents

HCl (BDH, cat. no. 10125).

Tris (Sigma, cat. no. 154563).

Ethylenediaminetetraacetic acid disodium salt (EDTA) (Sigma, e.g., cat. no. E5134).

Sodium dodecyl-sulfate (SDS) (Sigma, e.g., cat. no. L4390).

Proteinase K solution (Boehringer-Mannheim, cat. no. 161 519).

DNA-extraction buffer: 20 mM Tris–HCl, pH 8, 50 mM EDTA, 1 % w/v sodium dodecyl-sulfate (SDS) plus 0.5 μg/μl proteinase K.

0.5× TBE solution: made by diluting (1/20) 10× TBE buffer stock in H_2O.

Isopropanol (BDH, cat. no. 1133).

Long PCR™ kit (BD Advantage 2 PCR Kit, cat. no. 639207, BD Bioscience Clontech), or Expand 20 kb^{PLUS} PCR System (cat. no. 11811002001, Roche) (*see* **Note 1**).

Sequencing kit (Big Dye Chemistry version 3.1, ABI, part. no. 4337454-8) or 454 technology (GSX FLX or FLX Titanium, Roche) via commercial service provider.

Molecular grade agarose (Bio-Rad, cat. no. 161-3100).

6× loading dye (e.g., Promega, cat. no. G190A) for agarose gel electrophoresis.

1 kb DNA Ladder (=molecular size marker, Promega, cat. no. G5711).

10× TBE buffer (Bio-Rad, cat. no. 161-0741).

SYBR Gold™ stain (Invitrogen, cat. no. S11494).

2.2 Equipment

Standard pipettes (1 ml, 200 μl, and 20 μl).

Vortex.

RNase/DNase-free pipette tips (with filters; 1 ml, 200 μl, and 20 μl).

Disposable syringes (5 ml; Luer).

DNase/RNAse-free double-lock PCR tubes (0.6 ml) (e.g., Robbins Scientific, cat. no. 1048-01-0).

DNase/RNAse-free Eppendorf tubes (1.5 ml).

Promega vacuum suction system (Promega, cat. no. A7231) (optional).

Wizard™ DNA Clean-Up columns (Promega, cat. no. A7280); alternative columns can be used.

Wizard™ PCR-Preps columns (Promega, cat. no. A7170); alternative columns can be used.

Conventional heat blocks.

Incubator (25–50 °C).

Vacuum pump.

Microfuge (Beckman).

NanoDrop ND-1000 UV–VIS spectrophotometer version 3.2.1 (NanoDrop Technologies).

Conventional PCR thermal cycler (e.g., Perkin Elmer 480, 2400, or ABI2720).

Conventional agarose gel electrophoresis apparatus.

Transilluminator (Elchrom Scientific AG, DWT Dual wavelength, 220 V; product no. 2038).

Gel documentation setup (e.g., Gel Doc System, Bio-Rad).

Computer with programs Photoshop elements v4.0 and Microsoft Powerpoint (for image storage and labeling) and with access to the world wide web (www), to be able to access databases in the National Center for Biotechnology Information (http://www.ncbi.nlm.nih.gov/) as well as bioinformatics packages required for specific analyses of mt genome sequence data.

3 Methods

3.1 Isolation of Genomic DNA (See Note 2)

1. Place mid-body sections of individual worms, single larval stages or eggs in ~20–40 μl saline or H_2O) in 1.5 ml Eppendorf tube containing 100–150 μl of DNA-extraction buffer and incubate at 37 °C for 12–14 h. If genomic DNA is being isolated from organisms other than parasites, a packed volume of 50 μl (diced tissue or packed cells) should suffice (*see* **Note 3**).
2. Vortex tube, centrifuge at 12,000 × *g*, and purify the genomic DNA from supernatant using a mini-spin column (DNA Clean-Up, Wizard, Promega), according to the manufacturer's instructions.
3. For long-term storage at −20 or −70 °C, transfer the DNA to 0.6 ml double-lock PCR tubes. For short-term usage, DNA must be stored at +2 to +8 °C, since multiple freezing and thawing will degrade DNA (*see* **Note 4**).

3.2 Amplification by Long-Range PCR

1. PCR-amplify the whole mt genome in two fragments (each usually ~5–10 kb in size) from ~20 to 40 ng (quantified spectrophotometrically; *see* **Note 5**) of total genomic DNA purified from an individual nematode (or part thereof) by long-PCR using oligonucleotide (18–28-mer) primers to relatively conserved regions in, for example, the *nad*1, *rrn*L, *rrn*S, or *cox*1 genes using the following PCR master mix (BD Advantage 2 PCR kit) (*see* **Notes 6** and 7):

Reagents	Amount (μl)
H_2O	37 (41)[a]
10× BD Advantage 2 PCR Buffer	5
50× dNTP mix	1
50× BD Advantage 2 Polymerase Mix	1
Forward primer (10 mM)	1
Reverse primer (10 mM)	1
Total genomic DNA (5-w0 ng per μl)	4 (0)[a]
Total volume	50

[a]No template control

2. Aliquot PCR master mix (4 °C) into tubes, cool on ice and add 1–2 μl of genomic DNA (usually 5–20 ng) to each reaction, with or without mineral oil overlay (depending on the thermal cycler). Also include, "positive" (i.e., genomic DNA known to amplify) and "negative" (i.e., no-DNA and/or host DNA) control reactions in each PCR run.

3. Run PCR according to the following cycling conditions: 94 °C, 2 min, followed by 35 cycles of 94 °C, 30 s (denaturation); 50–60 °C, 30 s (annealing); 60–72 °C, 10 min (extension), followed by 60–72 °C for 10 min (final extension) (*see* **Note 8**).

3.3 Agarose Gel Electrophoresis and Examination of Amplicons (See Note 9)

1. To verify the quality of the amplicons and PCR conditions, add 5 μl of amplicon to 1 μl of 6× loading dye and load on to 1 % w/v agarose-TBE gel. Also load 5 μl of an appropriate DNA molecular size marker (e.g., 1 kb ladder) into a lateral lane.
2. Run the gel at 80 V for 2 h in 0.5× TBE buffer.
3. Stain the gel with ethidium bromide (0.5 μg/ml) or SYBR Gold™ (according to the manufacturer's instructions) in H_2O. Briefly destain in H_2O, and then photograph (*see* **Note 10**).

3.4 Sequencing from Amplicons (See Note 11)

1. Purify each amplicon using a PCR-Preps spin column (Wizard, Promega), according to the manufacturer's instructions, and elute in 30 μl H_2O.
2. Check ~2 μl of each purified amplicon on a 1 % agarose-TBE gel.
3. Subject 20–50 ng of each purified product directly to automated cycle sequencing (Big-Dye Chemistry, ABI, according to the manufacturer's protocol) using a primer walking strategy (*see* **Note 12**). Alternatively, both amplicons can be pooled and then sequenced directly using 454 technology (Genome Sequencer FLX; Roche), according to the manufacturer's recommended protocol.
4. Assembly: If using conventional sequencing, visually verify sequences and ensure that protein-coding genes have open reading frames (*see* **Note 13**); align sequences and compare with appropriate reference sequences. For NGS data, a consensus mt genome sequence may be assembled using one of a variety of proprietary or open-source software packages [19]. For 454 sequence data, a consensus mt genome sequence is assembled automatically using the Newbler program (Roche) and can be refined further using the program MIRA (*see* **Note 14**) from thousands of individual (300–1,000 bp) "reads" based on a majority rule threshold among all reads representing each contig.

3.5 Bioinformatic Annotation and Analysis of Sequence Data

Following the assembly, the genes and features of each mitochondrial genome from each worm are annotated.

3.5.1 Manual Annotation

1. Predict open reading frames (ORFs) for each protein-coding mt gene using ORFinder (accessible via http://www.ncbi.nlm.nih.gov/projects/gorf/).
2. Identify each predicted ORF by BLASTx comparison with the nonredundant nucleotide database (i.e., GenBank; accessible via www.ncbi.nlm.nih.gov).

3. Conduct pairwise manual alignments with related, published reference sequences (accessible via http://drake.physics.mcmaster.ca/ogre/) and define start and stop codons based on overall sequence length and the inferred peptide sequence.
4. Predict transfer RNAs using tRNAscan-SE [20], selecting the "Mito/Chloroplast" or "Nematode Mito" model.
5. Annotate the small and large subunits of the mitochondrial ribosomal RNA genes (*rrn*S and *rrn*L) and any tRNA genes missed by tRNAscan-SE (e.g., in our experience, the nematode tRNA-Serine is often not detected due to its unusual structure) by BLASTn comparison with any large, unannotated regions of the mt genome, employing the nonredundant nucleotide sequence database. Confirm additional tRNAs by manual folding.
6. Construct annotated general feature format (.gff) in SEQUIN (available via http://www.ncbi.nlm.nih.gov/Sequin/) to allow direct submission to GenBank and/or further refinements to the annotation.

3.5.2 Automated Annotation

An automated workflow system has been designed and evaluated for its accuracy of annotation for mt genomes of metazoans, using reference sequences from public databases [14] (*see* **Note 15**).

1. Construct coding gene and translated protein databases for each reference sequence to be used for automated annotation. Construct tRNA database for all related reference species, and group tRNAs based on amino acid sequence (*see* **Note 16**). Perl scripts are designed to automate all subsequent steps.
2. Identify each protein coding mt gene by local alignment comparison (performed in all six reading frames) using amino acid sequences, conceptually translated from corresponding genes from the most closely related reference sequence database (defined in **step 1**). In all annotations, *cox1* is annotated first, and the consensus sequence is "rotated" to ensure that the first position is the first nucleotide of *cox1*. In all instances, the optimum full-length alignment (i.e., highest % similarity, determined based on the Blosum 62 matrix) is chosen as the location for each gene (i.e., given the relative conservation of the mt genome content, it is presumed that all coding genes are present), with the start and stop codons defined based on the length of the reference sequence/s.
3. Refine annotation of the start and stop codons for each coding gene by step-wise assessment of neighboring codons (i.e., within ten codons of the start or stop codon, defined based on gene length) for a more appropriate start or stop designation (based on the known mitochondrial codon translation code).

4. Identify *rrn*S and *rrn*L by local alignment (i.e., using nucleotide sequence data), employing the same approach.
5. Detect and identify all transfer RNA (tRNA) genes using a three-step process. First, predict all possible tRNA genes present in each consensus sequence (from both strands) based on secondary structure (i.e., based on the known tRNA models for the reference group, predict all possible ~50–100 nucleotide sequences that might be folded into a three-armed (nematode) or four-armed tRNA structure). Cluster all predicted tRNA genes into groups based on amino acid encoded by their anti-codon sequence (*see* **Note 16**). Rank all predicted tRNAs for each encoded amino acid based on structural "strength" (as inferred by the number of mismatched nucleotide pairs in each stem), and then compare the 100 best-scoring structures for each group by BLASTn alignment with sequences in the tRNA databases (constructed in **step 1**). Identify the predicted tRNA structure with the highest sequence identity to a known tRNA in the reference database (constructed in **step 1**) for each anti-codon group (*see* **Note 16**).
6. Construct a summary of the annotated mt genome (in table format) according to the instructions for the program SEQUIN (available via http://www.ncbi.nlm.nih.gov/Sequin/) for final verification and submission to the GenBank database.

4 Notes

1. Although *Taq* polymerase might introduce nucleotide misincorporations in PCR, it is crucial to use high-fidelity *Taq* polymerase with proofreading activity to minimize artifacts.
2. Timing for this step: overnight (12–24 h) is convenient.
3. Using DNA Clean-Up columns (Wizard), genomic DNA can be purified effectively from single eggs, larvae, or adults of helminths or other pathogens, and can be used directly for enzymatic amplification. Also alternative minicolumns (e.g., Qiagen) can be used for the purification of genomic DNA, following the manufacturers' protocols.
4. Genomic DNA samples can be stored at −20 or −70 °C for months to years.
5. Use, for example, a Qubit 2.0 fluorometer (Invitrogen Life Science) if using minicolumns from Promega.
6. Although conventional PCR kits should not be used for the amplification of regions of ≥3.0 kb (because of insufficient fidelity and no proofreading), long-PCR kits, such as the BD Advantage 2 PCR Kit, are effective and reliable. It may be necessary to optimize the $MgCl_2$ concentration by serial titration

for efficient and specific amplification. However, the optimum $MgCl_2$ concentration for PCR will depend on the kit/reagents used. The primer and dNTPs concentrations in the PCR are critical to achieve optimum results, characterized by a single, discrete, and abundant amplicon, displaying no smearing upon agarose gel electrophoretic analysis. Various primers may need to be tested empirically, and, gradually, a panel of relatively conserved primers can be assembled for the species being studied.

7. Fresh PCR mix should be prepared just prior to use and not stored. A reaction volume of 50 μl is used (depending on amounts required for subsequent analyses).
8. These cycling conditions have been found to be effective for the amplification of mitochondrial DNA regions from nematodes (usually ~5–10 kb). However, cycling conditions may need to be optimized, depending on *size* and *A + T-content* of the mt genome. In particular, highly A + T-rich regions (including the "control" region) can be very challenging to amplify. The elongation temperature in the PCR might need optimization (usually reduced from 68–72 to 60 °C). This optimization usually enables reproducible and effective amplification as well as subsequent sequencing of the two regions (which each need to represent single bands on agarose gels). Some workers have encountered difficulties in amplifying across the variable non coding (control) region (VNR) by long-PCR from invertebrates, such as platyhelminths and insects [8, 21], possibly because of its tandem-repetitive nature and length (1.5–13 kb). Fortunately, the AT-rich regions of the mt genomes of nematodes are often relatively short [22–24], readily permitting their amplification. Another issue is that pseudogenes have the potential to cause problems when PCR-amplified or co-amplified together with authentic mt genomic sequences and should be considered if (a) an amplicon represents more than one band on an agarose gel or if "background bands" are detected, (b) sequencing is difficult because of ambiguities due to deletions/insertions, frameshifts, or unexpected stop codons, and/or (c) the sequences are markedly different from those expected based on comparative genome analyses [21]. Since nuclear integrations usually relate to short regions [21, 25], fortunately, such issues have not been encountered for nematodes using the present PCR-based sequencing approach.
9. Timing for this step: ~ 3 h.
10. For the amplification of products from circular mt genomes by long-range PCR, each amplicon should appear as one discrete band in the lane of the agarose gel, and there should be no smears or streaks. The expected yield per PCR reaction is usually 1–5 μg of DNA. The specificity of the PCR can be verified by direct sequencing of selected amplicons using the same primers

used for amplification or internal ones (if already available). Amplicons can be stored at 4 °C for days and at −20 or −70 °C for months or years.

11. Timing for this step: ~ 12 h
12. For conventional sequencing using primer walking, a panel of primers needs to be evaluated for specificity in a first phase of sequencing. Once mt sequence data are available, a large panel of primers can be assembled for widespread application.
13. If there are any significant problems reading a sequence, individual amplicons can be sequenced conventionally following cloning into, for example, the plasmid vector pGEM-T-Easy[T] (Promega). However, this should not be necessary, given the haplotypic nature of the mt genome sequence (from a single individual).
14. The optimal assembly program for NGS datasets is dependent on the platform used for sequencing; many open-source programs are available to assist assembly. In our hands, MIRA [26] performs well for the assembly of 454 data and can be used also for Sanger or short-read data (i.e., Illumina or SOLiD). However, for short sequence reads (i.e., <~150 bp), De Bruijn graph-based assemblers such as Velvet [27, 28] or SOAPdenovo [29] appear to yield the best assemblies and are thus recommended for such data.
15. Because the annotation process [14] is dependent on local and Smith-Waterman format alignments, accurate annotation requires a suitably related reference sequence to be used. For some taxonomic groups, for which few or no such sequences are available, this is not possible. In such instances, we advise that one representative mt genome sequence representing a larger dataset is manually annotated and then used as a reference for annotation.
16. It is important to construct separate anti-codon groups for duplicated tRNAs (e.g., two nematode tRNA-Serine genes), as appropriate.
17. A trouble-shooting guide is given in Table 1.

Acknowledgements

Our research has been supported largely through grants from the Australian Research Council (ARC) and the National Health and Medical Research Council. Other support from the Alexander von Humboldt Foundation, Australian Academy of Science, the Fulbright Commission, Melbourne Water Corporation, the Victorian Life Sciences Computation Initiative (VLSCI), and the IBM Collaboratory is gratefully acknowledged.

Table 1
Trouble-shooting checklist

Problem	Possible reason	Proposed solution
No amplicon	Not enough genomic DNA template	Isolate fresh genomic DNA
	Ineffective PCR reagents	Verify reagents and their concentrations
	Inefficient PCR	Optimize $MgCl_2$ concentration and cycling temperatures
	Genome is very AT rich	Decrease annealing and/or extension temperature/s in the PCR
	Primer mismatch	Redesign primers
	Self-annealing or hairpin loops in primers	Redesign primers
	Primers anneal elsewhere in genome(s)	Redesign primers
Multiple or smeary band(s)	Nonspecific amplification in PCR	Optimize PCR and cycling conditions
	Complex genome	Clone amplicon prior to sequencing
	Pseudogenes or insertion/deletion events	Clone amplicon prior to sequencing or modify PCR conditions
	Primers anneal at multiple sites in the mitochondrial or nuclear genome	Optimize PCR conditions ($MgCl_2$ concentration and annealing temperature) or redesign primers
Smeary band(s) conditions	Problem with agarose gel electrophoresis	Modify matrix, buffer and/or electrophoresis conditions
Unreadable sequence	Amplicon is not specific or complex	Optimize $MgCl_2$ concentration and annealing temperature in PCR
	Heteroplasmy (multiple sequences in amplicon)	Sequence following cloning of amplicon
	Sequencing primer anneals at different sites in the sequence of the amplicon	Redesign primer or modify sequencing conditions
	High A+T- or G+C-content in the sequence	Modify sequencing conditions
Short sequence reads	High A+T content in the sequence	Modify sequencing conditions
	Poly-A or poly-T tracts in the sequence	Modify sequencing conditions
		Sequence following conventional cloning (do not PCR-amplify from cloned DNA)

References

1. Hu M, Gasser RB (2006) Mitochondrial genomes of parasitic nematodes—progress and perspectives. Trends Parasitol 22:78–84
2. Jex AR, Littlewood DT, Gasser RB (2010) Toward next-generation sequencing of mitochondrial genomes—focus on parasitic worms of animals and biotechnological implications. Biotechnol Adv 28:151–159
3. Boore JL, Macey JR, Medina M (2005) Sequencing and comparing whole mitochondrial genomes of animals. In: Zimmer EA, Roalson X (eds) Molecular evolution: producing the biochemical data, part B. Elsevier, Burlington
4. Burger G, Lavrov DV, Forget L, Lang BF (2007) Sequencing complete mitochondrial and plastid genomes. Nat Protoc 2:603–614
5. Hu M, Chilton NB, Gasser RB (2004) The mitochondrial genomics of parasitic nematodes of socio-economic importance: recent progress, and implications for population genetics and systematics. Adv Parasitol 56: 133–212

6. Lang BF, Burger G (2007) Purification of mitochondrial and plastid DNA. Nat Protoc 2:652–660
7. Lavrov DV, Brown WM, Boore JL (2000) A novel type of RNA editing occurs in the mitochondrial tRNAs of the centipede *Lithbius forficatus*. Proc Natl Acad Sci U S A 97: 13738–13742
8. Le TH, Blair D, McManus DP (2000) Mitochondrial genomes of human helminths and their use as markers in population genetics and phylogeny. Acta Trop 77:243–256
9. Simison WB, Lindberg DR, Boore JL (2006) Rolling circle amplification of metazoan mitochondrial genomes. Mol Phylogenet Evol 39: 562–567
10. Gasser RB (2006) Molecular tools–advances, opportunities and prospects. Vet Parasitol 136:69–89
11. Hu M, Jex AR, Campbell BE, Gasser RB (2007) Long PCR amplification of the entire mitochondrial genome from individual helminths for direct sequencing. Nat Protoc 2: 2339–2344
12. Hu M, Chilton NB, Gasser RB (2003) The mitochondrial genome of *Strongyloides stercoralis* (Nematoda)—idiosyncratic gene order and evolutionary implications. Int J Parasitol 33:1393–1408
13. Mardis ER (2008) Next-generation DNA sequencing methods. Annu Rev Genomics Hum Genet 9:387–402
14. Jex AR, Hall RS, Littlewood DT, Gasser RB (2010) An integrated pipeline for next-generation sequencing and annotation of mitochondrial genomes. Nucleic Acids Res 38: 522–533
15. Jex AR, Hu M, Littlewood DT, Waeschenbach A, Gasser RB (2008) Using 454 technology for long-PCR based sequencing of the complete mitochondrial genome from single *Haemonchus contortus* (Nematoda). BMC Genomics 9:11
16. Jex AR, Waeschenbach A, Hu M, van Wyk JA, Beveridge I, Littlewood DT et al (2009) The mitochondrial genomes of *Ancylostoma caninum* and *Bunostomum phlebotomum*–two hookworms of animal health and zoonotic importance. BMC Genomics 10:79
17. Binladen J, Gilbert MT, Bollback JP, Panitz F, Bendixen C, Nielsen R et al (2007) The use of coded PCR primers enables high-throughput sequencing of multiple homolog amplification products by 454 parallel sequencing. PLoS One 2:e197
18. Glenn TC (2011) Field guide to next-generation DNA sequencers. Mol Ecol Resour 11:759–769
19. Horner DS, Pavesi G, Castrignano T, De Meo PD, Liuni S, Sammeth M et al (2010) Bioinformatics approaches for genomics and post genomics applications of next-generation sequencing. Brief Bioinform 11:181–197
20. Schattner P, Brooks AN, Lowe TM (2005) The tRNAscan-SE, snoscan and snoGPS web servers for the detection of tRNAs and snoRNAs. Nucleic Acids Res 33:W686–W689
21. Zhang DX, Hewitt GM (1996) Nuclear integrations: challenges for mitochondrial DNA markers. Trends Ecol Evol 11:247–251
22. Keddie EM, Higazi T, Unnasch TR (1998) The mitochondrial genome of *Onchocerca volvulus*: sequence, structure and phylogenetic analysis. Mol Biochem Parasitol 95:111–127
23. Lavrov DV, Brown WM (2001) *Trichinella spiralis* mtDNA: a nematode mitochondrial genome that encodes a putative ATP8 and normally structured tRNAS and has a gene arrangement relatable to those of coelomate metazoans. Genetics 157:621–637
24. Okimoto R, Macfarlane JL, Clary DO, Wolstenholme DR (1992) The mitochondrial genomes of two nematodes, *Caenorhabditis elegans* and *Ascaris suum*. Genetics 130:471–498
25. Nelson WS, Prodöhl PA, Avise JC (1996) Development and application of long-PCR for the assay of full-length animal mitochondrial DNA. Mol Ecol 5:807–810
26. Chevreux B, Pfisterer T, Drescher B, Driesel AJ, Muller WE, Wetter T et al (2004) Using the miraEST assembler for reliable and automated mRNA transcript assembly and SNP detection in sequenced ESTs. Genome Res 14:1147–1159
27. Zerbino DR, Birney E (2008) Velvet: algorithms for de novo short read assembly using de Bruijn graphs. Genome Res 18:821–829
28. Zerbino DR (2010) Using the Velvet de novo assembler for short-read sequencing technologies. Curr Protoc Bioinformatics Chapter 11:Unit 11.15
29. Li Y, Hu Y, Bolund L, Wang J (2010) State of the art de novo assembly of human genomes from massively parallel sequencing data. Hum Genomics 4:271–277

Chapter 4

A Beginners Guide to Estimating the Non-synonymous to Synonymous Rate Ratio of all Protein-Coding Genes in a Genome

Daniel C. Jeffares, Bartłomiej Tomiczek, Victor Sojo, and Mario dos Reis

Abstract

The ratio of non-synonymous to synonymous substitutions (*dN/dS*) is a useful measure of the strength and mode of natural selection acting on protein-coding genes. It is widely used to study patterns of selection on protein genes on a genomic scale—from the small genomes of viruses, bacteria, and parasitic eukaryotes to the largest eukaryotic genomes. In this chapter we describe all the steps necessary to calculate the *dN/dS* of all the genes using at least two genomes. We include a brief discussion on assigning orthologs, and of codon-aware alignment of orthologs. We then describe how to use the CODEML program of the PAML package for phylogenetic analysis to calculate the *dN/dS* and how to perform some statistical tests for positive selection. We then outline some methods for interpreting output and describe how one may use this data to make discoveries about the biology of your species. Finally, as a worked example we show all the steps we used to calculate *dN/dS* for 3,261 orthologs from six *Plasmodium* species, including tests for adaptive evolution (see worked_example.pdf).

Key words *dN/dS*, CODEML, PAML, Synonymous/non-synonymous rate ratio, Evolutionary rate, Adaptive evolution, *Plasmodium*, Malaria

1 Introduction

1.1 The dN/dS Ratio

With the production of a complete genome sequence a relatively routine task, the bottleneck is now the annotation, analysis, and understanding of this genome data. A particularly useful statistic for protein-coding genes is the ratio of non-synonymous to synonymous substitutions $\omega = dN/dS$ (non-synonymous substitutions are nucleotide changes that alter the protein sequence, synonymous substitutions do not). This ratio measures the strength and mode of natural selection acting on the protein genes, with $\omega > 1$

Electronic supplementary material: The online version of this chapter (doi:10.1007/978-1-4939-1438-8_4) contains supplementary material, which is available to authorized users.

Christopher Peacock (ed.), *Parasite Genomics Protocols*, Methods in Molecular Biology, vol. 1201, DOI 10.1007/978-1-4939-1438-8_4, © Springer Science+Business Media New York 2015

indicating positive (adaptive or diversifying) selection, $\omega = 1$ indicating neutral evolution, and $\omega < 1$ indicating negative (purifying) selection. The ω ratio summarizes the evolutionary rates of genes, and can be an informative feature, because it can identify which genes are the most (or least) conserved and also identify genes that may have undergone periods of adaptive evolution [1]. For parasite genomes, this can help to uncover genes that may be changing rapidly in the "evolutionary arms race" against the host's immune system [2, 3]. There is an extensive literature on the use of ω to study adaptive evolution (see for examples [1–5]).

1.2 Principles of Evolution in Protein-Coding Genes

To understand why ω measures the strength and mode of action of natural selection of genes, let's first consider a new mutation that appears in the genome of a single organism in a population. Over long evolutionary time scales, two outcomes are possible: the mutation may spread throughout the population, until all individuals carry the mutation, that is, the mutation becomes fixed in the population; or the mutation may be lost. The ultimate fate of the mutation (that is, whether it becomes fixed or lost) depends on the interplay between natural selection and random genetic drift. Population genetics classifies mutations as either neutral (having little effect on the organism), deleterious (bad for the organism), or advantageous (good for the organism). Neutral mutations will accumulate in the population at the same rate as the genomic mutation rate μ [6, 7]. On the other hand, deleterious mutations may still reach fixation due to drift, but will accumulate in the population at a slower rate $\mu_-(<\mu)$, while those that are advantageous will accumulate at a faster rate $\mu_+(>\mu)$.

Let's now consider only those mutations that occur at codon positions in protein-coding genes. Synonymous mutations are (mostly) neutral because they do not change the amino acid sequence of the protein encoded, and therefore the synonymous substitution rate will be the neutral rate $\mu_S = \mu$; on the other hand non-synonymous substitutions may be affected by selection and the non-synonymous substitution rate will be in general different to the neutral rate $\mu_N \neq \mu$. Therefore the ratio $\omega = \mu_N/\mu_S$ indicates the mode of selection acting at non-synonymous sites. In practice the rates μ_N and μ_S are not easy to estimate directly. However, the non-synonymous and synonymous distances, $dN = t\,\mu_N$ and $dS = t\,\mu_S$, among orthologous genes in a phylogeny can be estimated from a sequence alignment (with t being the time of divergence or branch length in the phylogeny) leading to the estimation of ω. Sometimes selection may also act at synonymous sites (since some codons may be suboptimal), but this is of main concern for highly expressed genes of fast-growing organisms, since selection on codon usage is in general very weak for most genes in most organism [8, 9]. Methods that explicitly model codon usage selection in the estimation of ω have been developed [10].

For an excellent account of the mathematical theory of ω, the reader can consult Bustamante [7].

Most non-synonymous changes in coding regions negatively alter the structure and function of the protein and are therefore deleterious, whereas most synonymous changes are nearly neutral. This will result in $\omega < 1$ for most genes. When there are strong structural constraints on a protein, purifying selection is strong and there is little or no accumulation of non-synonymous changes, such that the ω approaches zero. In this way, the ω estimate can be used to describe the degree of "selective constraint" (strength of purifying selection) in a gene. This can be a very informative value for describing sets of genes, which can aid in the interpretation of the functioning of the genome [11].

Of course positive selection does occur, if rarely. If positive selection has acted along many of the codons of a gene and throughout the entire phylogeny, then $\omega > 1$. In practice this seldom happens, because positive selection is usually only observed within a specific region of the protein (e.g.: a specific domain) and/or within one branch of the phylogeny (some but not all species). In this case the ω for the entire gene will be shifted (perhaps imperceptibly) towards 1. Models able to detect all these scenarios have been developed [12–16].

In this chapter we limit ourselves to describing how to estimate ω using the CODEML program from the Phylogenetic Analysis by Maximum Likelihood (PAML) package [17]. The CODEML program calculates dN and dS using the observed changes present in a multiple alignment of protein-coding gene sequences from several species in a phylogeny (i.e., given the phylogenetic tree). Statistical estimation of ω with CODEML uses maximum likelihood, employing sophisticated mathematical models to correct for multiple changes, accounting for the different numbers of non-synonymous and synonymous sites, among other complexities, as briefly described in later sections. We describe a few common tests of positive selection. We also describe how to prepare the necessary data for CODEML, that is, how to identify orthologs correctly and build an appropriate sequence alignment, and how to estimate the phylogeny (i.e., the tree topology and branch lengths). We show a real-life example of these methods by examining selection in a set of 3,269 one-to-one orthologs of six *Plasmodium* species.

2 Materials

2.1 Computer Resources

To implement the processes and run the examples described in this chapter you will need access to a computer running a UNIX-like operating system (such as Linux or Mac OS X). Although it is possible to run our examples (Subheading 5) on a typical desktop computer, many CPU hours will be required to process genome-scale data.

In particular, running CODEML for all genes of a genome could take considerable computational time, depending on the number of species and the number of genes in each species. For example, running CODEML on 3,261 *Plasmodium* orthologs took 10 h (average gene length 2.1 kb).

2.2 Software

We provide a list of recommended software in Table 1. We indicate which of these packages will require administration privileges and/or moderate knowledge of Unix to install. The software you will need depends on your data, so we strongly recommend that you read this chapter to the end, including the notes, before installing any necessary packages. All the software we recommend is free of charge for academic use. The essential software will include:

1. BioPerl (to process genome-scale data).
2. An alignment tool (depending on the proximity of your sequences, we recommend Clustal Omega [18] or $PRANK_C$ [19]).

Table 1
Software recommendations

Software	Function	URL	Refs.
BioPerl	Wrappers for automating running code and file I/O	http://www.bioperl.org/	
BLAST+	Ortholog assignment using RBB	Download ftp://ftp.ncbi.nlm.nih.gov/blast/executables/blast+/LATEST/ Manual http://www.ncbi.nlm.nih.gov/books/NBK1763/	[31]
Clustal omega	Protein alignment	http://www.clustal.org/omega/	[18]
PRANK	Protein alignment	http://www.ebi.ac.uk/goldman-srv/prank/src/prank/	[19, 33]
GUIDANCE	Alignment filtering	http://guidance.tau.ac.il/source.html	[38]
PAL2NAL	Protein-to-nucleotide alignment	http://www.bork.embl.de/pal2nal/	[40]
PAML	Calculating evolutionary rates (ω, etc.)	http://abacus.gene.ucl.ac.uk/software/paml.html	[17]
RAxML	Calculating phylogenetic trees	http://sco.h-its.org/exelixis/software.html	[20]
MACSE	De novo codon-based alignment	http://mbb.univ-montp2.fr/MBB/subsection/softExec.php?soft=macse	[41]
Custom perl scripts developed for this chapter	Various	http://www.danieljeffares.com/data	

Software that may require administrator privileges to install are in **bold underlined** text

3. A package for building phylogenetic trees (we recommend RAxML [20]).
4. The PAML package [17].

2.3 Input Files (Genomes and Annotations)

To calculate ω for all genes in your chosen genomes you will need the following:

1. An annotated genome for the species you are most interested in: This must include accurate protein-coding gene predictions.
2. An annotated genome for *at least* one related species (preferably more): To obtain reasonable sensitivity the additional species must be sufficiently closely related to be accurately aligned (*see* Subheading 3.3 and **Note 1**).

3 Methods

This section describes how to create the necessary files (with various options) for running CODEML and parsing results. The workflow for all these methods is shown in Fig. 1, and an example is provided in the file worked_example.pdf. Some guidance on interpreting results is also included. Throughout this chapter commands will be shown in monotype font, e.g.,

```
perl runscript.pl\
--input myinputfile\
[--parameter 100]\
> myoutputfile
```

Parameters (file names, etc.) that need to be defined by the user are italicized, and optional parameters are placed in square brackets. To display usage information and show what inputs the scripts require, all scripts described here can be run either with no options or using the -h flag (or its longer equivalent `--help`), e.g.,

```
perl runscript.pl -h
```

3.1 Generating (or Collecting) Input Files

For each species to be analyzed, obtain FASTA format sequences of all protein-coding genes, and their corresponding translations. Often, these can simply be downloaded from a variety of websites and servers; we explore this first.

3.1.1 Gathering Gene Sequences from a Database

If it is possible to download the annotation files for some/all of the genes in the genomes of the phylogeny you're analyzing, gathering the list of genes is trivial. We provide a script to extract coding sequences from a Genbank or Embl format file. This script is run like this:

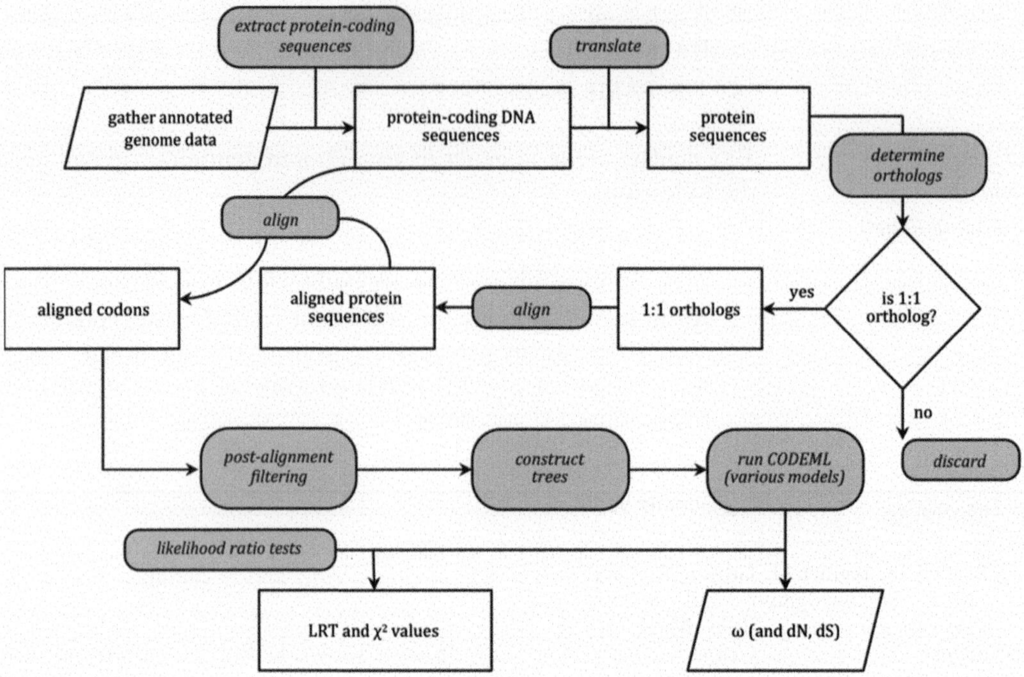

Fig. 1 A flow chart for calculating ω on a genome scale. Once you have obtained protein-coding sequences (DNA and protein), the steps to prepare data for analysis with CODEML will include ortholog assignment, alignment, possible post-alignment filtering, and tree construction. Finally running CODEML will produce ω values, as well as *dN* and *dS*. Running CODEML again with different models of evolution and then conducting likelihood ratio tests will give likelihood ratio test (LRT) values, which can then be tested for significance against χ^2 critical values. See main text for details

```
extract_genes_from_genome.pl\
-I "input_file1,input_file2, input_file3, etc"\
-s "species1,species2,species3"\
-t tag\
[-f genbank/embl (embl 175 default)]
```

The input_files will be Genbank or Embl format files, which contain both sequences and the start and end positions of all protein-coding exons. Species_name is used merely to name the output file, and the tag is the delimiter for gene names in the input file. This tag will differ depending on the species/input files. "systematic_id" is one example. We advise that you look into the Genbank or Embl file to determine this.

3.1.2 Generating Input Files from Contigs/Chromosomes and Corresponding Genome Annotations

Alternatively, it's possible to generate the necessary DNA and protein files from chromosome sequences or contigs (in FASTA format) and corresponding annotation files (in standard GFF format). We provide a script to gather all coding sequence (CDS) fragments from the genome and join together those corresponding to the same gene in the correct order, using gene coordinates from the GFF file. Protein sequences are then produced by translating these CDS sequences using BioPerl. If provided, the script will also compare its own translations to a set of corresponding protein sequences from an initial genome annotation and ignore those that do not match, or print a warning and take the translation. This script is run as follows:

```
perl get_cds_prot_from_gff_cont.pl -i input_folder -o output_folder -g annotations.gff -d contigs.fasta [-l list_of_desired_ids.csv]
```

The optional `-i` flag takes the address of a folder where the input files are to be found, and analogously for `-o` and the output files. `-g` is required; it takes an annotation file in GFF format (version 3 by default, although this can be changed). `-d` is also required, and it takes a DNA sequence file in which the sections that contain the genes specified in the annotations can be found. By using `-l` you can specify a list of desired IDs from the GFF file to process; if you don't provide such list, the script will simply process all the genes in the GFF file. You can additionally specify a file containing all protein sequences by using `-p`. As with all our scripts, a full list of options can be obtained by running it with no options specified, or using the `-h` (or `--help`) flag. The final output of this script consists of a pair of files for each gene, one for the combined CDS DNA sequences, and the other for the protein sequence, both identified by the same gene ID (`<gene_id>.dna.fasta` and `<gene_id>.prot.fasta`, respectively).

3.2 Identifying Orthologs

Once you have obtained the protein translations of each gene for each genome, the next step is grouping genes into sets of orthologs. These orthologous groups of genes will then be aligned and used as input for CODEML. "Orthologs" are homologous genes that were separated by speciation while "paralogs" are homologous genes that were separated by gene duplication [21]. For a review of the principles and complexities of orthology *see* ref. 22.

Many tools have been created to assign orthologs (for reviews, *see* refs. 23–26). The three main approaches use sequence similarity with graph clustering, phylogenetic trees, synteny, or a combination of several methods. The differences between the performance of the widely used methods on the same data appear to be fairly small—the major factor is the complexity of the proteomes involved (number of proteins, number and complexity of gene duplications, extent of multi-domain proteins, and domain-shuffling) [25–27]. It is important to appreciate that all orthology prediction methods will contain errors. These should be identified and removed where possible. We provide guidelines at later stages to account for these. The most common approach is to remove all orthology groups that have more than one ortholog per species (retaining only 1:1

Table 2
Useful websites

Contents/topic	URL
Guide to using BioPerl modules to run PAML and parse output	http://BioPerl.org/wiki/HOWTO:PAML
PAML discussion group	http://www.ucl.ac.uk/discussions/viewforum.php?f=54
Database of ortholog groups of eukaryotic proteins using InParanoid	http://inparanoid.sbc.su.se/
Database of ortholog groups of proteins using OrthoMCL	http://www.orthomcl.org
Clusters of orthologous groups of proteins from whole genomes	http://www.ncbi.nlm.nih.gov/COG/
OMA (Orthologous MAtrix) database of orthologs for complete genomes	http://omabrowser.org/
Treefam (tree families database)	http://www.treefam.org/
Gene ontology	http://www.geneontology.org/
A list of orthology databases	http://questfororthologs.org/orthology_databases

orthologs). This will remove the complication of paralogs that are due to gene duplications.

3.2.1 Obtaining Predefined Orthologs for the Species in Your Phylogeny

If a manually curated set of orthologs has been produced as part of a genome project, particularly any that use synteny, we encourage you to use these. Another alternative is that for published genomes, ortholog assignments may have been produced in one or more orthology prediction databases, such as OMA or OrthoMCL (*see* Table 2).

3.2.2 Reciprocal Best BLAST Hit Method to Assign Orthologs

If ortholog lists are not available we recommend the Reciprocal Best BLAST hit (RBB) method (with simple clustering for multiple species). This method is simple, fast, scalable, arguably as accurate as tree-based methods, and does not need extensive parameter optimization [27]. However, *see* ref. 28 for advice about BLAST options.

The (RBB) method assigns two proteins (genes) as orthologs if protein A from species 1 identifies protein A' from species 2 as its best hit, and vice versa [29, 30]. This requires that you run a BLAST search for each protein against each other genome in turn. The recommended BLAST parameters for RBB are a minimum BLASTP Evalue $\leq 1e^{-5}$ or $\leq 1e^{-6}$, and the combination of soft filtering with a Smith–Waterman final alignment (the -F "m S" -sT options in NCBI's BLASTP) [27, 28].

The pairwise RBB method may be extended into a clustering algorithm (cRBB) so that it can be applied to more than two genomes as follows [27]. For the genes A,B,C (from species *a, b, c*) to be clustered into an orthologous group gene B must be the reciprocal best hit (RBH) to gene A, and gene C must be the RBH to *either* gene A or gene B. If not, then gene C is not included in the ortholog group.

To determine cRBB you will require:

(a) FASTA format files of all protein sequences of all species considered (one file per species).

(b) An installation of BLAST (we advise BLAST+) [31].

(c) Wrapper script(s) to run BLAST searches and cluster best hits.

We provide a script to produce a list of orthologs from a full analysis of a set of genomes using this RBB approach. Note that this script is likely to take a few hours to run, due to the many BLAST searches performed:

```
perl prthologs_from_RBBH.pl -o orthologs.csv -i
"all_proteins_species_1.fasta,all_proteins_
species_2.fasta,all_proteins_species_3.fasta"
```

Multiple files can be specified in the usual UNIX way (e.g., `"*spec*.fasta"`). This script runs a Reciprocal Best BLAST hits method and produces a set of orthologous gene IDs separated by a comma and indexed by the first species. The output file consists of a list of such comma-separated sets. In the example above only lists with orthologs in all three species will be returned, but adding `-n 2` would return all sets with two or more orthologs.

3.3 Alignment of Orthologs

Alignments can be an important source of false-positive cases (incorrectly inferred adaptive evolution) [32–34], so the choice of an alignment tool is an important consideration. A particular concern in the context of this chapter is that alignment quality decreases with decreasing protein similarity [18]. So this places a limit on the degree of divergence that can be used in calculations of evolutionary rates, particularly when false positives are a concern (such as when the aim is to detect the few genes that are subject to positive selection). In general, reliable alignments can be produced with Clustal Omega (and several other tools) with proteins that have at least 70 % identity [18]. In this section we give an overview of the typical steps you will need to follow if you're writing your own scripts, followed by instructions for running a script we provide that integrates all the relevant tasks. Finally, we provide an alternative script when reliable translations of your coding sequences (CDS) are not available.

3.3.1 Considerations for Choosing a Multiple Alignment Tool

There are many alternative tools for aligning multiple protein sequences. At the present time, two of the most reliable programs are Clustal Omega [18] and PRANK [19]. Clustal Omega appears to be the most powerful (fast, accurate) for divergent proteins, whereas PRANK performs well on more closely related sequences. PRANK has been explicitly designed for evolutionary analysis and performs well under simulation [33, 35]. We do not advise using the older ClustalW which is an entirely different program to the newer Clustal Omega.

With any level of divergence that provides sufficient power to detect adaptive evolution, alignments will contain errors [18] that cause false positives and false negatives [33]. In particular insertions and deletions are a major source of false positives in the detection of adaptive evolution using CODEML [35]. Using the *cleandata* option in CODEML may reduce the false positives, but at the possible loss of interesting sites. There are various filtering methods available to remove potentially unreliable alignment columns or codons [36–38]. The use of these is equivocal; they appear to improve evolutionary analysis in some cases [39], but have negligible effects in others [33]. In practice, alignment filtering will produce a more conservative analysis—lowering both false positives and true positives. It is difficult to generalize about the cost/benefit of such approaches. For a detailed analysis, we recommend comparing results produced from filtered and unfiltered alignments. Based on two simulation studies [33, 39], we recommend the GUIDANCE tool for alignment filtering [38].

Programs that calculate dN/dS will require codon-based alignments of the DNA sequences of all genes in each ortholog group; therefore gaps should be positioned so as not to change the reading frame. If you aligned the CD sequences with PRANK using the translate option or using the empirical codon models you will already have codon-based DNA alignments. On the other hand, if your alignment program generated amino acid alignments, you will need to perform "reverse translation" to construct a codon alignment from the unaligned DNA sequence files and the aligned amino acid sequence files. This is best achieved by aligning the corresponding protein sequences, and then converting the protein alignment to a nucleotide alignment using the corresponding gene sequences. We recommend the conducting analysis tool for this task [40]. This software produces a codon alignment with options for removing gaps and in frame stop codons, as well as mismatched codons. The "native" format for the CODEML program is the PHYLIP format, with some small modifications. We suggest that you refer to the PAML manual before constructing DNA alignments (*see* Table 1).

In summary, care must be taken to obtain the best alignment, and we recommend particular care and skepticism for this stage in the analysis. This is particularly important when conducting analysis on a genome scale, because a few false positives could dominate any signal for adaptive evolution, or skew the ω estimate with a systematic bias to particular types of genes.

3.3.2 Producing Codon-Specific Alignments from an Ortholog List

We provide a script to automate the tasks described above for a genome-scale list of sets of orthologs. To run it, you will need:

1. A list of sets of orthologs in a comma-separated value (CSV) file, where each line has a set of related orthologous gene IDs separated by a comma (*see* **Note 2** for an example of the CSV file, and comments).
2. DNA and protein sequences for each desired gene, in FASTA format, identified by the same gene ID indicated in the orthologs CSV file, and named <gene_id>.dna.fasta and <gene_id>.prot.fasta.
3. Run the script:

```
perl align_orthologs.pl -l orthologs.csv -i input_folder -o output_folder -c -a
```

The `-l` option receives a list of orthologs in a CSV file, as described above. `-i` receives the location of the folder where your DNA and protein sequences reside, and you can also specify the folder where you want to put your output files via the `-o` option. `-c` tells the script to do the protein alignment using Clustal Omega, and -a indicates that you want to calculate PAL2NAL codon alignments. The script will print out warnings for any input files it cannot find, and it will only produce alignment files if it can find two or more of the orthologs in each set (i.e., each line of the CSV). You can send the list of any missing information to a file by adding the `-e` flag.

It is also possible in theory to align protein-coding sequences when reading frames are not known, using software such as MACSE [41] (*see* **Note 3**). If the goal of the analysis is to identify genes that are subject to positive selection, we advise caution when using such methods, because alignment inaccuracies increase the rate of false-positive results, as well increase (the already abundant) false negatives [35].

3.4 Estimating Phylogenetic Trees

Model testing with CODEML requires a phylogenetic tree of either the group of species or the gene concerned. When using CODEML "site tests" for positive selection are robust to tree topology, so in general the species tree should be used. This is the most common case (*see* below Subheading 3.5). If a species tree is not available, we recommend that you estimate one tree for the species you will analyze using a concatenation of all the aligned orthologs (see below). This should increase the power to detect cases where particular branches of the phylogeny have an increased evolutionary rate in a few orthologous groups. An exception is that in the unlikely scenario there is evidence for recombination between your species (i.e., if they are strains within a species or very recently separated species), then you may wish to estimate topologies for each gene. A concatenation of all aligned CDS sequences can be achieved with the concatenate_alignments.pl script that we provide (*see* Table 5).

There are many tools to estimate phylogenetic trees based on sequence data. We recommend the RAxML tool, which is fast and sufficiently accurate [20], run using the GTR gamma model. To obtain a phylogeny for ω estimation the command line required to run RAxML is

```
raxmlHPC -f a -x 12345 -p 12345 -# 100 -m GTRGAMMA
-s your_alignment_file.phy -n your_alignment_
prefix
```

3.5 Using CODEML to Calculate ω and Identify Positive Selection

3.5.1 Concepts in CODEML

PAML is a package of programs designed to analyze molecular sequences and estimate a variety of parameters of molecular evolution [17]. We concern ourselves here only with estimates of ω, and attempts to detect positive selection using the CODEML application in PAML. For more advice the PAML FAQ and PAML manual will be helpful, as will the PAML discussion group (*see* Table 2). The statistical theory of adaptive evolution is reviewed in [42]. We also recommend these more technical articles for further reading [15, 17, 32, 43]. There are of course other tools for calculating non-synonymous and synonymous evolutionary rates apart from CODEML. We recommend HyPhy, a particularly versatile tool for testing models of evolution [44]. While HyPhy allows the user to specify virtually any model for evolution, some expertise is needed to do this because this tool has its own batch language. The sitewise likelihood-ratio (SLR) software package is another alternative [45]. The SLR method makes less assumptions about how the strength of selection is distributed across sites and is considered complementary to PAML. As with PAML, the HyPhy and SLR tools are all in active development (as of 2012). We do not describe how to use HyPhy or SLR in this chapter.

Adaptive evolution seldom occurs in all species of a phylogeny, or over all sites in a gene, which makes it more difficult to locate the genes/sites concerned. The more likely scenario is that positive selection has occurred in some branches of the phylogeny, or in some specific sites in the gene or only in specific sites in some branches of the phylogeny. Each of these possibilities is formalized into a "model," so that possible processes of evolution can be tested for explicitly. The main classes of models used in CODEML are "branch models" (where ω can vary over different branches in the phylogeny), "site models" (where ω can vary at different sites in the gene), and "branch-site" models (where ω can vary in particular sites, in particular branches). In tests for adaptive evolution that use branch models, positive selection is detected along

Table 3
Models of adaptive evolution implemented in CODEML[a]

Model	Description
Site models	
M0 (one ratio)	One average ω for the gene Null model for testing if selected branches evolve with different rate than the background branches Specify using NSsites = 0, model = 0
M1a (nearly neutral)	One ω across all lineages, models only two classes of sites ($0 \leq \omega < 1$ and $\omega = 1$) Specify using NSsites = 1, model = 0
M2a (positive selection)	One ω across all lineages, models three classes of sites ($0 <= \omega < 1$, $\omega = 1$, and $\omega > 1$) Specify using NSsites = 2, model = 0
M7 (beta)	One ω across all lineages, ten classes of sites with $\omega <= 1$ Specify using NSsites = 7, model = 0
M8 (beta and ω)	One ω across all lineages, 11 classes of sites on all lineages, 10 with $\omega \leq 1$, 1 with $\omega > 1$ Specify using NSsites = 8, model = 0
Branch models	
Free-ratio model	Allows different ω for each branch of the tree. Specify using NSsites = 0, model = 1
Two-ratio model	Allows several ω values for a specified branch (the "foreground" branch, usually your species of interest). The user must specify which this "foreground" branch, and the other "background" branches Specify using NSsites = 0, model = 2
Branch-site models[b]	
Model A	Like site M1a, M2 site model, but marked branches are treated as foreground allowing three classes of sites ($0 < \omega < 1$, $\omega = 1$, $\omega => 1$), and others, as background with only two classes of sites ($\omega = 0$, $\omega = 1$) Specify using NSsites = 2, model = 2, fixomega = 0
Model A1	Null model, foreground branches allowing two classes of sites ($0 < \omega < 1$, $\omega = 1$), and others, as background with only two classes of sites ($0 < \omega < 1$, $\omega = 1$) Specify using NSsites = 2, model = 2, fixomega = 1

[a]In all these models $\omega < 1$ indicates purifying selection, $\omega \leq 1$ indicates selection in a purifying to nearly neutral range, $\omega = 1$ indicates neutral evolution, and $\omega > 1$ indicates adaptive evolution. The ω values can refer either to the entire gene or to some sites (codons) within a gene. In some cases the models allow for adaptive evolution ($\omega > 1$) in some sites within one branch of the tremanue ("site-branch" models)
[b]See the PAML manual (Version 4.6, March 2012) for how to direct CODEML to use these models. Note that Model B and Site model 3 are no longer recommended

the branches only if the average ω over all codons in the gene is larger than one. This is unlikely to occur, because even if a few sites in the protein are evolving fast along the branch the average ω may not be >1, because most of the sites in the protein will remain under purifying selection. However some authors managed to get positive results using this approach to detect adaptive evolution [4, 46]. A more realistic model is site models, which allow ω to vary only within specific sites of the gene, but for all species. Finally, branch-site models allow ω to vary both among sites and across the branches of the phylogeny. These are probably the most realistic models. The models that are currently recommended to test these alternatives are described below (*see* also Table 3 for summary of the models used in CODEML, and how to direct CODEML to use these models).

1. The *one-ratio model* (M0 in CODEML) calculates the average ω for the whole gene, over all branches in the phylogeny. This is useful to obtain the average ω value for the gene, but is not thought to be a sufficiently realistic model to detect adaptive evolution.
2. The *Nearly Neutral model* (M1a) classifies codon sites in a gene into two groups: one group has codons subjected to purifying selection ($\omega < 1$), and the other group has codons under neutral evolution ($\omega = 1$). There are no codons under positive selection ($\omega > 1$).
3. The *Positive Selection model* (M2a) as the NearlyNeutral model, but an extra class of codon sites subjected to positive selection ($\omega > 1$) is allowed.
4. The *beta model* (M7) uses the flexible beta distribution to describe ω variation among sites. The distribution of ω values can take a variety of shapes in the range from 0 to 1, so codons under positive selection are not allowed.
5. The *beta and ω model* (M8) is the same as the beta model, except that it allows for some sites to be subjected to positive selection ($\omega > 1$).

The application of maximum likelihood in CODEML allows these models of evolution to be described as mathematical summaries of the stochastic process of molecular evolution. CODEML uses a maximum likelihood approach to attempt to fit the observed data (the sequence alignment) to the model of evolution that you specify. This involves estimating parameters such as the branch lengths, the transition/transversion ratio, and the ω ratio (see the PAML manual for details). Once this is done CODEML provides the parameters that are its best fit to the data and a *likelihood value*. This *likelihood value* (L) is the probability of observing the data with parameters generated by the model. Likelihood values are provided in the natural log, $\ln L$ (*see* **Note 4** for further details).

To determine if positive selection has occurred in a gene, you will need to show that a model that includes positive selection (where $\omega > 1$ in some sites) fits the data better than one that does not include positive selection (i.e., no $\omega > 1$ sites). This is achieved as follows:

1. Run CODEML with a simple model that does not allow positive selection. CODEML will estimate ω and determine the $\ln L$ for each gene with this model (l_0).
2. Run CODEML with a more general model that allows positive selection. CODEML will estimate ω and determine the $\ln L$ for each gene with this model (l_1).
3. Determine which model is more likely for each gene using a likelihood ratio test (LRT, *see* **Note 4**). The LRT statistic = $2 \times (l_1 - l_0)$
4. You may reject the simpler model for any particular gene if the LRT statistic is greater than the critical χ^2 value with k degrees of freedom (*see* **Note 4**, and example in supplementary file worked_example.pdf).

We describe in the next sections a general schema for how to do this in practice. We also provide a detailed step-by-step example of this process in the supplementary file worked_example.pdf. This example shows how we calculated ω for all the 1:1 orthologs in six *Plasmodium* species, and detected some statistically supported cases of adaptive evolution.

3.5.2 Estimating ω for All Genes Using the Simple One Ratio Model

Once you have a codon-specific alignment and a phylogenetic tree, the next step is to run CODEML with a null model (usually model M0). The models are specified in the CODEML control file (usually with a .ctl extension). The control file also specifies which sequence file, the tree file, and other parameters that CODEML should use. A detailed explanation of this file and all the options available is given in the PAML manual, and we provide an example CODEML-M0.ctl in supplementary material. The most important parameters to note are:

```
seqfile = myfile.paml
```

The sequence alignment file (containing all gene alignments).

```
treefile = tree.txt
```

The plain text file containing the phylogenetic tree of the species, in Newick format.

```
outfile = M0-output.txt
```

The name of the output file.

```
ndata = N
```

Where N is the number of alignments to be analyzed.

```
CodonFreq = 2
```

Which specifies which positions to use to calculate the nucleotide frequencies.

```
model = 0
```

This specifies whether to allow ω to vary among lineages in the phylogeny.

```
NSsites = 0
```

Specifies whether the model CODEML uses ω to vary among sites of a gene.

Once you have edited your control file, you should run CODEML:

```
codeml codeml-M0.ctl
```

Expect CODEML to run for many hours (our example of 3,261 *Plasmodium* orthologs took 10 h). The output will be contained in the file you specified (M0-output.txt above). It is simple to extract the ω values from CODEML's output file with grep:

```
grep omega M0-output.txt > M0-omega.txt
```

3.5.3 Estimating ω and the lnL for All Genes Using Alternative Models

To evaluate whether the data for a particular gene fits an alternative evolutionary model you will need to run CODEML again, specifying another model. This is done by modifying the control file (saving it with a new name), sometimes modifying the tree file, and running CODEML again. The most common use for this is to examine whether each gene better fits a model that includes *some sites* that have adaptive evolution (where $\omega > 1$). Remember that it is unlikely that the average ω for the *entire gene* will be >1. Once this is done CODEML will produce a log likelihood estimate (lnL), which you can use for likelihood ratio tests. To determine

Table 4
Recommended tests of selection in CODEML

Models		k	Hypothesis tested
Site models:	M2a vs. M1a	2[a]	Does adding a third class of sites with $\omega > 1$ (adaptive evolution) fit the data better than a model with two classes $\omega < 1$, $\omega = 1$?
	M8 vs. M7	2[a]	Does adding an extra class of sites with $\omega > 1$ (adaptive evolution) fit the data better than a model with ten classes with flexible normalized non-synonymous ratio distribution?
Branch models:	*Free-ratio* vs. *one-ratio* model	2*s*–4	For a tree of *s* species, is ω different among lineages?
	Two-ratio vs. *one-ratio* model	1	Are the foreground branches that you specify more likely to have different ω from background branches?
Branch-site models:	MA($\omega > 1$) vs. MA($\omega = 1$)	1[b]	Is the defined "foreground branch" more likely to contain sites with $\omega > 1$

[a] In these models the regularity conditions are not met and the asymptotic distribution of the LRT statistic is not known. Using χ^2 with the given *k* degrees of freedom possible makes the test conservative (Yang and dos Reis [32])

[b] In the branch-site test, the asymptotic distribution of the LRT statistic is a 1:1 mixture of point mass zero and χ^2 with $k = 1$ (Yang and dos Reis [32])

which model to use as the null and alternatives, consult Table 4. For example, comparing the model M2a (which allows some sites to have $\omega > 1$) against the null model M1a (which doesn't allow this) examines whether the data better fits a model with some adaptive evolution. See below for more detail about the likelihood ratio tests.

1. *To specify a site model*: Modify your control file to include these lines (deleting the previous settings). The "NSsites = 0 1 2 7 8" text instructs CODEML to determine the lnL with several models. Note that site models allow you to predict which sites have been subject to selection.

   ```
   model = 0
   NSsites = 0 1 2 7 8
   outfile = site-models-output.txt
   ```

2. *To specify a branch model*: Modify your tree file to mark the branch that you wish to test. This is done by adding a hash tag (e.g., "#1") to the branch: e.g.:

   ```
   (((2,(3, 1)),6 #1),5,4)
   ```

   For clarity, save your tree file with a new name. Then modify your control file to include these lines:

   ```
   treefile = marked-tree.txt
   model = 2
   NSsites = 0
   outfile = branch-model-output.txt
   ```

3. *To specify a "branch-site" model*: Modify your control file to include these lines:

   ```
   treefile = marked-tree.txt
   model = 2
   NSsites = 2
   outfile = branch-site-model-output.txt
   fix_omega = 1
   omega = 1
   ```

3.5.4 Likelihood Ratio Tests (LRT) of Positive Selection

Once you have run CODEML with a null model and an alternative model, you will then use a likelihood ratio test to see if the data are a significantly better fit to the alternative model (*see* Table 3 or models and Table 4 for which null and alternative models to test). Note that adaptive evolution is usually rare in genomes, so the no-selection model is usually the null. The steps to take to perform a likelihood ratio test are the following:

1. A log likelihood (ln*L*) value for each gene has been calculated by CODEML for a null and alternative model.
2. Calculate the value of the LRT statistic (twice the difference of the log-likelihood between the null model and alternative model).
3. Determine the degrees of freedom (*k*) for your test. This is calculated as $k = p_1 - p_0$, where p_1 is the number of parameters estimated in the alternative model, and p_0 is the number of parameters in the null model. For simplicity, we list the degrees of freedom in Table 4.
4. Compare the LRT statistic with the critical value from χ^2 distribution, with the appropriate degrees of freedom (*k*) and the significance level that you want (α), which is usually 0.05 or 0.01.
5. (a) If the value of the LRT statistic is greater than the critical value, you reject the null hypothesis, which means that there are sites (or branches, depending on the test) that have undergone adaptive evolution.

   (b) Alternatively, the *p*-value can be calculated from the cumulative distribution function of the χ^2 statistic where appropriate (*see* **Note 5**). These models are described in the following references [14, 47, 48].

Table 5
Scripts we provide with this chapter

Script	Function
`genes_from_genome.pl`	Extracts CDS sequences and proteins from a Genbank or Embl file
`gff_cds_proteins_processor.pl`	Extracts CDS sequences from a FASTA nucleotide file according to GFF coordinates, translates
`orthologs_from_RBBH.pl`	Assigns orthologs using the clustered Reciprocal Best Blast (cRBB) approach
`align_orthologs.pl`	Aligns proteins, generates codon-aware nucleotide alignment
`multiple_sequence_splitter.pl`	Splits a FASTA file with many sequences into one file per sequence
`concatenate_alignments.pl`	Takes a list of alignment files and outputs a single concatenated alignment file
`codeml_simple.pl`	Runs CODEML for all genes in a list
`codeml_site_models.pl`	Runs CODEML for all genes in a list. Performs the likelihood ratio tests for site models
`codeml_branch_models.pl`	Runs CODEML for all genes in a list. Performs the likelihood ratio tests for branch models

When testing for adaptive evolution on thousands of genes, a method to correct for multiple testing is desirable. We recommend using the false discovery rate approach described by Benjamini et al. [49].

For large sets of data you can perform all CODEML calculations and the tests using our perl wrapper scripts (*see* Table 5).

3.5.5 Detecting Particular Sites That Have Been Subject to Selection

Genes that were identified as containing sites under selection can be investigated further to determine the probability that each codon has been subject to adaptive evolution ($\omega > 1$). CODEML will already have performed a Bayesian identification of these sites (as described in [50, 51], which is presented in the main output file (e.g., branch-site-model-output.txt above). We provide more detail about how to examine this output in the supplementary data file worked_example.pdf.

4 Interpreting Results on a Genome Scale

A genome-scale evolutionary analysis of protein-coding genes can be very useful for describing features of the genome. Two approaches to describing genomes using evolutionary values are (a) to plot and correlate evolutionary parameters (ω, etc.) with other quantitative features of genes (e.g., expression levels) and (b) to group genes by various methods (e.g., Gene Ontology) and then look for groups of gene with significantly higher/lower evolutionary parameters.

4.1 Correlating Evolutionary Parameters with Other Features of Genes

It is most often found that ω correlates with the expression "breadth" (the number of tissues it is present in) or expression level of a gene, for example [11], but other correlating features of genes with ω (or dN or dS) could also reveal new features of genomes. It is important to appreciate in these analyses that many aspects of genes are correlated [52], so further analysis will be required to determine which aspect(s) of the gene causes the correlation [53, 54]. A balanced analysis should take into account that statistically significant *p*-values can be obtained with large data sets, even when the strength of the effect is very weak (i.e., high *p*-value, but low correlation coefficient *rho*). Plotting data and reporting only the strongest effects will help to distinguish biologically meaningful results from those that are very weak effects that produce very statistically significant *p*-values merely because of the large number of observations.

4.2 Comparing Evolutionary Parameters Between Group of Genes

There are a variety of ways to group genes that can be revealing. The use of Gene Ontology (GO) is common [1, 11, 55]. Within gene ontology both biological function, biochemical function, and the cellular location aspects can be revealing. The PANTHER software system for inferring the functions of genes based on their evolutionary relationships [56] is another alternative. Clustering genes by their similarity of expression or by principal tissue (or life cycle stage for parasites) they are expressed in can also reveal salient patterns [11, 57]. Genetic or protein interaction maps can also be used to group genes.

Once genes have been grouped, the approach is to show the extent to which different groups of genes differ in their evolutionary features by comparing evolutionary parameters (such as ω, dN, dS, the LRT statistic, or the p-value from LRT tests) between groups of genes. Simply sorting groups by their median values and plotting can be sufficient, for example [5]. To test whether specific groups of genes have more/less constraint a Mann–Whitney test is most often used because it is nonparametric (does not assume that the data have any particular distribution, e.g., normal distribution). In this case one might test whether the genes in a particular group have a different distribution to another group, or differs to all other genes. To locate particular groups of genes that are evolving adaptively, then the likelihood ratio test (LRT) statistics of the genes can be compared with Mann–Whitney tests. Another alternative is to count the number of genes in a group that pass a meaningful LRT significance value, and use a Fisher's exact test to determine if the group is enriched for positively selected genes.

Regardless of the methods used it is important to use a method to correct for multiple comparisons. The Bonferonni correction is in common use, but other methods are available [58, 59]. Finally, we suggest some healthy scepticism about genes that appear to have undergone adaptive evolution (e.g., high ω). If possible manual checks of the alignments and orthology may aid in rejecting false positives.

5 Final Comments

The methods we have described should enable you to calculate ω and detect possible cases of adaptive evolution for all the genes in your genome. We advise care with all steps, particularly collecting sufficient data to have good power (more genomes is better), alignments, and CODEML model testing. Keep in mind that ω is not the only test for non-neutral evolution; some other methods are described in [60], which may require different data types. The methods described are, to the best of our knowledge, up to date when this chapter was written. However, things change, so we encourage readers to post comments on CiteULike at www.citeulike.org/user/danieljeffares/publications.

Supplementary data will be available on http://www.danieljeffares.com/data.

6 Notes

1. To calculate *dN*/*dS* accurately from alignments of orthologous genes the sequences must not be too closely related, or too distant. If sequences are too closely related (e.g., all sequences >95 % identity on the DNA level) there will be little power to accurately estimate *dN* and *dS*, since there will be too few observed changes. Because CODEML (and other tools) estimate parameters such as these from the observed genetic changes, the power of the analysis increases when there are more changes observed. Increased power can be attained in two ways. First, by choosing species that are sufficiently divergent. This approach is helpful up to the point where orthologs cannot be assigned correctly or DNA mutations (substitutions) at fast-evolving sites are saturated. The PAML FAQ also states that the method is reasonable if the synonymous distance over all branches of the tree is >0.5; this approximate figure is supported by simulations [33]. In practice, this means that when looking at an alignment of a protein-coding gene most synonymous sites have a change in one or more of the species. Secondly, the power increases with increasing number of orthologs (species) in the alignment. The PAML FAQ recommends that the absolute minimum number of species is 4 or 5 and that 10 is good, but 20 would be better. Simulations show that good estimates of ω can be obtained with six species, while detection of adaptive evolution has relatively low power with this many taxa [33]. In practice of course, it is nontrivial to add another genome to your analysis after data have been gathered for a project, but it is an important consideration if accurate and sensitive evolutionary analysis is a desired outcome.
2. The ortholog list file that the `ortholog_processor_aligner.pl` script requires should be in this format:

   ```
   GENE_1_SPECIES1,GENE1_SPECIES2,GENE1_SPECIES3
   GENE_2_SPECIES1,GENE2_SPECIES2,GENE2_SPECIES3
   ```

   Since the scripts use each gene ID to find corresponding files, sequences, and annotations, it is crucial to use exactly the same spelling all across (however, note that our scripts can get rid of most non-word characters like "_" or "#"). In case you already have files containing the sets of orthologous DNA sequences, provide a list with only one ID per line, corresponding to the base name of each of the files that contain the orthologs, which in turn should be named `<gene_id>.orthologs.dna.fasta` and, optionally, `<gene_id>.orthologs.prot.fasta`

3. When annotations of protein-coding sequences are unreliable or absent it is possible to produce "de novo" codon-based nucleotide alignments with packages such as MACSE [41]. This software can generate multiple-sequence alignments accounting for disruptions in the reading frame (stop codons or frame shifts arising either from sequencing errors or biological deviations) without knowing the reading frame, or any corresponding amino acid translation, in advance. MACSE recognizes the reading frame and produces the alignment in the FASTA format. Any possible frameshifts and stop codons are detected and the nucleotides are aligned in a way that any alignment gaps are more likely inserted as a multiplication of three. This results in higher quality of the codon-specific alignment for coding regions than could be achieved using alignments tools that are not codon aware.

   Our experience has shown that it is important to adjust the penalty parameters of MACSE depending on the type of data you are using. Transcriptome data (such as assembled RNASeq data) can be aligned with the default parameters since the occurrence of the stop codon in the middle of the gene is less likely than the sequencing error. While exon sequences extracted from a genome may contain single base "overhangs," so should be given a higher penalty for frameshifts. We advise caution, since inappropriately tuned parameters may result in alignment errors that will affect downstream results.

   The MACSE java application is invoked like so:

```
java -jar macse_v0.8b2.jar -i your_orthologs_file.fa -o your_output_prefix
```

4. A full discussion of maximum likelihood, likelihood values, and likelihood ratio tests in molecular evolution is beyond the scope of this chapter. For the purposes of using PAML, the important principles are that *CODEML* and other programs in the PAML package use ***maximum likelihood*** to try to find parameters that best fit the observed data to the model you have specified (e.g.: model M0). The ***likelihood function*** is a function of the parameters of the model: the ***likelihood*** of a set of parameter values given the observed data is the probability of observing the data given the parameter values. It is not necessary to fully understand the theory of maximum likelihood to use PAML effectively. The main point to appreciate is that the log likelihood (lnL) is a probability of data fitting the model. The aim with testing various models is to find a model that better fits the data. The ***likelihood ratio test*** is used to determine if one data-model fit is significantly better than another. We recommend excellent Wikipedia articles as a short primer on these topics and [42, 61] for further reading.

5. This can be achieved, for example, using the function pchisq in R. Only for the branch test the LRT follows the χ^2 (chi-square) distribution. For the site test or the branch site test the LRT does not follow a χ^2 distribution, but using this distribution in both cases will make your *p*-values conservative [32].
6. A tutorial about using CODEML to calculate ω for 3,261 orthologs from six Plasmodium species is given as supplementary file worked_example.pdf. The data used are as described in [57]. All the files required to follow through this example are provided in the supplementary file worked_example_files.zip. All supplementary data will be available on http://www.danieljeffares.com/data.

Acknowledgements

We thank Ziheng Yang, Caroline Biagosch, and Sanne Nygaard for comments on the manuscript.

References

1. Kosiol C, Vinar T, da Fonseca RR, Hubisz MJ, Bustamante CD, Nielsen R, Siepel A (2008) Patterns of positive selection in six Mammalian genomes. PLoS Genet 4:e1000144
2. Yang W, Bielawski JP, Yang Z (2003) Widespread adaptive evolution in the human immunodeficiency virus type 1 genome. J Mol Evol 57:212–221
3. Lefébure T, Stanhope MJ (2009) Pervasive, genome-wide positive selection leading to functional divergence in the bacterial genus Campylobacter. Genome Res 19:1224–1232
4. Yang Z (1998) Likelihood ratio tests for detecting positive selection and application to primate lysozyme evolution. Mol Biol Evol 15: 568–573
5. Clark AG, Eisen MB, Smith DR, Bergman CM, Oliver B, Markow TA, Kaufman TC, Kellis M, Gelbart W, Iyer VN, Pollard DA, Sackton TB, Larracuente AM, Singh ND, Abad JP, Abt DN, Adryan B, Aguade M, Akashi H, Anderson WW, Aquadro CF, Ardell DH, Arguello R, Artieri CG, Barbash DA, Barker D, Barsanti P, Batterham P, Batzoglou S, Begun D, Bhutkar A, Blanco E, Bosak SA, Bradley RK, Brand AD, Brent MR, Brooks AN, Brown RH, Butlin RK, Caggese C, Calvi BR, de Carvalho AB, Caspi A, Castrezana S, Celniker SE, Chang JL, Chapple C, Chatterji S, Chinwalla A, Civetta A, Clifton SW, Comeron JM, Costello JC, Coyne JA, Daub J, David RG, Delcher AL, Delehaunty K, Do CB, Ebling H, Edwards K, Eickbush T, Evans JD, Filipski A, Szlig SF, Freyhult E, Fulton L, Fulton R, Garcia ACL, Gardiner A, Garfield DA, Garvin BE, Gibson G, Gilbert D, Gnerre S, Godfrey J, Good R, Gotea V, Gravely B, Greenberg AJ, Griffiths-Jones S, Gross S, Guigó R, Gustafson EA, Haerty W, Hahn MW, Halligan DL, Halpern AL, Halter GM, Han MV, Heger A, Hillier L, Hinrichs AS, Holmes I, Hoskins RA, Hubisz MJ, Hultmark D, Huntley MA, Jaffe DB, Jagadeeshan S, Jeck WR, Johnson J, Jones CD, Jordan WC, Karpen GH, Kataoka E, Keightley PD, Kheradpour P, Kirkness EF, Koerich LB, Kristiansen K, Kudrna D, Kulathinal RJ, Kumar S, Kwok R, Lander E, Langley CH, Lapoint R, Lazzaro BP, Lee S-J, Levesque L, Li R, Lin C-F, Lin MF, Lindblad-Toh K, Llopart A, Long M, Low L, Lozovsky E, Lu J, Luo M, Machado CA, Makalowski W, Marzo M, Matsuda M, Matzkin L, McAllister B, McBride CS, McKernan B, McKernan K, Mendez-Lago M, Minx P, Mollenhauer MU, Montooth K, Mount SM, Mu X, Myers E, Negre B, Newfeld S, Nielsen R, Noor MAF, O'Grady P, Pachter L, Papaceit M, Parisi MJ, Parisi M, Parts L, Pedersen JS, Pesole G, Phillippy AM, Ponting CP, Pop M, Porcelli D, Powell JR, Prohaska S, Pruitt K, Puig M, Quesneville H, Ram KR, Rand D, Rasmussen MD, Reed LK, Reenan R, Reily A, Remington KA, Rieger TT, Ritchie MG, Robin C, Rogers Y-H, Rohde C, Rozas J, Rubenfield MJ, Ruiz A, Russo S, Salzberg SL, Sanchez-Gracia A,

Saranga DJ, Sato H, Schaeffer SW, Schatz MC, Schlenke T, Schwartz R, Segarra C, Singh RS, Sirot L, Sirota M, Sisneros NB, Smith CD, Smith TF, Spieth J, Stage DE, Stark A, Stephan W, Strausberg RL, Strempel S, Sturgill D, Sutton G, Sutton GG, Tao W, Teichmann S, Tobari YN, Tomimura Y, Tsolas JM, Valente VLS, Venter E, Venter JC, Vicario S, Vieira FG, Vilella AJ, Villasante A, Walenz B, Wang J, Wasserman M, Watts T, Wilson D, Wilson RK, Wing RA, Wolfner MF, Wong A, Wong GK-S, Wu C-I, Wu G, Yamamoto D, Yang H-P, Yang S-P, Yorke JA, Yoshida K, Zdobnov E, Zhang P, Zhang Y, Zimin AV, Baldwin J, Abdouelleil A, Abdulkadir J, Abebe A, Abera B, Abreu J, Acer SC, Aftuck L, Alexander A, An P, Anderson E, Anderson S, Arachi H, Azer M, Bachantsang P, Barry A, Bayul T, Berlin A, Bessette D, Bloom T, Blye J, Boguslavskiy L, Bonnet C, Boukhgalter B, Bourzgui I, Brown A, Cahill P, Channer S, Cheshatsang Y, Chuda L, Citroen M, Collymore A, Cooke P, Costello M, D'Aco K, Daza R, De Haan G, DeGray S, DeMaso C, Dhargay N, Dooley K, Dooley E, Doricent M, Dorje P, Dorjee K, Dupes A, Elong R, Falk J, Farina A, Faro S, Ferguson D, Fisher S, Foley CD, Franke A, Friedrich D, Gadbois L, Gearin G, Gearin CR, Giannoukos G, Goode T, Graham J, Grandbois E, Grewal S, Gyaltsen K, Hafez N, Hagos B, Hall J, Henson C, Hollinger A, Honan T, Huard MD, Hughes L, Hurhula B, Husby ME, Kamat A, Kanga B, Kashin S, Khazanovich D, Kisner P, Lance K, Lara M, Lee W, Lennon N, Letendre F, Levine R, Lipovsky A, Liu X, Liu J, Liu S, Lokyitsang T, Lokyitsang Y, Lubonja R, Lui A, MacDonald P, Magnisalis V, Maru K, Matthews C, McCusker W, McDonough S, Mehta T, Meldrim J, Meneus L, Mihai O, Mihalev A, Mihova T, Mittelman R, Mlenga V, Montmayeur A, Mulrain L, Navidi A, Naylor J, Negash T, Nguyen T, Nguyen N, Nicol R, Norbu C, Norbu N, Novod N, O'Neill B, Osman S, Markiewicz E, Oyono OL, Patti C, Phunkhang P, Pierre F, Priest M, Raghuraman S, Rege F, Reyes R, Rise C, Rogov P, Ross K, Ryan E, Settipalli S, Shea T, Sherpa N, Shi L, Shih D, Sparrow T, Spaulding J, Stalker J, Stange-Thomann N, Stavropoulos S, Stone C, Strader C, Tesfaye S, Thomson T, Thoulutsang Y, Thoulutsang D, Topham K, Topping I, Tsamla T, Vassiliev H, Vo A, Wangchuk T, Wangdi T, Weiand M, Wilkinson J, Wilson A, Yadav S, Young G, Yu Q, Zembek L, Zhong D, Zimmer A, Zwirko Z, Alvarez P, Brockman W, Butler J, Chin C, Grabherr M, Kleber M, Mauceli E, MacCallum I (2007) Evolution of genes and genomes on the Drosophila phylogeny. Nature 450:203–218

6. Kimura M (1983) The neutral theory of molecular evolution. Cambridge University Press, Cambridge, 1968. ISBN 0-521-23109-4
7. Bustamante CD (2005) Population genetics of molecular evolution. In: Nielsen R (ed) Statistical methods in molecular evolution. Springer, New York
8. Sharp PM, Bailes E, Grocock RJ, Peden JF, Sockett RE (2005) Variation in the strength of selected codon usage bias among bacteria. Nucleic Acids Res 33:1141–1153
9. dos Reis M, Wernisch L (2009) Estimating translational selection in eukaryotic genomes. Mol Biol Evol 26:451–461
10. Yang Z, Nielsen R (2008) Mutation-selection models of codon substitution and their use to estimate selective strengths on codon usage. Mol Biol Evol 25:568–579
11. Jeffares DC, Pain A, Berry A, Cox AV, Stalker J, Ingle CE, Thomas A, Quail MA, Siebenthall K, Uhlemann A-C, Kyes S, Krishna S, Newbold C, Dermitzakis ET, Berriman M (2007) Genome variation and evolution of the malaria parasite Plasmodium falciparum. Nat Genet 39:120–125
12. Ziheng Y, Bielawski JP (2000) Statistical methods for detecting molecular adaptation. Trends Ecol Evol 15:496–503
13. Swanson, W. J., Z. Yang, M. F. Wolfner, and C. F. Aquadro. 2001. Positive Darwinian selection drives the evolution of several female reproductive proteins in mammals. Proc. Natl. Acad. Sci. USA 98:2509-2514
14. Anisimova M, Bielawski JP, Yang Z (2001) Accuracy and power of the likelihood ratio test in detecting adaptive molecular evolution. Mol Biol Evol 18:1585–1592
15. Anisimova M, Bielawski JP, Yang Z (2002) Accuracy and power of bayes prediction of amino acid sites under positive selection. Mol Biol Evol 19:950–958
16. Yang Z, Nielsen R (2002) Codon-substitution models for detecting molecular adaptation at individual sites along specific lineages. Mol Biol Evol 19:908–917
17. Yang Z (2007) PAML 4: phylogenetic analysis by maximum likelihood. Mol Biol Evol 24:1586–1591
18. Sievers F, Wilm A, Dineen D, Gibson TJ, Karplus K, Li W, Lopez R, McWilliam H, Remmert M, Söding J, Thompson JD, Higgins DG (2011) Fast, scalable generation of high-quality protein multiple sequence alignments using Clustal Omega. Mol Syst Biol 7:539
19. Löytynoja A, Goldman N (2008) Phylogeny-aware gap placement prevents errors in

sequence alignment and evolutionary analysis. Science 320:1632–1635

20. Stamatakis A, Ludwig T, Meier H (2005) RAxML-III: a fast program for maximum likelihood-based inference of large phylogenetic trees. Bioinformatics 21:456–463
21. Fitch WM (1970) Distinguishing homologous from analogous proteins. Syst Zool 19:99
22. Koonin EV (2005) Orthologs, paralogs, and evolutionary genomics. Annu Rev Genet 39: 309–338
23. Kuzniar A, van Ham RCHJ, Pongor S, Leunissen JAM (2008) The quest for orthologs: finding the corresponding gene across genomes. Trends Genet 24:539–551
24. Altenhoff AM, Dessimoz C (2012) Inferring orthology and paralogy. Methods Mol Biol 855:259–279
25. Trachana K, Larsson TA, Powell S, Chen W-H, Doerks T, Muller J, Bork P (2011) Orthology prediction methods: a quality assessment using curated protein families. Bioessays 33: 769–780
26. Kristensen DM, Wolf YI, Mushegian AR, Koonin EV (2011) Computational methods for gene orthology inference. Brief Bioinform 12:379–391
27. Salichos L, Rokas A (2011) Evaluating ortholog prediction algorithms in a yeast model clade. PLoS One 6:e18755
28. Moreno-Hagelsieb G, Latimer K (2008) Choosing BLAST options for better detection of orthologs as reciprocal best hits. Bioinformatics 24:319–324
29. Bork P, Dandekar T, Diaz-Lazcoz Y, Eisenhaber F, Huynen M, Yuan Y (1998) Predicting function: from genes to genomes and back. J Mol Biol 283:707–725
30. Tatusov RL, Koonin EV, Lipman DJ (1997) A genomic perspective on protein families. Science 278:631–637
31. Camacho C, Coulouris G, Avagyan V (2009) BLAST+: architecture and applications. BMC Bioinformatics 10:421
32. Yang Z, dos Reis M (2011) Statistical properties of the branch-site test of positive selection. Mol Biol Evol 28:1217–1228
33. Jordan G, Goldman N (2012) The effects of alignment error and alignment filtering on the sitewise detection of positive selection. Mol Biol Evol 29:1125–1139
34. Markova-Raina P, Petrov D (2011) High sensitivity to aligner and high rate of false positives in the estimates of positive selection in the 12 Drosophila genomes. Genome Res 21: 863–874
35. Fletcher W, Yang Z (2010) The effect of insertions, deletions, and alignment errors on the branch-site test of positive selection. Mol Biol Evol 27:2257–2267
36. Castresana J (2000) Selection of conserved blocks from multiple alignments for their use in phylogenetic analysis. Mol Biol Evol 17:540–552
37. Notredame C, and Abergel C (2003) Using Multiple Alignment Methods to Assess the Quality of Genomic Data Analysis, in Bioinformatics and Genomes: Current Perspectives, M. Andrade, Editor. 2003, Horizon Scientific Press. p. 30–50
38. Penn O, Privman E, Landan G, Graur D, Pupko T (2010) An alignment confidence score capturing robustness to guide tree uncertainty. Mol Biol Evol 27:1759–1767
39. Privman E, Penn O, Pupko T (2012) Improving the performance of positive selection inference by filtering unreliable alignment regions. Mol Biol Evol 29:1–5
40. Suyama M, Torrents D, Bork P (2006) PAL2NAL: robust conversion of protein sequence alignments into the corresponding codon alignments. Nucleic Acids Res 34: W609–W612
41. Ranwez V, Harispe S, Delsuc F, Douzery EJP (2011) MACSE: multiple alignment of coding SEquences accounting for frameshifts and stop codons. PLoS One 6:e22594
42. Yang Z (2006) Computational molecular evolution. Oxford University Press, UK
43. Yang Z, Nielsen R, Goldman N (2009) In defense of statistical methods for detecting positive selection. Proc Natl Acad Sci U S A 106:E95–E95
44. Sergei L, Pond S, Frost S (2005) HyPhy: hypothesis testing using phylogenies. Bioinformatics 21:676–679, Advance Access published on March 1, 2005
45. Massingham T, Goldman N (2005) Detecting amino acid sites under positive selection and purifying selection. Genetics 169:1753–1762
46. Messier W, Stewart CB (1997) Episodic adaptive evolution of primate lysozymes. Nature 385:151–154
47. Swanson WJ, Nielsen R, Yang Q (2003) Pervasive adaptive evolution in mammalian fertilization proteins. Mol Biol Evol 20:18–20
48. Wong WSW, Yang Z, Goldman N, Nielsen R (2004) Accuracy and power of statistical methods for detecting adaptive evolution in protein coding sequences and for identifying positively selected sites. Genetics 168:1041–1051
49. Benjamini Y, Hochberg Y (1995) Controlling the false discovery rate—a practical and power-

ful approach to multiple testing. J Roy Stat Soc B Met 57:289–300

50. Yang Z, Wong W (2005) Bayes empirical Bayes inference of amino acid sites under positive selection. Mol Biol Evol 22:1107–1118
51. Zhang J, Nielsen R, Yang Z (2005) Evaluation of an improved branch-site likelihood method for detecting positive selection at the molecular level. Mol Biol Evol 22:2472–2479
52. Krylov DM, Wolf YI, Rogozin IB, Koonin EV (2003) Gene loss, protein sequence divergence, gene dispensability, expression level, and interactivity are correlated in eukaryotic evolution. Genome Res 13: 2229–2235
53. Fraser HB, Hirsh AE (2004) Evolutionary rate depends on number of protein-protein interactions independently of gene expression level. BMC Evol Biol 4:13
54. Park S, Choi S (2010) Expression breadth and expression abundance behave differently in correlations with evolutionary rates. BMC Evol Biol 10:241
55. Lindblad-Toh K, Garber M, Zuk O, Lin MF, Parker BJ, Washietl S, Kheradpour P, Ernst J, Jordan G, Mauceli E, Ward LD, Lowe CB, Holloway AK, Clamp M, Gnerre S, Alföldi J, Beal K, Chang J., Clawson H, Cuff J, Di Palma F, Fitzgerald S, Flicek P, Guttman M, Hubisz MJ, Jaffe DB, Jungreis I, Kent WJ, Kostka D, Lara M, Martins AL, Massingham T, Moltke I, Raney BJ, Rasmussen MD, Robinson J, Stark A, Vilella AJ, Wen J, Xie X, Zody MC, Broad Institute Sequencing Platform and Whole Genome Assembly Team, Baldwin J, Bloom T, Chin CW, Heiman D, Nicol R, Nusbaum C, Young S, Wilkinson J, Worley KC, Kovar CL, Muzny DM, Gibbs RA, Baylor College of Medicine Human Genome Sequencing Center Sequencing Team, Cree A, Dihn HH, Fowler G, Jhangiani S, Joshi V, Lee S, Lewis LR, Nazareth LV, Okwuonu G, Santibanez J, Warren WC, Mardis ER, Weinstock GM, Wilson RK, Genome Institute at Washington University, Delehaunty K, Dooling D, Fronik C, Fulton L, Fulton B, Graves T, Minx P, Sodergren E, Birney E, Margulies EH, Herrero J, Green ED, Haussler D, Siepel A, Goldman N, Pollard KS, Pedersen JS, Lander ES, Kellis M (2011) A high-resolution map of human evolutionary constraint using 29 mammals. Nature 478:476–482
56. Mi H, Dong Q, Muruganujan A, Gaudet P, Lewis S, Thomas PD (2010) PANTHER version 7: improved phylogenetic trees, orthologs and collaboration with the gene ontology consortium. Nucleic Acids Res 38:D204–D210
57. Essien K, Hannenhalli S, Stoeckert CJ (2008) Computational analysis of constraints on non-coding regions, coding regions and gene expression in relation to Plasmodium phenotypic diversity. PLoS One 3:e3122
58. Holm, S. (1979). "A simple sequentially rejective multiple test procedure". *Scandinavian Journal of Statistics* 6 (2): 65–70
59. Abdi H (2007) Chapter Bonferroni and Sidak corrections for multiple comparisons. In: Salkind NJ (ed) Encyclopedia of measurement and statistics. Sage, Thousand Oaks, CA
60. Nielsen R (2005) Molecular signatures of natural selection. Annu Rev Genet 39:197–218
61. Nielsen R (2005) Statistical methods in molecular evolution. Springer, New York, NY

Chapter 5

Exploiting Genetic Variation to Discover Genes Involved in Important Disease Phenotypes

Paul Capewell, Anneli Cooper, Caroline Clucas, Willie Weir, Heli Vaikkinen, Liam Morrison, Andy Tait, and Annette MacLeod

Abstract

Elucidating the underlying genetic determinants of disease pathology is still in the early stages for many pathogenic parasites. There have, however, been a number of advances in which natural genetic diversity has been successfully utilized to untangle the often complex interactions between parasite and host. In this chapter we discuss various methods capable of exploiting this natural genetic variation to determine genes involved in phenotypes of interest, using virulence in the pathogenic parasite *Trypanosoma brucei* as a case study. This species is an ideal system to benefit from such an approach as there are several well-characterized laboratory strains; the parasite undergoes genetic exchange in both the field and the laboratory, and is amenable to efficient reverse genetics and RNAi.

Key words Genetics, Parasites, Virulence, Trypanosome, Genetic variation, Forward genetics

1 Introduction

The degree of pathogenicity or virulence of an infection is the primary method to assess the severity of a disease. It has been studied extensively in trypanosome infections, although largely at the level of the host. The virulence of a trypanosome infection can vary greatly between parasite strains, even when hosts of identical genetic backgrounds are infected. Understanding the virulence factors that confer this variability may bring the option to control or alleviate symptoms. If we define virulence as the growth of the parasite in vivo or in vitro (assayed by parasitemia) and by the ability to cause a measurable disease phenotype, candidate genes that are "knocked out" or their expression "knocked down" via reverse genetics will often have reduced growth and pathology and so will be less virulent. While ablating gene function is an ideal mechanism for investigating virulence in the laboratory, applying conclusions generated in such an artificial setting to natural populations can be problematic. Many genes can affect parasite growth and

Christopher Peacock (ed.), *Parasite Genomics Protocols*, Methods in Molecular Biology, vol. 1201,
DOI 10.1007/978-1-4939-1438-8_5,

virulence when disrupted using reverse genetics, although most are unlikely to be directly controlling parasite growth or disease pathology in the real world. In order to identify the genes that control these processes one can look at natural populations, where allelic variation is likely to be the main determinant on which selection and evolution act. Consequently, when investigating phenotypes such as virulence in the parasite's natural environment, it is necessary to identify the alleles of genes associated with phenotypic variation. Uncovering the genes in which allelic differences greatly affect the disease phenotype will reveal the determinants of virulence. Once the key determinants of virulence have been identified, it will then be possible to develop novel intervention strategies based on limiting virulence and disease pathology.

1.1 African Trypanosomes

African trypanosomes are significant disease-causing pathogens found throughout sub-Saharan Africa, although three species in particular have the greatest effect on humans and their livestock: *Trypanosoma congolense*, *T. vivax*, and *T. brucei*. These three species infect a varied range of mammals and are predominantly transmitted from host to host by a tsetse insect vector. Livestock infection causes the most significant losses to productivity (an economic cost of approximately $1.3 billion per year to the continent), with some 20 million cattle, significant numbers of small ruminant livestock, and working equines infected [1]. Additionally, two *T. brucei* subspecies (*T. brucei gambiense* and *T. brucei rhodesiense*) are pathogenic to humans and cause approximately 70,000 cases of Human African Trypanosomiasis per year [2, 3]. However this number is likely a substantial underestimate by virtue of the difficulties in accessing the appropriate diagnosis and treatment in the isolated rural communities most affected by the disease [4, 5]. These human infective subspecies persist as geographically discrete foci across Africa with *T. b. rhodesiense* restricted to East and Southern Africa and *T. b. gambiense* to West and Central Africa [6]. Matters are complicated in that isoenzyme markers and phenotype analysis has identified two distinct types of *T. b. gambiense* [7–9]. Group 1 *T. b. gambiense* is the predominant type, found throughout Western and Central Africa while the less prevalent group 2 *T. b. gambiense* is limited to foci in Cote d'Ivoire and Burkina Faso and appears to be freely breeding with local *T. b. brucei* [8].

The genetic relationship between the *T. brucei* subspecies has been much studied and remains a subject of debate in the research community. Past evidence suggested that *T. b. rhodesiense* is host range variants of *T. b. brucei* while group 1 *T. b. gambiense* is a true subspecies [7–14]. A recent analysis using sequence data from mitochondrial genes and microsatellite polymorphisms shows that this initial interpretation was correct [15] and that group 2 *T. b. gambiense* isolates genetically cluster more closely to *T. b. brucei* and *T. b. rhodesiense* than to group 1 *T. b. gambiense*. However this may be a consequence of using East African *T. b. brucei* strains as a more recent study has shown evidence that group 2 *T. b. gambiense* and local *T. b. brucei* are a freely mating population [8].

1.2 The Genetic System

Before using genetic diversity to ask biological questions, one must first understand how *T. brucei* generates genetic diversity. Trypanosomes are a unicellular diploid protozoa [16] that have relatively small but complex genomes of approximately 35 Mb per haploid genome [17, 18]. Like many parasites, *T. brucei* has a complex life cycle involving both a mammalian host and an insect vector, during which it undergoes a series of morphological and biochemically distinct stages associated with adaptation or transmission to several disparate environments (i.e. the glucose rich mammalian bloodstream or the nutrient-poor tsetse gut).

Allele frequency analysis from multilocus enzyme electrophoresis of a population of *T. b. brucei* isolates [19], and the observation of hybrid-like enzyme profiles in *T. brucei* field stocks [12], provided early evidence that some degree of mating may be occurring in *T. brucei*. However, it was not until 1986 that genetic crosses were first performed between two different strains in a laboratory setting [20]. Although the conditions permissive for mating in *T. brucei* are still not fully understood, a number of experimental laboratory crosses, assisted in recent years by the use of fluorescently-tagged trypanosomes [21, 22], have enabled several important features of the process to be understood, culminating in the recent description of short-lived promastigote-like haploid gametes [23]. The F1 progeny that are formed in the salivary glands suggest that only a single round of mating occurs, with allelic segregation ratios that do not differ significantly from those predicted by a diploid Mendelian genetic system and meiosis [16, 24–26]. An extra consideration with *T. brucei* is that, unlike many other protozoa, mating is not an obligatory part of the life cycle and crosses usually result in large numbers of parental metacyclic clones in addition to F1 progeny [27, 28]. Furthermore, there is also evidence of self-fertilization within parental strains, both in the presence of other cross-mating strains [29, 30], and, more recently, during the transmission of a single strain [31]. However these self-fertilization events appear to occur at a low prevalence, suggesting that there may be some intrinsic mechanism that prevents self-fertilization in the parasite [29, 31].

Several crosses have been performed in a laboratory setting that show that genetic exchange can occur between different strains of *T. b. brucei*, group 2 *T. b. gambiense*, and *T. b. rhodesiense* [26, 28, 32]. There appear to be limited barriers to mating in *T. brucei* under laboratory conditions, even between different subspecies. A notable exception is that no cross has been successfully demonstrated with group 1 *T. b. gambiense*, supporting the conclusion from field studies that genetic exchange may not occur regularly in this subspecies [8, 33, 34]. Nevertheless, successful genetic crosses have allowed the formulation of genetic maps for strains of both *T. b. brucei* [35] and group 2 *T. b. gambiense* [36]. These were constructed using over 120 microsatellite markers distributed throughout the genome and formed linkage groups that could be aligned to the physical map of the *T. b. brucei* genome sequence [17]. The formulation and analysis of these genetic maps have also revealed that trypanosome genetics has both "crossing over" between pairs of homologous chromosomes with a recombination rate consistent with other organisms of a similar genome size and that chromosomes possess regions of high and low recombination (hot and cold spots). To summarize, *T. brucei* has a diploid, Mendelian genetic system similar to many other eukaryotes. Although trypanosome genetics has some unique features, there are enough similarities to other well-understood genetic systems that standard methods can be used to identify genetic determinants of important phenotypes such as virulence [37].

1.3 Genetic Diversity

In order to use genetic or population genomics to identify genes involved in important phenotypes, the organism being studied must possess both genotypic and phenotypic diversity. Fortunately, trypanosomes are an ideal case study as significant genotypic strain diversity has been shown in field populations [38, 39]. Some phenotypic diversity between strains of *T. brucei* has also been described, although less comprehensively than for genotypic diversity. These include drug resistance [40], tsetse transmission [41], and virulence [42]. It is this virulence phenotype that we will consider in our case study to harness genotypic variance to discover potential genes of interest.

1.4 The Virulence Phenotype

Virulence in African trypanosomes can be defined in several possible ways that take into account the various dynamics of the host and parasite. One method is to gauge the mortality of the infection, in that strains that are more virulent cause considerable mortality despite low parasitemia while less virulent strains that exhibit a higher growth rate have little impact on the host. A reverse genetics approach utilizing techniques such as RNAi and gene ablation may be able to identify many nonessential genes that contribute to virulence in the field but are unlikely to detect essential genes that cannot be ablated. In such cases, adopting a forward genetics strategy incorporating naturally occurring field phenotypes can become a powerful tool due to the fact that most variance in the field is likely to be due to allelic variation rather than gene loss. However, when using forward genetics approaches, there is usually limited initial information on the genes involved and it is therefore unclear whether the phenotype measured is due to a single determinant or multiple interacting determinants. Applying a combination of both reverse and forward genetics can resolve this issue.

A pertinent consideration when investigating virulence in field populations is the potential for complex interactions between the parasite and individual hosts that will have different genetic backgrounds that can influence the parasite phenotype. Furthermore, this can be made infinitely more complex when you consider that the parasite and host are embedded within a complex ecology that adds a large range of unknown variables to any study. For example, in many livestock animals it is common to find several species of trypanosome co-infecting a single host, in addition to numerous other pathogens [43–45]. Due to these complex interactions, host variation is usually minimized in studies of virulence by using inbred mice strains with limited genetic diversity. Unfortunately, as a consequence, this may impact on the biological relevance of such studies.

1.5 Species and Subspecies Variation

Despite the issues of working with genetically diverse field isolates, there are some well-defined variations in human disease severity between the different subspecies of *T. brucei* that can be exploited. After geography, one of the most defining differences between the two human infective subspecies is that *T. b. rhodesiense* causes an acute disease while *T. b. gambiense* causes a more chronic infection that can persist for decades [6]. Looking in more detail, within each human infective subspecies there is also a range of clinical pathologies. For example *T. b. rhodesiense* infections can be delineated into "mild" and "severe" forms of East African sleeping sickness that appear to be segregated somewhat by geography [46–48]. The parasites causing the two forms of the disease possess the same gene responsible for human infectivity (*SRA*) [46], so it may be that different parasite genotypes are affecting the severity of the disease. However, the alternative hypothesis cannot be discounted that there is a host trait that differentiates the patients living in the different disease foci that leads to the differences in disease progression and outcome. In *T. b. gambiense* infections, there is also a range of clinical features of the human disease, although less thoroughly studied. It is known, however, that group 1 *T. b. gambiense* infections follow a much more long-term and chronic infection course than group 2 *T. b. gambiense*. An interesting facet that has recently emerged is the discovery of asymptomatic individuals with detectable *T. b. gambiense* parasites and high antibody titers, who show no signs of clinical progression [49]. While there is some evidence that variation in virulence of the parasites may cause the differences in disease phenotype [49, 50], it is also possible that the differences in symptoms are due to host variation [51–53].

1.6 Strain-Specific Diversity

Although defining and measuring virulence between the different species, subspecies, and clades of *T. brucei* is a vital first step to understanding the disease profile, it provides little information as to the genetic basis for the variation. There are three possible

Table 1
Measures of trypanosome virulence

Phenotype	Parasite species	Host	Reference
Survival	*T. b. gambiense* and *T. congolense*	Mouse	[56]
Prepatent period	*T. b. gambiense* and *T. congolense*	Mouse	[56]
Maximum parasitemia	*T. b. gambiense* and *T. congolense*	Mouse	[56]
Anaemia (PCV)	*T. congolense* and *T. b. brucei*	Mouse	[42]
Organomegaly	*T. b. brucei*	Mouse	[42]
Reticulocytosis	*T. b. brucei*	Mouse	[42]
Macrophage activation	*T. b. brucei* and *T. b. gambiense*	Mouse and human	[56, 57]
Blood–brain barrier	*T. b. brucei* and *T. b. gambiense*	In vitro	[65]
Stage 1/2 progression	*T. b. brucei* and *T. b. rhodesiense*	Human	[48]
Asymptomatic	*T. b. brucei* and *T. b. gambiense*	Human	[50]

methods to elucidate the genes involved: firstly, to use biochemical and molecular analysis of candidate genes in vitro and in in vivo models of the disease process; secondly, using a genetic mapping approach to investigate inheritance of the phenotype in crosses of different strains; and thirdly, to examine population genomics and associate alleles with a phenotype. It is also important to screen for inheritable differences in gene expression as variation in cis-acting elements is responsible for a large component of phenotypic variation in several different organisms, including humans [54, 55]. Gene expression-based analysis can also be incorporated with genetic analysis to conduct an expression quantitative trait loci (e-QTL) approach. For analysis of inheritance during crosses and population genomics associations, it is necessary to first identify strain variation in the virulence phenotype within a species or subspecies. As previously highlighted, clinical studies suggest that such strain-specific variation occurs, although it is likely to be also influenced by other factors, such as host susceptibility.

Most studies of virulence have used laboratory studies in rodents and so raise the question of the relevance to the situation in the field. It is therefore important that when genetic or other mechanisms determining a phenotype are identified, appropriate field studies are used to validate these experimental findings. Several measures of virulence have been defined (Table 1) and involve a broad set of phenotypes in the mouse model and natural hosts. There have been a number of studies demonstrating significant differences between *T. brucei* strains with many of these different measures of virulence. For example, when sets of strains of *T. b. gambiense* from Cote d'Ivoire are inoculated individually into mice, a range of strain-specific differences in infectivity, mortality, prepatent period, and parasitemia are found [56]. The two most widely different strains, used in these studies, were shown to have highly statistically significant differences in the prepatent period, maximum parasitemia, and survival time [56]. There was also a marked difference in the ability to stimulate arginase activity in isolated macrophages between the two strains, suggesting a possible mechanism for the differences in virulence [56]. Strain-specific differences in virulence have also been demonstrated in *T. b. brucei* infections, involving organomegaly, reticulocytosis, and anaemia, although there were limited differences in mortality or maximum parasitemia [57]. There is therefore good evidence for strain-related variation in virulence in *T. brucei* using survival time, prepatent period, anaemia (PCV), maximum parasitemia, organomegaly, and reticulocytosis as proxies for virulence. These traits are liable to be inheritable and have a genetic basis, although it is unlikely that a single gene determines such complex pathology.

1.7 Genetic Basis of Virulence

While a possible approach to finding genes involved in virulence is a reverse genetic analysis of candidate genes, this is hardly an efficient method, especially when the complexity of interactions and the number of potentially independent phenotypes is considered. A more efficient approach is to exploit any phenotypic variation found in the target species. As an added benefit, genes determining this natural variation will be more relevant to virulence in the field and give a better understanding of the evolution of virulence. The observed diversity in virulence allows genomic approaches to identify genes that determine the phenotype. Two approaches have been previously utilized in such research: firstly crosses between strains that differ in virulence phenotypes with differing inheritance patterns (for example quantitative trait analysis) and secondly, population genomics, using association analysis between phenotype and genotype. Due to the inherent difficulties in isolating and characterizing sufficient numbers of field strains, the latter has not yet been applied to African trypanosomes.

1.8 Quantitative Trait (QTL) Analysis

The large number of different methods to define virulence and the multitude of measurable phenotypes (Table 1) indicates that virulence is likely to involve many genes, making reverse genetics difficult and costly. Using a forward genetic analysis that can work holistically on the whole genome may be a more efficient approach to identify loci of interest by using the segregation of the phenotype in crosses and treating the phenotypes as quantitative traits (QTL analysis). A summary of the approach used in *T. brucei* is illustrated in Fig. 1. In brief, analysis is based on phenotypic characterization of F1 progeny derived from a *T. brucei* cross between two strains which differ in the phenotype of interest, followed by linkage analysis using the microsatellite-based genetic map [35].

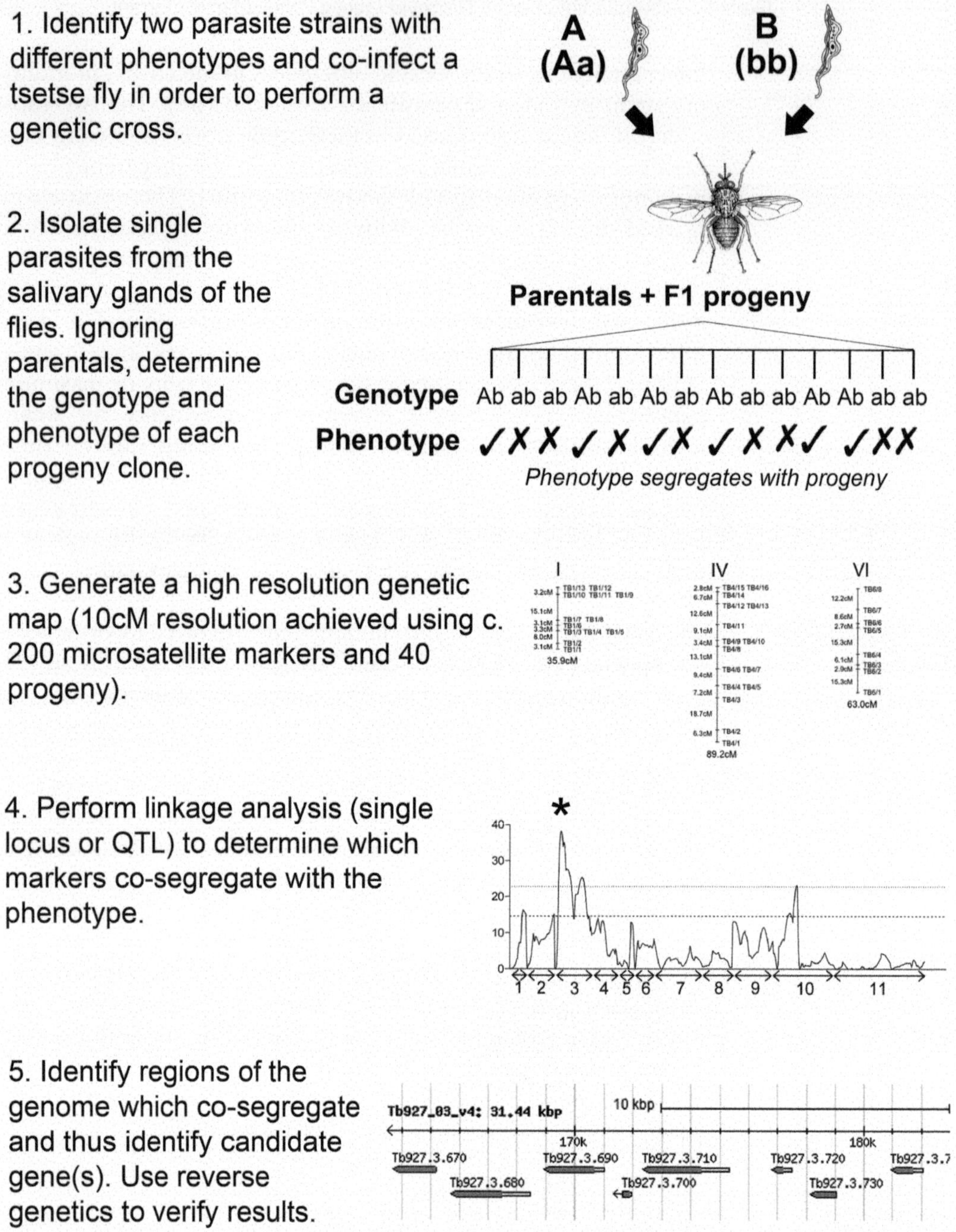

Fig. 1 Diagram of a trypanosome cross and the strategy used to map genes determining phenotypes such as virulence. Two strains of parental trypanosomes, one of which displays the phenotype of interest, are used to co-infect tsetse flies. Once the infection has developed to the salivary gland stage, indicated by the presence in the salivary glands of trypanosomes with genetic markers derived from both parents, individual trypanosomes can be expanded vegetatively. This may occur by direct isolation of individual metacyclic trypanosomes from dissected salivary glands, which are then expanded in mice, or by allowing the tsetse fly to transmit the uncloned population of metacyclics to a mouse through a blood meal followed by cloning of individual bloodstream trypanosomes after expansion. These "clones" are then genotyped with microsatellite markers and phenotyped by infection of mice. The phenotype is treated as being determined by more than one locus and the segregation of the quantitative trait used to map the loci on the genetic map

By comparing inheritance of markers on the genetic map with the inheritance of the phenotype in the progeny, markers that co-segregate with the phenotype can be identified, allowing an estimate of the likelihood of linkage to the phenotype to be made [58]. A statistical value is determined for each marker that indicates the logarithm of the odds (LOD) value of the likelihood that there is genetic linkage at that marker relating to the phenotype. The statistically significant LOD value used for diploid organisms is usually 3, which suggests that the odds are a thousand to one in favor of genetic linkage to the phenotype ($p < 0.001$). Several software packages can be used to assess linkage of inherited progeny haplotypes to a phenotype, although two in particular were used in the study by Morrison et al. [42]; MapManager QTX [59] and QTL Express [60]. Although both use a similar methodology to assign linkage, they differ in the protocols designed to assess significance. These programs were therefore used in tandem to ensure that identified loci were robust.

Unlike classical QTL analysis that uses highly inbred parents, this case study [42] used two *T. b. brucei* strains that had only recently been derived from the field so as to maximize the likelihood that observed phenotypes will be relevant to field populations. The *T. b. brucei* strains used were TREU927 (the genome strain) originally derived from a tsetse fly in Kenya, and STIB247 isolated from a Hartebeest in Tanzania. An important factor that determined the choice of these strains was that they cause different pathology in the host based on several measures of virulence and that they could be crossed. Due to using outbred parasite strains, the F1 progeny from the mating experiment only permitted the linkage analysis of loci that were heterozygous in one of the parents (equivalent to F2 progeny in classical studies) but did not allow the analysis of loci that were homozygous. The investigation was focused on mapping the loci that determine strain-specific differences in the measures of virulence that are more pertinent to *T. b. brucei*, namely organomegaly, reticulocytosis, and anaemia [42]. Individual progeny from the cross were inoculated into inbred mice and the phenotype of each parameter were quantified. As to be expected from a multi-gene effect, the measures of virulence in the progeny segregated in a semiquantitative manner. This suggested that allelic variation at several loci were determinants of virulence, so genetic linkage analysis was a valid approach to defining such loci.

Both splenomegaly and hepatomegaly showed evidence for a highly significant QTL (LOD scores >7) on chromosome 3 accounting for 66 % and 64 %, respectively, of the phenotypic variance [42]. Although it remains difficult to assess the number of genes that may be involved in the phenotypic variance, this locus appears to be the major contributor to these two phenotypes. The region of interest is still quite large, containing more than 300 genes, but this number is significantly reduced from the nearly

10,000 putative open reading frames identified in the *T. brucei* genome [17]. Having identified the region of interest, the addition of more markers, using RFLPs for example, allows more detailed mapping to the region using existing progeny. Alternatively, adding more progeny clones from new mating events would increase the chance that crossovers occurred within the locus and allow it to be mapped more finely. This would, however, be a considerable amount of work as laboratory crossing through the tsetse is still inefficient. Finally, analysis of gene expression across the region would allow the number of possible candidates to be more finely tuned. It would be expected that any virulence-associated gene would be expressed in the bloodstream stage of the parasite to affect the mammalian host allowing insect-stage-specific genes to be discounted [42]. When the number of genes is reduced to a manageable level, a directed reverse genetics approach can be utilized to confirm which alleles or genes are responsible for the phenotype. In addition to the locus on chromosome 3, other significant QTLs were identified for reticulocytosis, anaemia, and organomegaly on several chromosomes [42]. This case study has shown that a genetic analysis can be used to map the loci determining natural variation in virulence with no prior knowledge of the genes involved, which is an extremely powerful tool. The study can also be extended to investigate phenotypes that do not differ between TREU927 and STIB247 by undertaking further crosses between other parasite strains or by further phenotyping of the parental and progeny clones from other previously generated crosses [61]. Indeed, effectively any phenotype can be investigated that has a measurable difference between the parental strains.

2 Potential Association Studies

With the recent development of rapid and more economical genomic sequencing technology, a second genetic approach using population genomics to identify genes determining important phenotypes has become a possibility. Although not yet utilized for trypanosome research, the approach has successfully been applied to identify drug resistance genes in *Plasmodium falciparum* [62], virulence factors in *Toxoplasma gondii* [63], and genes involved in a wide range of phenotypes in *Saccharomyces cerevisiae* [64]. If applied to trypanosomes, this approach would require a large collection of phenotyped strains from one or more parasite populations. Examining the virulence phenotype by this method would involve measuring various virulence parameters in inbred mice and then whole-genome sequencing each strain using next-generation sequencing technology. These strains would then be subdivided into different phenotypes classes and any SNPs or haplotypes that characteristically differed between classes would be candidate gene regions determining virulence factors. This approach is summarized

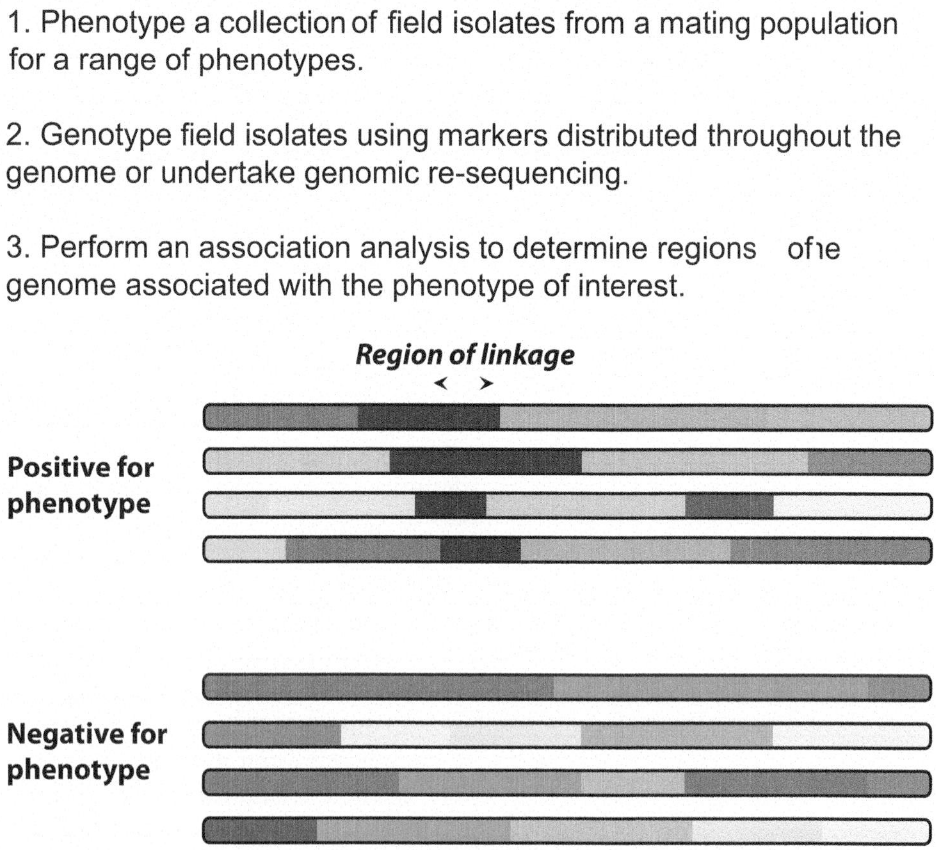

Fig. 2 Population genomic association analysis. The diagram illustrates how strains are phenotyped for one or more measures of virulence and then divided into classes of similar virulence (in this example for highly virulent and avirulent). The isolates are then genotyped using markers distributed over the whole genome or by whole-genome sequencing. In this example haplotypes of each strain are illustrated just using a single chromosome (for simplicity). The black haplotype appears to be associated with high virulence and is not present in the avirulent class. Other haplotypes are randomly distributed across the virulence classes and so show no association

in a hypothetical experiment (Fig. 2) in which the highlighted region shared between all members of the virulent phenotypic class is linked to a virulence locus.

Once a locus had been identified, narrowing it down to a specific gene or genes would then be undertaken in a similar way to that previously outlined for QTL analysis, for example, eliminating genes that are not expressed in the bloodstream form of the disease and those not having SNPs. Again, when the number of candidates is reduced to a manageable number they can be investigated further using reverse genetics. There are several limitations to using such a genomics-based approach, especially given the unknowns of using field populations. The level of mating or the amount of polymorphism in a population would significantly affect the number of samples required for statistical significance, so such studies would ideally have to be preceded by a large strain-sequencing project of a discrete parasite population. Despite these caveats, a major benefit to the approach is that it allows study of any phenotype that exhibits diversity rather than just the phenotypes that differ between two parental strains in laboratory crosses.

3 Biochemical and Cell Biological Analysis

An additional approach that does not rely on strain variation is to use in vitro models of host functions that could be modulated by parasite strains of different virulence (in vivo), such as macrophage arginase activation. Parasite protein extracts that produce phenotypic effects on macrophages can be purified by fractionation and the active proteins identified by mass spectrometry. Candidate genes can also be tested by expressing them as recombinant proteins or alternatively by knocking down expression in the trypanosome to investigate whether they have a phenotypic effect. An example of an in vitro system that uses macrophage arginase activation to assess possible virulence factors has recently been described using macrophages co-cultured with *T .b. gambiense* [56]. Two strains of *T .b. gambiense* with different in vivo virulence profiles induced different levels of macrophage arginase expression. By using conditioned culture medium with the parasites removed, it was shown that the protein effectors are secreted components and fractionation of these components will either elucidate the individual protein involved or at least reduce the list of candidates to one suitable for a reverse genetics screen. Another well-characterized in vitro model is the penetration of brain microvascular endothelial cells by *T. brucei* which is used as a proxy for invasion of the blood–brain barrier [65]. Inhibitors of the parasite cysteine proteinase brucipain are able to prevent invasion, suggesting that this gene is essential for this particular marker of virulence. For this invasion phenotype, there is also evidence of diversity between strains indicating that a human infective *T. b. rhodesiense* strain is better able to cross the endothelial layer compared to *T. b. brucei* strains. The isolate also has considerably higher amounts of brucipain activity, which coupled with the inhibitor experiments suggests that this protein is a key determinant of the endothelial layer invasion phenotype and a virulence factor [66]. It is, however, important to note that these effects have not yet been demonstrated in vivo and so may not be relevant outside of the model.

4 Conclusions and Future Prospects

Although research on the genetic basis of trypanosome virulence is at a relatively early stage, the existence of both interspecies and intraspecies variation can be exploited to identify candidate genes for various phenotypes. Molecular and biochemical analysis has shown that the cysteine proteinase brucipain is involved in the invasion of parasites across an in vitro endothelial layer and may have relevance to invasion of the blood–brain barrier, a key virulence phenotype [65]. This is also notable in that a human infective strain that possesses more brucipain than the *T. b. brucei* used in the experiment was better able to cross the endothelial layer. RNAi knockdown of brucipain expression affects virulence in vivo, with decreased mortality and splenomegaly [67], although the in vivo effect on penetration of the blood–brain barrier was not tested.

Two strains of *T. b. gambiense* that differ in virulence and in their secreted proteins have been shown to differentially activate macrophage arginase, suggesting that secreted proteins are involved in this measure of virulence [56]. In a separate experiment the differences in virulence between two different strains of *T. b. brucei* show that the differences in phenotype are associated with the activation of the innate immune response in the spleen, also partially involving arginase. These two facts may be related although how is unclear. It is also unknown if these virulence measures are due to brucipain. As outlined in the case study, a major locus determining splenomegaly has successfully been mapped [42], but it does not include the brucipain gene. The same study also identified a further locus on chromosome 2 that contributes to splenomegaly and reticulocytosis although these again do not contain the brucipain gene. These separate measures of virulence, and the evidence for brucipain, suggest that there are at least three loci or genes that have a role as virulence factors in *T. brucei* infections, however this is likely a large underestimate. With at least three different genes and allelic variants identified to be involved with virulence, it is evidently a complex phenotype that is consistent with the many different phenotypes observed during infection (Table 1).

All the studies discussed here rely on either in vitro or mouse models of virulence, so it is important to consider whether the identified genes and loci are relevant to the field. Unfortunately study of field infections is complicated by both the sheer practical issues of working in the field but also the significant difference in genetic diversity in the natural hosts compared to inbred mice. Any parasite genes or alleles that have been identified in model systems should be examined in field populations to test for association with virulence in the natural hosts. With livestock trypanosomiasis, experimental infections with different strains of parasites is feasible and could provide an approach to analyse virulence. Such experimentation is obviously not applicable in the human disease, so these studies would have to rely on natural infections. It is important to restate that the studies described here examine naturally occurring virulence variation, so any identified genes are more likely to have relevance to virulence in the field.

In addition to virulence factors found within the parasite there is a growing body of research investigating variation of host response to infection and the genetic basis of tolerance or "resistance" [68–71]. For example, mapping studies of *T. congolense* infections in strains of mice with varying susceptibility to infection have identified three major QTL regulating survival time [68]. Two QTL regions were refined, generating strong candidates for this phenotype using next-generation re-sequencing and array-based comparative genomic hybridization [72]. Similarly, genetic analysis in cattle has been undertaken to identify "trypanotolerance" loci that are major determinants of disease outcome [73, 74]. In addition there are several association studies looking at human disease that identified specific host genetic variants that affect the response to infection [69]. Interestingly, there are also recently described cases of asymptomatic patients of *T. b. gambiense* [49] and a range of disease severities in *T. b. rhodesiense* infections, suggesting that trypanotolerance may also occur in humans. This demonstrates that there is a significant genetic component in the host that determines how virulent an infection will be. The degree of interaction between the host and parasite determinants of virulence is still unknown and will need complex experimental design to tease apart the relative roles of host and parasite in the virulence phenotype. For example, what would be the result of a virulent parasite infecting a highly tolerant host—which phenotype would be the determinant? Recent advances in methodologies, tools, and technologies will allow us to readily identify host and parasite loci involved in many phenotypes, including virulence, with little prior knowledge. Once the loci and genes are identified it will be possible to elucidate the interactions between parasite and host and fully define the differences between symptomatic and asymptomatic infection. While this would be important in understanding virulence and pathogenesis to predict prognosis, it also raises the opportunity to intervene in these interactions to prevent or minimize the pathological consequences of infection.

References

1. Shaw A (2004) Economics of African Trypanosomiasis. The trypanosomiases. CABI, London, UK
2. Simarro PP, Diarra A, Postigo JAR, Franco JR, Jannin JG (2011) The human African trypanosomiasis control and surveillance programme of the World Health Organization 2000–2009: the way forward. PLoS Negl Trop Dis 5:e1007. doi:10.1371/journal.pntd.0001007
3. World Health Organization (2006) Weekly epidemiological record: relevé épidémiologique hebdomadaire
4. Odiit M, Coleman PG, Liu W-C, McDermott JJ, Fèvre EM et al (2005) Quantifying the level of under-detection of *Trypanosoma brucei rhodesiense* sleeping sickness cases. Trop Med Int Health 10:840–849. doi:10.1111/j.1365-3156.2005.01470.x
5. Fèvre EM, Wissmann B, Welburn SC (2008) The burden of human African trypanosomiasis. PLoS Negl Trop Dis 2:e333
6. Hoare CA (1972) The trypanosomes of mammals. A zoological monograph. Blackwell Scientific Publications, NJ

7. Beadell JS, Balmer O, Gibson W, Caccone A (2011) Phylogeography and taxonomy of *Trypanosoma brucei*. PLoS Negl Trop Dis 5:e961
8. Capewell P, Cooper A, Duffy CW, Tait A, Turner CM et al (2013) Human and animal trypanosomes in Côte d'Ivoire form a single breeding population. PLoS One 8:e67852. doi:10.1371/journal.pone.0067852
9. Gibson W (1986) Will the real *Trypanosoma b. gambiense* please stand up. Parasitol Today 2:255–257
10. Mehlitz D, Zillmann U, Scott CM (1982) Epidemiological studies on the animal reservoir of Gambiense sleeping sickness. Part III. Characterization of trypanozoon stocks by isoenzymes and sensitivity to human serum. Tropenmed Parasitol 33:113–118
11. Tait A, Babiker EA, Le Ray D (1984) Enzyme variation in *Trypanosoma brucei* ssp. I. Evidence for the sub-speciation of Trypanosoma brucei gambiense. Parasitology 89:311–326
12. Gibson W, Marshall DC, Godfrey DG (1980) Numerical analysis of enzyme polymorphism: a new approach to the epidemiology and taxonomy of trypanosomes of the subgenus *Trypanozoon*. Adv Parasitol 18:175–246
13. Godfrey DG, Scott CM, Gibson WC, Mehlitz D (1987) Enzyme polymorphism and the identity of *Trypanosoma brucei gambiense*. Parasitology 94:337–347
14. Capewell P, Veitch NJ, Turner CMR, Raper J, Berriman M et al (2011) Differences between *Trypanosoma brucei gambiense* groups 1 and 2 in their resistance to killing by trypanolytic factor 1. PLoS Negl Trop Dis 5:e1287. doi:10.1371/journal.pntd.0001287
15. Balmer O, Beadell JS, Gibson W, Caccone A (2011) Phylogeography and taxonomy of *Trypanosoma brucei*. PLoS Negl Trop Dis 5:e961. doi:10.1371/journal.pntd.0000961
16. Turner CMR, Sternberg J, Buchanan N, Smith E, Hide G et al (1990) Evidence that the mechanism of gene exchange in *Trypanosoma brucei* involves meiosis and syngamy. Parasitology 106:209–214
17. Berriman M (2005) The genome of the African trypanosome *Trypanosoma brucei*. Science 309:416–422. doi:10.1126/science.1112642
18. Jackson AP, Sanders M, Berry A, McQuillan J, Aslett MA et al (2010) The genome sequence of *Trypanosoma brucei gambiense*, causative agent of chronic human African trypanosomiasis. PLoS Negl Trop Dis 4:e658
19. Tait A (1980) Evidence for diploidy and mating in trypanosomes. Nature 287:536–538
20. Jenni L, Marti S, Schweizer J, Betschart B, Le Page RWF et al (1986) Hybrid formation between African trypanosomes during cyclical transmission. Nature 322:173–175. doi:10.1038/322173a0
21. Bingle LE, Eastlake JL, Bailey M, Gibson W (2001) A novel GFP approach for the analysis of genetic exchange in trypanosomes allowing the in situ detection of mating events. Microbiology 147:3231–3240
22. Peacock L, Ferris V, Bailey M, Gibson W (2008) Fly transmission and mating of *Trypanosoma brucei brucei* strain 427. Mol Biochem Parasitol 160:100–106
23. Peacock L, Bailey M, Carrington M, Gibson W (2014) Meiosis and haploid gametes in the pathogen *Trypanosoma brucei*. Curr Biol 24:181–186. doi:10.1016/j.cub.2013.11.044
24. Gibson W, Garside L (1991) Genetic exchange in *Trypanosoma brucei brucei*: variable chromosomal location of housekeeping genes in different trypanosome stocks. Mol Biochem Parasitol 45:77–89
25. Macleod A, Tweedie A, McLellan S, Taylor S, Cooper A et al (2005) Allelic segregation and independent assortment in *T. brucei* crosses: proof that the genetic system is Mendelian and involves meiosis. Mol Biochem Parasitol 143:12–19
26. Tait A, Macleod A, Tweedie A, Masiga D, Turner CMR (2007) Genetic exchange in *Trypanosoma brucei*: evidence for mating prior to metacyclic stage development. Mol Biochem Parasitol 151:133–136. doi:10.1016/j.molbiopara.2006.10.009
27. Schweizer J, Tait A, Jenni L (1988) The timing and frequency of hybrid formation in African trypanosomes during cyclical transmission. Parasitol Res 75:98–101. doi:10.1007/BF00932707
28. Gibson W, Peacock L, Ferris V, Williams K, Bailey M (2008) The use of yellow fluorescent hybrids to indicate mating in *Trypanosoma brucei*. Parasit Vectors 1:4. doi:10.1186/1756-3305-1-4
29. Tait A, Buchanan N, Hide G, Turner CMR (1996) Self-fertilisation in *Trypanosoma brucei*. Mol Biochem Parasitol 76:31–42. doi:10.1016/0166-6851(95)02528-6
30. Gibson W, Winters K, Mizen G, Kearns J, Bailey M (1997) Intraclonal mating in *Trypanosoma brucei* is associated with outcrossing. Microbiology 143:909–920
31. Peacock L, Ferris V, Bailey M, Gibson W (2009) Intraclonal mating occurs during tsetse transmission of *Trypanosoma brucei*. Parasit Vectors 2:43. doi:10.1186/1756-3305-2-43

32. Gibson W, Stevens J (1999) Genetic exchange in the *Trypanosomatidae*. Adv Parasitol 43: 1–46. doi:10.1016/S0065-308X(08)60240-7
33. Koffi M, De Meeûs T, Bucheton B, Solano P, Camara M et al (2009) Population genetics of *Trypanosoma brucei gambiense*, the agent of sleeping sickness in Western Africa. Proc Natl Acad Sci U S A 106:209–214
34. Morrison LJ, Tait A, McCormack G, Sweeney L, Black A et al (2008) *Trypanosoma brucei gambiense* type 1 populations from human patients are clonal and display geographical genetic differentiation. Infect Genet Evol 8:847–854
35. MacLeod A (2005) The genetic map and comparative analysis with the physical map of *Trypanosoma brucei*. Nucleic Acids Res 33:6688–6693
36. Cooper A, Tait A, Sweeney L, Tweedie A, Morrison L et al (2008) Genetic analysis of the human infective trypanosome *Trypanosoma brucei gambiense*: chromosomal segregation, crossing over, and the construction of a genetic map. Genome Biol 9:R103
37. Tait A, Masiga D, Ouma J, Macleod A, Sasse J et al (2002) Genetic analysis of phenotype in *Trypanosoma brucei*: a classical approach to potentially complex traits. Philos Trans R Soc Lond B Biol Sci 357:89–99. doi:10.1098/rstb.2001.1050
38. MacLeod A, Tweedie A, Welburn SC, Maudlin I, Turner CM et al (2000) Minisatellite marker analysis of *Trypanosoma brucei*: reconciliation of clonal, panmictic, and epidemic population genetic structures. Proc Natl Acad Sci 97:13442–13447
39. Agbo EC, Majiwa PAO, Claassen HJHM, te Pas MFW (2002) Molecular variation of *Trypanosoma brucei* subspecies as revealed by AFLP fingerprinting. Parasitology 124: 349–358
40. Kibona SN, Matemba L, Kaboya JS, Lubega GW (2006) Drug-resistance of *Trypanosoma b. rhodesiense* isolates from Tanzania. Trop Med Int Health 11:144–155. doi:10.1111/j.1365-3156.2005.01545.x
41. Welburn SC, Maudlin I, Milligan PJ (1995) *Trypanozoon*: infectivity to humans is linked to reduced transmissibility in tsetse. I. Comparison of human serum-resistant and human serum-sensitive field isolates. Exp Parasitol 81:404–408. doi:10.1006/expr.1995.1131
42. Morrison LJ, Tait A, McLellan S, Sweeney L, Turner CMR et al (2009) A major genetic locus in *Trypanosoma brucei* is a determinant of host pathology. PLoS Negl Trop Dis 3:e557. doi:10.1371/journal.pntd.0000557
43. Pinchbeck GL, Morrison LJ, Tait A, Langford J, Meehan L et al (2008) Trypanosomosis in the gambia: prevalence in working horses and donkeys detected by whole genome amplification and PCR, and evidence for interactions between trypanosome species. BMC Vet Res 4:7. doi:10.1186/1746-6148-4-7
44. Bronsvoort B, Wissmann BV, Fèvre EM, Handel IG, Picozzi K et al (2010) No gold standard estimation of the sensitivity and specificity of two molecular diagnostic protocols for *Trypanosoma brucei* spp. in Western Kenya. PLoS One 5:e8628
45. Cox AP, Tosas O, Tilley A, Picozzi K, Coleman P et al (2010) Constraints to estimating the prevalence of trypanosome infections in East African zebu cattle. Parasit Vectors 3:82. doi:10.1186/1756-3305-3-82
46. MacLean L, Chisi JE, Odiit M, Gibson WC, Ferris V et al (2004) Severity of human African trypanosomiasis in East Africa is associated with geographic location, parasite genotype, and host inflammatory cytokine response profile. Infect Immun 72:7040–7044
47. Sternberg JM, MacLean L (2010) A spectrum of disease in human African trypanosomiasis: the host and parasite genetics of virulence. Parasitology 137:2007–2015. doi:10.1017/S0031182010000946
48. MacLean L, Chisi JE, Odiit M, GIBSON WC, Ferris V et al (2004) Severity of human african trypanosomiasis in East Africa is associated with geographic location, parasite genotype, and host inflammatory cytokine response profile. Infect Immun 72:7040–7044. doi:10.1128/IAI.72.12.7040-7044.2004
49. Jamonneau V, Ilboudo H, Kaboré J, Kaba D, Koffi M et al (2012) Untreated Human Infections by *Trypanosoma brucei gambiense* Are Not 100 % fatal. PLoS Negl Trop Dis 6:e1691. doi:10.1371/journal.pntd.0001691
50. Jamonneau V, Ravel S, Garcia A, Koffi M (2004) Characterization of *Trypanosoma brucei* sl infecting asymptomatic sleeping-sickness patients in Cote d'Ivoire: a new genetic group? Ann Trop Med Parasitol 98:329–337
51. Garcia A, Courtin D, Solano P, Koffi M, Jamonneau V (2006) Human African trypanosomiasis: connecting parasite and host genetics. Trends Parasitol 22:405–409. doi:10.1016/j.pt.2006.06.011
52. Bucheton B, MacLeod A, Jamonneau V (2011) Human host determinants influencing the outcome of *Trypanosoma brucei gambiense* infections. Parasite Immunol 33:438–447. doi:10.1111/j.1365-3024.2011.01287.x
53. Kaboré J, Koffi M, Bucheton B, Macleod A, Duffy C et al (2011) First evidence that

parasite infecting apparent aparasitemic serological suspects in human African trypanosomiasis are *Trypanosoma brucei gambiense* and are similar to those found in patients. Infect Genet Evol 11:1250–1255. doi:10.1016/j.meegid.2011.04.014

54. Göring HHH, Curran JE, Johnson MP, Dyer TD, Charlesworth J et al (2007) Discovery of expression QTLs using large-scale transcriptional profiling in human lymphocytes. Nat Genet 39:1208–1216. doi:10.1038/ng2119
55. Stranger BE, Nica AC, Forrest MS, Dimas A, Bird CP et al (2007) Population genomics of human gene expression. Nat Genet 39: 1217–1224. doi:10.1038/ng2142
56. Holzmuller P, Biron DG, Courtois P, Koffi M, Bras-Gonçalves R et al (2008) Virulence and pathogenicity patterns of *Trypanosoma brucei gambiense* field isolates in experimentally infected mouse: differences in host immune response modulation by secretome and proteomics. Microbes Infect 10:79–86. doi:10.1016/j.micinf.2007.10.008
57. Morrison LJ, McLellan S, Sweeney L, Chan CN, MacLeod A et al (2010) Role for parasite genetic diversity in differential host responses to *Trypanosoma brucei* Infection. Infect Immun 78:1096–1108. doi:10.1128/IAI.00943-09
58. Doerge RW (2002) Mapping and analysis of quantitative trait loci in experimental populations. Nat Rev Genet 3:43–52. doi:10.1038/nrg703
59. Manly KF, Cudmore RH, Meer JM (2014) Map manager QTX, cross-platform software for genetic mapping. Mamm Genome 12:930–932. doi:10.1007/s00335-001-1016-3
60. Seaton G, Haley CS, Knott SA, Kearsey M (2002) QTL express: mapping quantitative trait loci in simple and complex pedigrees. Bioinformatics 18:339–340
61. MacLeod A, Turner C, Tait A (2007) The system of genetic exchange in Trypanosoma brucei and other trypanosomatids. Trypanosomes: after the genome. Horizon, Linton, UK
62. Mu J, Myers RA, Jiang H, Liu S, Ricklefs S et al (2010) *Plasmodium falciparum* genome-wide scans for positive selection, recombination hot spots and resistance to antimalarial drugs. Nat Genet 42:268–271. doi:10.1038/ng.528
63. Reid AJ, Vermont SJ, Cotton JA, Harris D, Hill-Cawthorne GA et al (2012) Comparative genomics of the apicomplexan parasites *Toxoplasma gondii* and *Neospora caninum*: coccidia differing in host range and transmission strategy. PLoS Pathog 8:e1002567. doi:10.1371/journal.ppat.1002567
64. Liti G, Carter DM, Moses AM, Warringer J, Parts L et al (2009) Population genomics of domestic and wild yeasts. Nature 458:337–341. doi:10.1038/nature07743
65. Grab DJ, Nikolskaia O, Kim YV, Lonsdale-Eccles JD, Ito S et al (2004) African trypanosome interactions with an in vitro model of the human blood–brain barrier. J Parasitol 90:970–979. doi:10.1645/GE-287R
66. Nikolskaia OV (2006) Blood–brain barrier traversal by African trypanosomes requires calcium signaling induced by parasite cysteine protease. J Clin Invest 116:2739–2747. doi:10.1172/JCI27798
67. Abdulla M-H, O'Brien T, Mackey ZB, Sajid M, Grab DJ et al (2008) RNA interference of *Trypanosoma brucei* cathepsin B and L affects disease progression in a mouse model. PLoS Negl Trop Dis 2:e298. doi:10.1371/journal.pntd.0000298
68. Kemp SJ, Iraqi F, Darvasi A, Soller M, Teale AJ (1997) Localization of genes controlling resistance to trypanosomiasis in mice. Nat Genet 16:194–196. doi:10.1038/ng0697-194
69. Courtin D, Berthier D, Thevenon S, Dayo G-K, Garcia A et al (2008) Host genetics in African trypanosomiasis. Infect Genet Evol 8:229–238. doi:10.1016/j.meegid.2008.02.007
70. Iraqi F, Clapcott SJ, Kumari P, Haley CS, Kemp SJ et al (2014) Fine mapping of trypanosomiasis resistance loci in murine advanced intercross lines. Mamm Genome 11:645–648. doi:10.1007/s003350010133
71. Nganga JK, Soller M, Iraqi FA (2010) High resolution mapping of trypanosomosis resistance loci Tir2 and Tir3 using F12 advanced intercross lines with major locus Tir1 fixed for the susceptible allele. BMC Genomics 11:394. doi:10.1186/1471-2164-11-394
72. Goodhead I, Archibald A, Amwayi P, Brass A, Gibson J et al (2010) A comprehensive genetic analysis of candidate genes regulating response to *Trypanosoma congolense* infection in mice. PLoS Negl Trop Dis 4:e880. doi:10.1371/journal.pntd.0000880
73. Hanotte O, Ronin Y, Agaba M, Nilsson P, Gelhaus A et al (2003) Mapping of quantitative trait loci controlling trypanotolerance in a cross of tolerant West African N'Dama and susceptible East African Boran cattle. Proc Natl Acad Sci U S A 100:7443–7448. doi:10.1073/pnas.1232392100
74. Hill EW, O'Gorman GM, Agaba M, Gibson JP, Hanotte O et al (2005) Understanding bovine trypanosomiasis and trypanotolerance: the promise of functional genomics. Vet Immunol Immunopathol 105:247–258. doi:10.1016/j.vetimm.2005.02.004

Chapter 6

Identification and Analysis of Ingi-Related Retroposons in the Trypanosomatid Genomes

Frédéric Bringaud, Matthew Rogers, and Elodie Ghedin

Abstract

Transposable elements (TE), defined as discrete pieces of DNA that can move from one site to another site in genomes, represent significant components of eukaryotic genomes, including trypanosomatids. Up to 5 % of the trypanosomatid genome content is composed of retroposons of the ingi clade, further divided into subclades and subfamilies ranging from short extinct truncated elements (SIDER) to long active elements (ingi). Important differences in ingi-related retroposon content have been reported between trypanosomatid species. For instance, *Leishmania* spp. have expanded and recycled a whole SIDER family to fulfill an important biological pathway, i.e., regulation of gene expression, while trypanosome genomes are primarily composed of active elements. Here, we present an overview of the computational methods used to identify, annotate, and analyze ingi-related retroposons for providing a comprehensive picture of all these TE families in newly available trypanosomatid genome sequences.

Key words Transposable element, Ingi-related retroposon, Trypanosomatid, Computational methods, Identification, Annotation, Classification, Evolution, Consensus sequence, (Sub)family

1 Introduction

Mobile genetic elements, also called transposable elements (TEs), can be defined as DNA fragments that can move into new locations in the host genome by excision or replication of an existing copy. Despite their abundance in most genomes (over 40 % of the human genome [1]), TEs are often called "junk," "selfish," or "parasitic" DNA, because they appear as functionless DNA sequences replicating themselves into as many copies as possible. This view began to change in the early 1990s and now tends to be replaced by a "functionalist" view of TE biology. This is supported by a rapidly increasing number of reports describing domestication or exaptation of TE to play a role in cellular function, such as transcriptional regulation, and contribution to protein-coding regions (for a recent review *see* [2]). Since most of these are degenerate and extinct elements that are barely detectable as TEs [3–5],

Christopher Peacock (ed.), *Parasite Genomics Protocols*, Methods in Molecular Biology, vol. 1201,
DOI 10.1007/978-1-4939-1438-8_6, © Springer Science+Business Media New York 2015

developing bioinformatics tools and approaches to identify highly degenerate TEs is a major challenge. The aim of this chapter is to provide a bioinformatics approach (including a workflow) to identify and annotate active, as well as extinct, degenerate TEs using as example the trypanosomatid ingi retroposon family, which contains highly degenerate sequences involved in the regulation of gene expression [5].

The trypanosomatid family includes some of the most important protist parasites of humans in the genera *Leishmania* and *Trypanosoma*, as well as other species parasitic in a wide variety of vertebrates, invertebrates, ciliates, and plants [6]. The genomes of nine trypanosomatids have been sequenced and published to date, i.e., *T. brucei* [7], *T. b. gambiense* [8], *T. cruzi* [9], *L. major* [10], *L. braziliensis* [11], *L. infantum* [11], *L. mexicana* [12], *L. donovani* [13], and *L. tarentolae* [14, 15, 16]. During edition of this chapter the genome of two plant-infecting (*Phytomonas*) and two endosymbiont-bearing (Angomonas deanei and Strigomonas culicis) trypanosomatid species have been sequenced. In addition, a number of ongoing trypanosomatid genome projects have data available online (http://tritrypdb.org/tritrypdb), such as for *Crithidia fasciculata*, and two other African trypanosomes, *T. congolense* and *T. vivax*. All these genomes contain active and/or traces of inactive TEs (for reviews *see*: [17, 18]).

All TEs described to date in trypanosomatids belong to four groups: the VIPER LTR retrotransposons, the site-specific SLACS/CZAR retroposons (also named non-LTR retrotransposons), the ingi/L1Tc retroposons, and TATE, for which the TE class is unknown (for a recent review *see* [17, 18]). Retroposons of the ingi clade, which will be further considered herein, include two categories of active elements (*see* Fig. 1): (1) the long (4,736–5,419 bp) and autonomous elements originally characterized in *T. brucei* (Tbingi) [19, 20] and *T. cruzi* (L1Tc) [21], and subsequently described in *T. vivax* (Tvingi) [22] and *T. congolense* (Tcoingi and L1Tco) [22], and (2) the short (260–1,030 bp) and non-autonomous elements identified in *T. brucei* (TbRIME) [23], *T. cruzi* (NARTc) [24], and *T. vivax* (TvRIME) [22]. The short elements (TbRIME, TvRIME, and NARTc) are truncated versions that are mobilized by the retrotransposition machinery of the corresponding long elements (Tbingi, Tvingi, and L1Tc, respectively) [22, 25, 26]. Consequently, the Tbingi/TbRIME, Tvingi/TvRIME, and L1Tc/NARTc associations are considered as pairs of retroposons akin to the human LINE1/Alu pairs [27, 28].

Trypanosome and *Leishmania* genomes also contain highly degenerate elements related to retroposons of the ingi clade, named DIREs for "degenerate ingi-related elements" [29, 30]. Tbingi/TbRIME, L1Tc/NARTc, and DIREs share the first 76–79 residues, which constitute the hallmark of trypanosomatid retroposons ("76–79 bp signature"). Recently, small degenerate

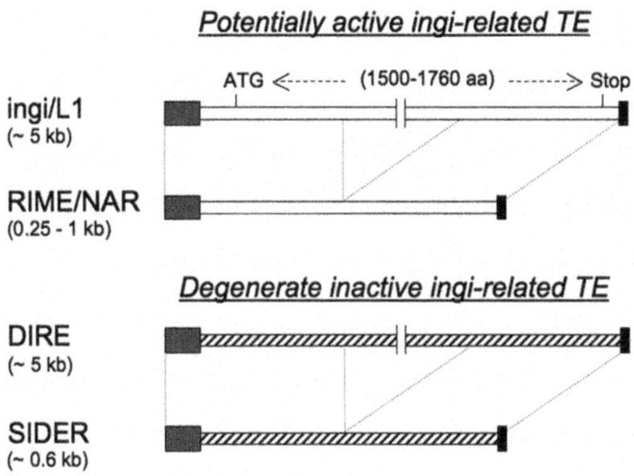

T.b.	*T.co.*	*T.v.*	*T.cr.*	*L.m.*	*L.b.*
Tbingi (4)	Tcoingi (6) L1Tco (1)	Tvingi (6)	L1Tc (1)	×	×
TbRIME (4)	×	TvRIME (6)	NARTc (1)	×	×
TbDIRE (4,5,6)	TcoDIRE (1,3,5,6)	TvDIRE (6)	TcrDIRE (1,4)	LmDIRE (2)	LbDIRE (1,2)
TbSIDER1 (6) TbSIDER2 (4)	TcoSIDER1 (6)	TvSIDER1 (6)	?	LmSIDER (2)	LbSIDER (2)

Fig. 1 Schematic representation of ingi-related retroposons identified in the trypanosomatid genomes. The *left panel* represents potentially active (ingi/L1 and RIME/NAR) and degenerate (DIRE and SIDER) retroposons. The conserved "76–79 bp signatures" and the poly(dA) tails are showed by *grey* and *black boxes*, respectively. The approximate size of these mobile elements is indicated in *brackets*. The long and autonomous ingi/L1 elements are the only retroposons coding for a protein (from 1,500 to 1,760 aa long) responsible for retrotransposition of themselves or other ingi-related retroposons. The *right panel* shows the name of each ingi-related family in the genome of *T. brucei* (*T.b.*), *T. congolense* (*T.co.*), *T. vivax* (*T.v.*), *L. major* (*L.m.*), and *L. braziliensis* (*L.b.*), as well as the ingi subclade(s) they belong to (from 1 to 6 in *brackets*). A *cross* (x) and an *interrogation mark* (?) mean that no such TE have been identified or their presence has not been investigated, respectively

retroposons (~0.55 kb) named LmSIDERs (for "short interspersed degenerate retroposons") containing this motif have been identified in the genomes of *L. major* [5], *L. infantum*, and *L. braziliensis* [5, 31]. LmSIDER constitutes the largest retroposon family described so far in trypanosomatids; members are located in the 3′-UTR of genes, where they play a role in the regulation of gene expression [5, 32, 33].

2 Materials

Computational TE analyses can be performed on a local desktop machine with Internet access. However, large-scale studies require a local software installation, typically in a UNIX environment (*see* **Note 1**), a working knowledge of the UNIX language, and the ability to install applications in a Linux environment. Also essential is a basic knowledge of genome annotation and sequence viewers (while there are many out there, the authors of this chapter are partial to Artemis). Basic knowledge of PERL is necessary for some of the pipelines proposed. For the identification and analysis of mobile elements, the following software materials will be necessary.

1. *Rapid Annotation Transfer Tool (RATT)*: freely available from its own sourceforge site (http://ratt.sourceforge.net/), or as

part of the Pagit package (http://www.sanger.ac.uk/resources/software/pagit/).

2. *Exonerate*: for alignment of full-length ingi peptides against reference genomes (www.ebi.ac.uk/~guy/exonerate/).
3. *Tabix*: used for indexing and retrieving rows from a tabular file format, such as a General Feature Format File (GFF); it is freely available as part of the Samtools package (http://sourceforge.net/projects/samtools/files/).
4. *Water*: used for performing Smith-Waterman local alignments; freely available as part of the EMBOSS package (http://emboss.sourceforge.net/download/).
5. *Artemis*: or a similar sequence viewer program able to handle GFF format files (http://www.sanger.ac.uk/resources/software/artemis/).
6. *Blastall*: or a similar multifunctional Blast package installed locally; freely available from NCBI (ftp://ftp.ncbi.nlm.nih.gov/blast/executables/blast+/LATEST/).
7. A web browser with access to major online genome databases (*see* http://tritrypdb.org/tritrypdb/).
8. Other recommended software for DNA and protein analysis (ClustalX, DNA strider, etc.).

3 Methods

3.1 Identification of Ingi-Related Retroposons

Identification of ingi-related sequences in all the investigated trypanosomatid genomes was primarily based on Blast searches using tBlastN and BlastN approaches, followed by extensive manual annotation and curation. Here we propose a workflow that will help jumpstart curation of the elements by replacing a significant portion of what was originally done as manual work.

Ingi sequences are identifiable by BlastN searches within species (e.g., Tbingi), but are not conserved at the nucleotide sequence level across species (e.g., Tbingi versus L1Tc) and are poorly conserved at the amino acid level (Tbingi and L1Tc peptides are only 23.8 % identical). Further complicating prediction of ingi elements across trypanosomatids is the phylogenetic diversity of ingi elements, with identification of six different subclades [22]; for example ingi6 (Tcoingi) and ingi1 (L1Tco) in *T. congolense*. For this reason, BlastN searches alone will not suffice in making predictions in distantly related trypanosomatid genomes. We propose a pipeline relying on peptide to DNA alignments using representative members of each ingi subclade as a query. It is important to mention at this stage that the proposed electronic annotation pipeline is useful to identify and perform a pre-annotation of transposable elements in a given genome; manual detailed curation and annotation will however be required to perform in-depth analyses.

At the core of our ingi prediction pipeline is the European Bioinformatic Institute's (EBI) sequence alignment program Exonerate [34]. We use Exonerate to query representative full-length peptides from the Tbingi, Tcoingi, Tvingi, L1Tc, and L1Tco families (accession numbers: JQ917146, JQ917147, JQ917148, JQ917144, and JQ917145, respectively—*see* **Note 2**) against a reference genome. Custom Perl scripts are then employed to select the highest scoring alignment from regions where multiple ingi family members have been aligned (Fig. 2), and to generate CDS features for each predicted element. We have designed a wrapper for exonerate called Ingihelper.pl (https://sites.google.com/a/nyu.edu/ghedin-lab/tools), which accepts as input a fasta file of peptide sequences (preferably one for each family of ingi) as query, and a whole-genome sequence as subject. This script calls exonerate to perform a protein2DNA alignment, returning a GFF file of target alignments, and then searches across the GFF file for overlapping alignments, selecting the highest scoring alignment when this occurs. This frequently occurs due to the similarity of all ingi family members at the peptide level. These results are outputted in a modified GFF format where the gene model and corresponding coding sequence features can be viewed for each ingi prediction, allowing a quick assessment of the degeneracy in each element (Fig. 2). Our goal with this method is not to provide a fully automated prediction pipeline for ingi-related elements in trypanosomatids, but to rapidly provide a mostly (>90 %) complete set of ingi predictions in a format that can then be easily viewed and combined with other predictions (e.g., Blast) by annotators familiar with the structure of these elements. In most cases, a human eye will be required to determine the absolute 5′ and 3′ ends of these elements and assess whether neighboring ingi alignments belong to the same element. A major advantage of Exonerate is the multiple output options available. Ingi predictions can be outputted not just in GFF format for viewing and further manual curation, but can also be outputted in Fasta format for multiple alignments, allowing for identification of ingi subclade using phylogenetic methods. We recommend using Artemis to view Exonerate and Blast outputs, and to annotate ingi-related retroposons. Although there are other freely available sequence viewers, the authors of this chapter favor the use of Artemis for its versatility in both sequence annotation and analysis, its ability to read and output sequence data in a variety of formats (EMBL, GenBank, and GFF), and its legacy as a tool in trypanosomatid genome annotation.

We have tested this method against manual predictions that have been performed on the *T. vivax* and *T. congolense* genomes. The *T. congolense* genome is a challenge for ingi prediction as it

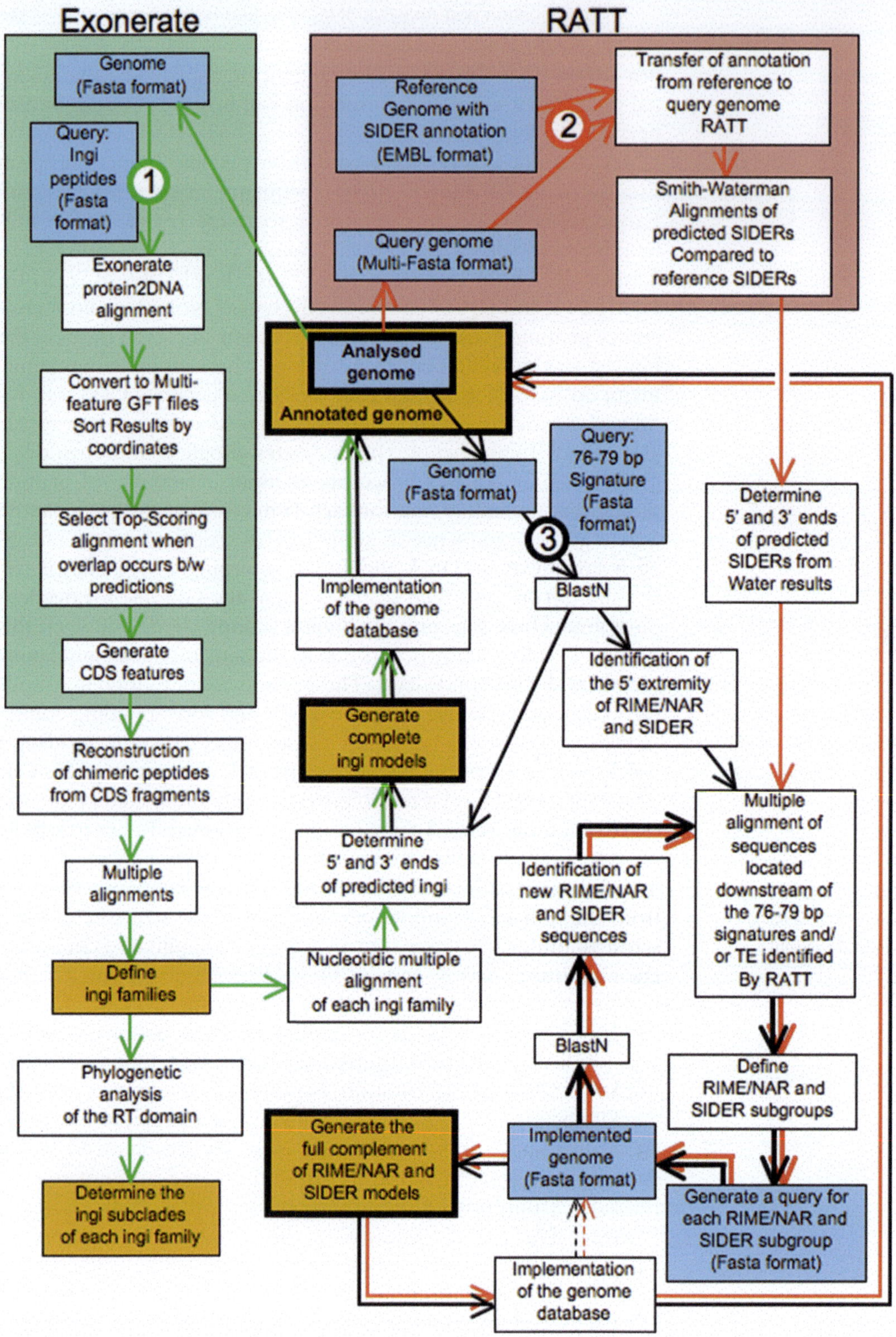

Fig. 2 Flow diagram of the annotation of ingi-related retroposons in trypanosomatid genomes. The Exonerate pipeline and the Rapid Annotation Transfer Tool (RATT) approach are shown on *green* and *red* backgrounds, respectively. *Blue* and *orange boxes* show input and output files, respectively. Identification and annotation of ingi/L1Tc-coding sequences are performed first (①), followed by identification and annotation of short non-autonomous ingi-related sequences (RIME/NAR and SIDER) by RATT (②) and BlastN approaches (③)

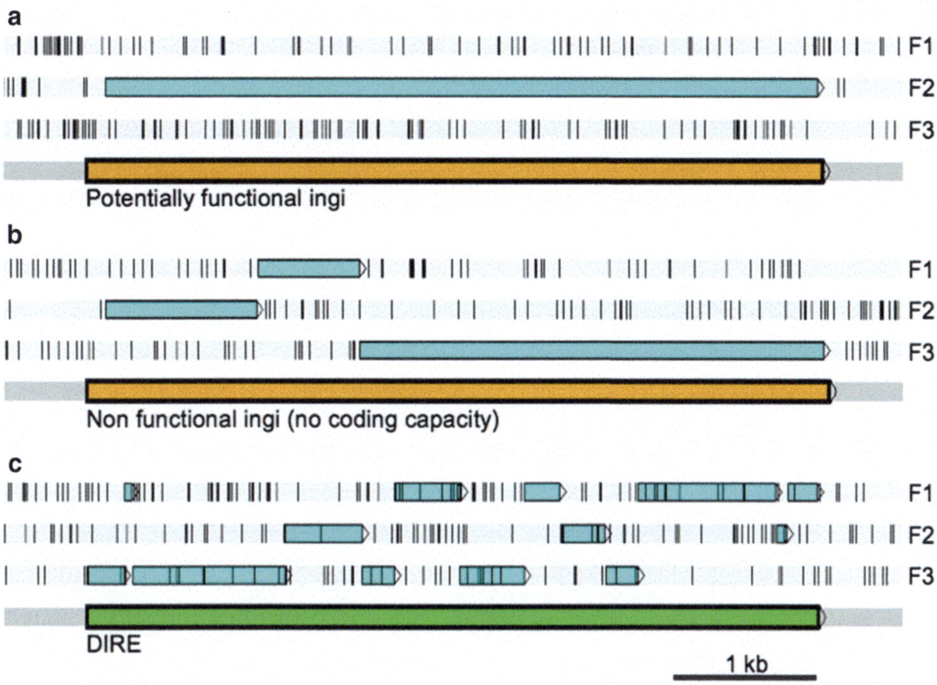

Fig. 3 Annotation of potentially functional ingi (**a**), nonfunctional ingi (**b**), and DIRE (**c**) using Exonerate. Partial (**b** and **c**) or complete (**a**) gene models generated by Exonerate are shown in *blue* as they appear in Artemis. *Vertical bars* represent stop codons found in each of the three frames (F1–F3). Potentially functional ingi contain a single long ORF (**a**). Only few frameshifts and/or stop codons are present in the coding sequence of nonfunctional ingi (**b**), while numerous frameshifts or stop codons characterize DIRE sequences (**c**). Manual curation is required to generate a complete gene model for nonfunctional ingi and DIRE, as well as to determine the 5′- and 3′-extremities of all these ingi-related TE

contains ingi elements of the ingi1 (L1Tco) and ingi6 (Tcoingi) subclades, which are distantly related at both extremities of the ingi phylogenetic tree [22]. The *T. vivax* genome is a challenge due to the sheer number of ingi elements present, and these are frequently found on contiguated unordered contigs where TE elements may be truncated at contig boundaries. We have used earlier versions of both of these genomes in order to compare previous manual annotations [22] to Exonerate's automated predictions.

The resulting Exonerate predictions are summarized in Fig. 3. Using the full-length peptide sequences for Tbingi, Tcoingi, Tvingi, L1Tc, and L1Tco (accession numbers: JQ917146, JQ917147, JQ917148, JQ917144, and JQ917145, respectively—*see* **Note 2**) as queries, Exonerate predicts 98.7 % of Tcoingi elements and 100 % of L1Tco elements (Table 1). Of the predicted Tcoingi elements, 13 % are split into 2 or more alignments for Tcoingi, and 8.3 % for L1Tco. In most cases, these occur where the coding sequence is interrupted by a large gap. A smaller proportion (85.3 %) of TcoDIRE elements are predicted, and a larger

Table 1
Performance of exonerate based computational predictions compared to manual annotation of TE

Genome	TE	Manual annotation (ref: [22])	Prediction overlap	% Correct	% Split
T. congolense	Tcoingi	78	77	98.7	12.0
T. congolense	L1Tco	12	12	100.0	8.3
T. congolense	TcoDIRE	170	145	85.3	19.3
T. vivax	Tvingi	741	701	94.6	3.7
T. vivax	TvDIRE	107	97	90.6	11.3
T. vivax	TvRIME	58	51	87.9	76.5

proportion of them (19.3 %) are split into multiple alignments because of their high degree of degeneracy. The Exonerate method performs similarly well against the *T. vivax* genome (Table 1). Here, 94.6 % of Tvingi elements are predicted, and only 3.7 % of these are split into multiple alignments. TvDIREs and TvRIMEs are also predicted in the *T. vivax* genome (90.6 % and 87.9 %, respectively); however, none of the TvSIDER1 elements are detected. Analysis of these two trypanosome genomes confirms that this approach can detect short truncated ingi-related elements, such as TvRIME (1,030 bp), as long as they contain fragment(s) of ingi-coding sequences [22]. Identifying degenerate short ingi-related families that do not have significant matches to ingi peptides, such as TcoSIDER1 and TvSIDER1 [35], is not possible with this approach. In addition to predictions that overlap with manual annotations, Exonerate makes 43 more predictions for *T. congolense* and 27 for *T. vivax*. Most of these are, however, short alignments (median length of 384/298 bp) with low scores.

Although very powerful, Exonerate cannot predict with accuracy start and end coordinates using ingi peptides as queries, since all ingi families identified so far contain 5′-untranslated sequences ranging from 9 to 193 bp (Tbingi and L1Tco) and 3′-untranslated sequences ranging from 15 to 271 bp (L1Tc and Tbingi). Because of their lack of site specificity for insertion, ingi are flanked by non-conserved sequences. Consequently, a simple multiple alignment of all nucleotide sequences flanking the single long ingi ORF of the same family will allow to determine both 5′- and 3′-extremities. To validate the exact ingi boundaries, it is essential to remember that all ingi families start with the conserved 76–79 bp ingi signature and end with the poly(A) stretch (retroposon hallmark). Blast searches with full-length nucleotide sequences of newly identified ingi families will be necessary to complete gene model determination and curation. Taken together, Exonerate performs well in detecting ingi elements and produces a rapid survey of ingi elements in a new trypanosomatid genome. However, an exhaustive annotation of all ingi elements will require combining Exonerate gene models with other search results (e.g., tBlastN, BlastN). Furthermore, a degree of manual annotation is necessary to correct models that are split into multiple alignments and to determine the exact ingi boundaries. Despite its shortcomings, the use of this method in generating models of ingi elements should greatly increase the speed at which these elements are called in trypanosomatid genomes.

3.2 Phylogenetic Analyses of Ingi Families and DIREs

The method described above also attempts to assign the family/subclade of ingi from the query that produces the highest scoring alignment. To conclusively assign an ingi element to a family or a subclade, a phylogenetic analysis should also be performed. Among the three ingi-coding domains, i.e., apurinic/apyrimidinic endonuclease, RNAseH, and reverse transcriptase (RT), the latter is the most conserved and ideal for phylogenetic tree reconstruction [22, 30]. It corresponds to the amino acid positions 492–770 of Tbingi and 527–822 of L1Tc. These phylogenetic analyses can include potentially active ingi families, including Tbingi, Tvingi, Tcoingi, L1Tc, and L1Tco, as well as DIRE, the degenerate ingi-related sequences. Since members of potentially active ingi families are highly conserved at the nucleotide level, it is recommended to generate for each family a consensus sequence from multiple alignments of full-length elements or of only the RT domain. Online tools available from EBI (http://srs.ebi.ac.uk/) can be used to generate the consensus (CosN). In contrast, DIREs are typically unique ingi degenerate sequences containing a number of frameshifts and stop codons in their nonfunctional coding sequences. Since phylogenetic analyses require amino acid sequences, frameshifts were removed manually from the DNA sequences using the Exonerate output to tentatively reconstitute proteins from the analyzed DIREs. This approach generated a pseudogene for each DIRE element, encoding a single ingi-like sequence, which in most cases contained numerous stop codons. Alternatively, the Blast-Extend-Repraze (BER) algorithm developed at the J. Craig Venter Institute (formerly the Institute for Genomic Research) can be used to localize frameshifts (*see* [30]). So far, RT-based phylogenetic analyses identified six ingi subclades, three of them containing potentially active ingi families [22].

3.3 Identification and Analysis of Short Ingi-Related Sequences (RIME/NAR and SIDER)

3.3.1 Identification by BlastN Searches

As mentioned above, certain short truncated ingi-related families cannot be detected by tBlastN using ingi peptides as query, implying that a BlastN-based approach is more appropriate. The "76–79 bp signature" shared by all retroposons of the ingi clade is the only query sequence that can be used to detect such elements in new trypanosomatid genomes. Several rounds of multiple alignments and BlastN searches with conserved sequences downstream of identified "76–79 bp signatures" are in fact required to annotate the full complement of the analyzed ingi-related family. For example, BlastN with the "76–79 bp signature" detected 108 significant matches in the *L. major* genome. After 8 consecutive comparisons/BlastN cycles, 1,858 related, but highly divergent, LmSIDER sequences were identified in this genome [5]. In contrast, only a single comparison/BlastN cycle was necessary to identify and annotate all of the 70 TcoSIDER1 sequences contained in the *T. congolense* genome after identification of their "76–79 bp signature" [35].

This approach leads to the generation of a list of SIDERs that are then aligned. Because of the degenerate nature of these elements, a manual alignment is first required. A profile alignment of new SIDER elements can then be done using the initial manual alignment, leading to the identification of more SIDERs. In [31], HMM profiles of SIDER sequences were generated by optimizing parameters and maximum likelihood estimators allowing a high-quality alignment of LmSIDER1 comparable to that published for LmSIDER2. But this approach is difficult and requires genome-specific parameter optimization. For example, we used HMMER 2.32 with default parameters to identify SIDER elements in the *L. mexicana* genome using a profile generated from a manual alignment of *L. major* SIDER1 elements. Only 562 SIDERs were predicted in the *L. mexicana* genome, which we assume is on the low side considering that the *L. major* genome has 1,858 SIDERS identified.

Once the complete, or near-complete, set of SIDERs is identified, 5′- and 3′-extremities can be determined by multiple alignments of all the members of a given family, as described above for ingi/L1Tc elements.

3.3.2 Identification by Transfer of Annotation

A second proposed method for predicting SIDERs in *Leishmania* genomes relies on the transfer of annotation between genomes (Fig. 2). *Leishmania* genomes are highly syntenic with few to no breaks in gene order among the old world *Leishmania* species. Even the divergent *Leishmania Viannia* clade (represented solely by the *L. braziliensis* genome) shares obvious blocks of synteny compared to members of the *Leishmania leishmania* clade. Recent annotation of both the *L. donovani* and *L. mexicana* genomes has relied heavily on automatic predictions of gene models using synteny (in the form of Nucmer matches between genomes) and the Rapid Annotation Transfer Tool (RATT) [36]. The majority (98.3 %) of predicted *L. infantum* genes could be transferred to *L. donovani* [13] in this manner, and slightly fewer (93 %) gene models were transferred from *L. major* to *L. mexicana* [12].

As RATT will indiscriminately transfer any sequence feature, we have attempted to transfer manual predictions of SIDER elements in *L. major* [5] to *L. mexicana* for which no SIDER predictions have been made to date. This method works moderately well with 1,223 of the 1,885 manual predictions (66 %) being transferred to *L. mexicana* from *L. major*. We generated Smith-Waterman alignments for each of these and, compared to its *L. major* orthologue, revealed that 800 of these share the same 5′ end. The benefit of this method is that it is easy to use and relatively rapid, although moderately sensitive. Drawbacks of this method are that it will only predict SIDER elements known in another species, and requires somewhat closely related *Leishmania* genomes (ideally > 95 % identity). RATT also assumes orthology and will not transfer the same model more than once, thus failing in cases where segmental duplications have resulted in multiple SIDERs of recent descent. Absence of synteny between SIDERs is probably the main reason why approximately one-third of *L. mexicana* SIDERs failed to be identified by this method. This is consistent with the identification of slightly more than 100 SIDERs in the *L. infantum* and *L. braziliensis* genomes, when more than 1,900 could be found in each genome by HMM [31].

3.3.3 Evolutionary Analyses of Ingi-Related TE Families

Once a consistent alignment of a given TE family is obtained, statistical analyses can be performed to gain insight into its evolutionary dynamics. This analysis is based on the determination of the degree of divergence within members of an ingi family. To study the extent of this divergence, the percentage of divergence between the consensus sequence deduced from the alignment and each TE copy aligned is determined. Since the consensus sequence is assumed to approximate the element's original sequence at the time of insertion, the percentage of substitutions from the consensus sequence is correlated to the age of a given element (the age corresponds to the time of retrotransposition). In other words, the younger the family, the more conserved the sequences.

Online tools available from EBI are very useful to generate the consensus sequence (CosN) of a given alignment and to calculate the percentage of divergence from the consensus sequences (InfoalignN). The values obtained can be expressed as the number of elements as a function of their divergence from their consensus sequence, to calculate the median divergence value. The higher the value, the older and degenerate the family analyzed, such as TbSIDER2, TbSIDER1, TcoSIDER, and LmSIDER2 (11, 16, 16, and >20, respectively), while low median values reflect youth and possible functionality of a TE family, as observed for TvRIME, NARTc, and TbRIME (1, 2, and 4, respectively).

3.4 Conclusion

The whole strategy developed to identify and annotate the full complement of ingi-related sequences in trypanosomatid genomes is presented in Fig. 2. This approach developed to identify protein-coding and noncoding ingi-related retroposons can be adapted to transposable element families from other organisms, with as ultimate goal to identify highly degenerate family members potentially recycled by the host genome to fulfill housekeeping cellular functions.

4 Notes

1. While UNIX is typically stated as a requirement, many of the tools commonly used also work under the UNIX-based Macintosh OS X operating system, and also under Microsoft Windows with environments like Cygwin or MSYS.
2. These entries correspond to consensus nucleotide sequences of 63 Tbingi (JQ917146), 27 Tcoingi (JQ917147), and 48 L1Tc (JQ917144) full-length elements, 46 Tvingi copies larger than 3.5 kb (JQ917148) or 8 L1Tco copies larger than 300 bp (JQ917145) [21]. Each of the five consensus sequences corresponds to a potentially functional retroposon encoding a full-length protein.

Acknowledgement

FB is supported by the Centre National de la Recherche Scientifique (CNRS), the Université Bordeaux Segalen, and the Laboratoire d'Excellence (LabEx) ParaFrap ANR-11-LABX-0024.

Ingihelper.pl is available at https://sites.google.com/a/nyu.edu/ghedin-lab/tools.

References

1. International-Human-Genome-Sequencing-Consortium (2001) Initial sequencing and analysis of the human genome. Nature 409: 860–921
2. Biemont C, Vieira C (2006) Genetics: junk DNA as an evolutionary force. Nature 443: 521–524
3. Bejerano G, Lowe CB, Ahituv N, King B, Siepel A, Salama SR, Rubin EM, Kent WJ, Haussler D (2006) A distal enhancer and an ultraconserved exon are derived from a novel retroposon. Nature 441:87–90
4. Santangelo AM, de Souza FS, Franchini LF, Bumaschny VF, Low MJ, Rubinstein M (2007) Ancient exaptation of a CORE-SINE retroposon into a highly conserved mammalian neuronal enhancer of the proopiomelanocortin gene. PLoS Genet 3:1813–1826
5. Bringaud F, Muller M, Cerqueira GC, Smith M, Rochette A, El-Sayed NM, Papadopoulou B, Ghedin E (2007) Members of a large retroposon family are determinants of post-transcriptional gene expression in *leishmania*. PLoS Pathog 3:e136
6. Stevens JR, Noyes HA, Schofield CJ, Gibson W (2001) The molecular evolution of Trypanosomatidae. Adv Parasitol 48:1–56
7. Berriman M, Ghedin E, Hertz-Fowler C, Blandin G, Renauld H, Bartholomeu DC, Lennard NJ, Caler E, Hamlin NE, Haas B et al (2005) The genome of the African trypanosome *Trypanosoma brucei*. Science 309: 416–422
8. Jackson AP, Sanders M, Berry A, McQuillan J, Aslett MA, Quail MA, Chukualim B, Capewell P, MacLeod A, Melville SE et al (2010) The genome sequence of *Trypanosoma brucei gambiense*, causative agent of chronic human African trypanosomiasis. PLoS Negl Trop Dis 4:e658
9. El-Sayed NM, Myler PJ, Bartholomeu DC, Nilsson D, Aggarwal G, Tran AN, Ghedin E, Worthey EA, Delcher AL, Blandin G et al (2005) The genome sequence of *Trypanosoma cruzi*, etiologic agent of Chagas disease. Science 309:409–415
10. Ivens AC, Peacock CS, Worthey EA, Murphy L, Aggarwal G, Berriman M, Sisk E, Rajandream MA, Adlem E, Aert R et al (2005)

The genome of the kinetoplastid parasite, *Leishmania major*. Science 309:436–442

11. Peacock CS, Seeger K, Harris D, Murphy L, Ruiz JC, Quail MA, Peters N, Adlem E, Tivey A, Aslett M et al (2007) Comparative genomic analysis of three *Leishmania* species that cause diverse human disease. Nat Genet 39:839–847
12. Rogers MB, Hilley JD, Dickens NJ, Wilkes J, Bates PA, Depledge DP, Harris D, Her Y, Herzyk P, Imamura H et al (2011) Chromosome and gene copy number variation allow major structural change between species and strains of *Leishmania*. Genome Res 21:2129–2142
13. Downing T, Imamura H, Decuypere S, Clark TG, Coombs GH, Cotton JA, Hilley JD, de Doncker S, Maes I, Mottram JC et al (2011) Whole genome sequencing of multiple *Leishmania donovani* clinical isolates provides insights into population structure and mechanisms of drug resistance. Genome Res 21:2143–2156
14. Raymond F, Boisvert S, Roy G, Ritt JF, Legare D, Isnard A, Stanke M, Olivier M, Tremblay MJ, Papadopoulou B et al (2012) Genome sequencing of the lizard parasite *Leishmania tarentolae* reveals loss of genes associated to the intracellular stage of human pathogenic species. Nucleic Acids Res 40:1131–1147
15. Porcel BM, Denoeud, F, Opperdoes F, Noel B, Hammarton TC, Field MC, Da Silva C, Couloux A, Poulain J, Katinka M et al (2014) The Streamlined genome of *Phytomonas* spp. relative to human pathogenic kinetoplastids reveals a parasite tailored for plants. PLoS Genetics 10:e1004007
16. Motta MC, Martins AC, de Souza SS, Catta-Preta CM, Silva R, Klein CC, de Almeida LG, de Lima Cunha O, Ciapina LP, Brocchi M et al (2013) Predicting the proteins of Angomonas deanei, Strigomonas culicis and their respective endosymbionts reveals new aspects of the trypanosomatidae family. PLoS One 8:e60209
17. Bringaud F, Ghedin E, El-Sayed NM, Papadopoulou B (2008) Role of transposable elements in trypanosomatids. Microbes Infect 10:575–581
18. Thomas MC, Macias F, Alonso C, Lopez MC (2010) The biology and evolution of transposable elements in parasites. Trends Parasitol 26:350–362
19. Kimmel BE, Ole-MoiYoi OK, Young JR (1987) *Ingi*, a 5.2-kb dispersed sequence element from *Trypanosoma brucei* that carries half of a smaller mobile element at either end and has homology with mammalian LINEs. Mol Cell Biol 7:1465–1475
20. Murphy NB, Pays A, Tebabi P, Coquelet H, Guyaux M, Steinert M, Pays E (1987) *Trypanosoma brucei* repeated element with unusual structural and transcriptional properties. J Mol Biol 195:855–871
21. Martin F, Maranon C, Olivares M, Alonso C, Lopez MC (1995) Characterization of a non-long terminal repeat retrotransposon cDNA (L1Tc) from *Trypanosoma cruzi*: homology of the first ORF with the ape family of DNA repair enzymes. J Mol Biol 247:49–59
22. Bringaud F, Berriman M, Hertz-Fowler C (2009) Trypanosomatid genomes contain several families of ingi-related retroposons. Eukaryotic Cell 8:1532–1542
23. Hasan G, Turner MJ, Cordingley JS (1984) Complete nucleotide sequence of an unusual mobile element from *Trypanosoma brucei*. Cell 37:333–341
24. Bringaud F, Garcia-Perez JL, Heras SR, Ghedin E, El-Sayed NM, Andersson B, Baltz T, Lopez MC (2002) Identification of non-autonomous non-LTR retrotransposons in the genome of *Trypanosoma cruzi*. Mol Biochem Parasitol 124:73–78
25. Bringaud F, Biteau N, Zuiderwijk E, Berriman M, El-Sayed NM, Ghedin E, Melville SE, Hall N, Baltz T (2004) The *ingi* and RIME non-LTR retrotransposons are not randomly distributed in the genome of *Trypanosoma brucei*. Mol Biol Evol 21:520–528
26. Bringaud F, Bartholomeu DC, Blandin G, Delcher A, Baltz T, El-Sayed NM, Ghedin E (2006) The *Trypanosoma cruzi* L1Tc and NARTc non-LTR retrotransposons show relative site-specificity for insertion. Mol Biol Evol 23:411–420
27. Jurka J (1997) Sequence patterns indicate an enzymatic involvement in integration of mammalian retroposons. Proc Natl Acad Sci U S A 94:1872–1877
28. Dewannieux M, Esnault C, Heidmann T (2003) LINE-mediated retrotransposition of marked *Alu* sequences. Nat Genet 35:41–48
29. Ghedin E, Bringaud F, Peterson J, Myler P, Berriman M, Ivens A, Andersson B, Bontempi E, Eisen J, Angiuoli S et al (2004) Gene synteny and evolution of genome architecture in trypanosomatids. Mol Biochem Parasitol 134:183–191
30. Bringaud F, Ghedin E, Blandin G, Bartholomeu DC, Caler E, Levin MJ, Baltz T, El-Sayed NM (2006) Evolution of non-LTR retrotransposons in the trypanosomatid genomes: *Leishmania major* has lost the active elements. Mol Biochem Parasitol 145:158–170

31. Smith M, Bringaud F, Papadopoulou B (2009) Organization and evolution of two SIDER retroposon subfamilies and their impact on the *Leishmania* genome. BMC Genomics 10:240
32. Boucher N, Wu Y, Dumas C, Dube M, Sereno D, Breton M, Papadopoulou B (2002) A common mechanism of stage-regulated gene expression in *Leishmania* mediated by a conserved 3′-untranslated region element. J Biol Chem 277:19511–19520
33. McNicoll F, Muller M, Cloutier S, Boilard N, Rochette A, Dube M, Papadopoulou B (2005) Distinct 3′-untranslated region elements regulate stage-specific mRNA accumulation and translation in *Leishmania*. J Biol Chem 280:35238–35246
34. Slater GS, Birney E (2005) Automated generation of heuristics for biological sequence comparison. BMC Bioinformatics 6:31
35. Bringaud F, Berriman M, Hertz-Fowler C (2011) TSIDER1, a short and non-autonomous Salivarian trypanosome-specific retroposon related to the ingi6 subclade. Mol Biochem Parasitol 179:30–36
36. Otto TD, Dillon GP, Degrave WS, Berriman M (2011) RATT: Rapid Annotation Transfer Tool. Nucleic Acids Res 39:e57

Chapter 7

Approaches for Studying mRNA Decay Mediated by SIDER2 Retroposons in *Leishmania*

Barbara Papadopoulou, Michaela Müller-McNicoll, and Prasad K. Padmanabhan

Abstract

Regulated mRNA turnover is a highly important process in the control of gene expression in *Leishmania* and related trypanosomatid protozoa, as these organisms lack control at the level of transcription initiation. A large number of *Leishmania* transcripts harbor in their 3′UTRs two phylogenetically distinct subfamilies of extinct Short Interspersed DEgenerate Retroposons (SIDER1 and SIDER2) that are involved in posttranscriptional regulation of gene expression. We have shown recently that members of the SIDER2 subfamily promote mRNA destabilization and that degradation of SIDER2-containing mRNAs is initiated by site-specific endonucleolytic cleavage within the second 79-nt SIDER2 signature sequence without prior shortening of the poly(A) tail. Here, we describe experimental procedures for studying the mechanism of SIDER2-mediated mRNA decay. These include RNase protection assays to identify in vivo-generated mRNA decay intermediates following endonucleolytic cleavage, primer extension analysis to precisely map the site(s) of cleavage within SIDER2, and deadenylation assays to assess the polyadenylation state of unstable SIDER2-containing mRNAs in *Leishmania*.

Key words *Leishmania*, mRNA decay, SIDER2 retroposons, Endonucleolytic cleavage, RNase protection, Primer extension, Deadenylation assay

1 Introduction

Most eukaryotic mRNAs are degraded through two alternative pathways, each of which is initiated by the removal of the poly(A) tail (deadenylation) by a variety of deadenylases. Subsequently, the cap (5′-m7GpppN) structure is removed by the decapping enzymes DCP1/DCP2 and mRNAs are degraded by 5′ to 3′ exonucleases [1]. Alternatively, deadenylated mRNAs can be degraded from their 3′-ends by the exosome, a multimeric protein complex possessing 3′ to 5′ exoribonuclease activity [2]. In addition to these pathways, a small number of mRNAs are targeted for decay via endonucleolytic cleavage of a specific sequence within their 3′UTR by sequence-specific endoribonucleases [3–5]. Messenger RNAs

Christopher Peacock (ed.), *Parasite Genomics Protocols*, Methods in Molecular Biology, vol. 1201, DOI 10.1007/978-1-4939-1438-8_7, © Springer Science+Business Media New York 2015

that are degraded through endonucleolytic cleavage are generally short-lived and highly regulated, and in many cases degradation does not involve prior shortening of the poly(A) tail [4–7].

We have recently identified a new class of widespread extinct retroposons in *Leishmania* termed *S*hort *I*nterspersed *DE*generate *R*etroposons (SIDER1/SIDER2) that are predominantly located within 3′UTRs and play a role in posttranscriptional regulation [8, 9]. We have demonstrated that members of the SIDER2 subfamily promote mRNA destabilization through endonucleolytic cleavage without prior deadenylation [10]. Endonucleolytic digestion products from SIDER2-bearing mRNAs were detected in vivo by different methods (e.g., RNase protection, primer extension, northern blotting, and reverse ligation-mediated PCR). The most prevalent endonucleolytic cleavage site was mapped within the second conserved 79-nt signature II sequence of SIDER2 retroposons [10].

Regulated mRNA turnover is a highly important process in the control of gene expression in *Leishmania* and related trypanosomatid protozoa, as these organisms lack control at the level of transcription initiation, and regulation takes place almost exclusively at the posttranscriptional level [11, 12]. The current model for mRNA degradation in trypanosomatids involves at least two pathways: a regulated pathway that is rapid and seems to be deadenylation independent [10, 13, 14] and a constitutive pathway that is initiated with a progressive shortening of poly(A) tails and operates at a slower kinetics during the degradation of stable mRNAs [8, 11, 12].

Here, we describe assays to identify in vivo-generated mRNA decay intermediates following endonucleolytic cleavage (e.g., RNase protection assay) and to precisely map the site(s) of cleavage within the target RNA (e.g., primer extension). We also describe a deadenylation assay to assess the polyadenylation state of unstable mRNAs in *Leishmania*.

1.1 RNase Protection Assay (RPA) to Detect In Vivo Degradation Cleavage Products

In general, endonucleolytic cleavage products cannot be detected in vivo because they are rapidly degraded by exoribonucleases. However, specific endonucleolytic cleavage intermediates for some abundant transcripts have been visualized using RNase protection assays [15]. RNase protection assay (RPA) is a highly sensitive method to detect, map, and quantify RNA degradation products in total cellular RNA. We established an RPA protocol using *Leishmania* transfectants overexpressing reporter gene constructs harboring SIDER2 retroposon elements in their 3′UTR in order to detect degradation cleavage products generated in vivo (Fig. 1a). Total RNA (free of DNA) is hybridized with specific antisense RNA probes that are complementary to the predicted cleavage region within the 3′UTR (e.g., SIDER2) of reporter transcripts. To generate antisense RNA probes, DNA fragments are cloned into the pCR2.1 vector containing a T7 promoter and in vitro

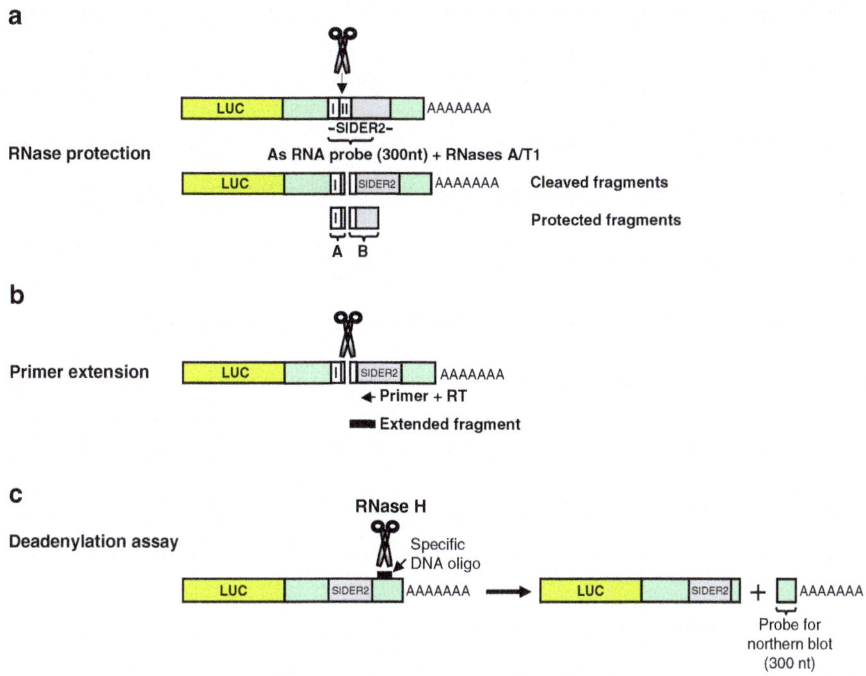

Fig. 1 Approaches for studying mRNA decay mediated by SIDER2 retroposons in 3′UTRs in *Leishmania* through site-specific endonucleolytic cleavage without prior shortening of the poly(A) tail. (**a**) Detection of in vivo-generated cleavage products derived from SIDER2-containing mRNAs by RNase protection assays. *Leishmania* transfectants expressing a luciferase (*LUC*) transcript under the control of a SIDER2-containing 3′UTR. Degradation of SIDER2-harboring transcripts occurs via endonucleolytic cleavage within the second conserved signature of SIDER2 retroposons (II) [10]. Total RNA is independently mixed with an in vitro-transcribed radiolabeled antisense (As) SIDER2 RNA probe of 300 nt and thereafter subjected to RNase A/T$_1$ treatment to detect protected double-stranded RNA fragments by northern blot hybridization. (**b**) Mapping of cleavage site(s) in a SIDER2-containing mRNA by primer extension analysis. *Leishmania* RNA is mixed with reverse primers at 100–200 nt from the cleavage region and reverse transcriptase (RT) enzyme. The size of the extended fragments with RT is precisely calculated together with the nucleotide sequence of the cleavage site(s) following migration on SDS-PAGE in the presence of a dideoxy-sequencing ladder. (**c**) Schematic representation of the deadenylation assay. *LUC* transcripts harboring a SIDER2 element in the 3′UTR are specifically cleaved at 300 nt from the poly(A) tail using oligonucleotide-directed RNase H cleavage. The resulting 3′-products containing the poly(A) tail are visualized by northern blot using a probe complementary to the last 300 nucleotides of the *LUC*-SIDER2 transcript

transcribed with T7 RNA polymerase. This permits the inclusion of additional sequences that will not hybridize to the target mRNA, so that undigested probe can be differentiated from the probe that is protected by hybridization to the intact RNA. Gel-purified radiolabeled RNA probes are annealed to the denatured *Leishmania* total RNA and then digested with a mix of RNases A/T1 to remove all single-stranded RNA sequences. Following inactivation of the RNases, the protected double-stranded RNA products are precipitated, resolved on a denaturing 8 % polyacrylamide/urea gel, and visualized by autoradiography. A radiolabeled probe mixed with unspecific RNA (e.g., yeast tRNA) and not treated with A/T1 RNases usually serves as a positive control showing the size of the expected full-length fragment. The same sample treated with A/T1 RNases serves as a negative control to assess whether the RNase treatment was complete. To determine the exact size of the protected fragments, a labeled size marker should be included. Protected fragments whose sizes sum up to the length of the full-length probe correspond to specific degradation cleavage intermediates.

1.2 Primer Extension

Primer extension is an alternative sensitive method for detecting and mapping the 5′-end of in vivo endonucleolytic cleavage products [16]. This approach was adapted to map cleavage RNA products in *Leishmania* derived from highly expressed unstable reporter transcripts (Fig. 1b). First, a 30-mer antisense primer located approximately 100–200 nt downstream of the putative cleavage region of the target RNA is designed (*see* **Notes 1** and **2**). This primer is 5′-end-labeled using [γ-^{32}P]ATP and T4 polynucleotide kinase and mixed with denatured total *Leishmania* RNA to allow hybridization (*see* **Note 3**). AMW reverse transcriptase (RT) is then added to the mix to extend RNA through cDNA synthesis until the RT runs off the template when it reaches the site of endonucleolytic cleavage within the RNA. The products are analyzed in an 8 % SDS denaturing polyacrylamide gel and visualized by autoradiography. To map precisely the location of the cleavage site, it is crucial to load a labeled dideoxy sequencing ladder on the same gel. This can be easily prepared using the same 5′-labeled primer and a purified PCR fragment. Because secondary structures within an mRNA can result in pausing of reverse transcriptase [17], a control of an in vitro-transcribed RNA of similar size than the RNA under study must be included in the assay (*see* **Note 4**). This RNA is treated at the same time and loaded on the same gel.

1.3 Deadenylation Assay

To study changes in the length of the poly(A) tail of unstable mRNAs, the original RNase H deadenylation assay protocol [18] was adapted in our laboratory for *Leishmania* mRNAs [13] (Fig. 1c). For this, parasites are treated with actinomycin D (ActD) to arrest transcription, and samples are taken from appropriate time points to cover the onset of mRNA degradation and to measure transcript's half-life. At each time point, total RNA is isolated and mixed with a specific 20-mer DNA oligonucleotide complementary to a region on the 3′UTR located approximately 300 nt upstream of the poly(A) tail addition site. The RNA samples are then subjected to digestion with RNase H, an enzyme that cleaves specifically RNA:DNA hybrids. One sample is treated with oligo(dT), which cleaves off the poly(A) tail and serves as a control for completely deadenylated RNA species. Another sample is not treated with RNase H to distinguish between specific and unspecific cleavage products. The resulting RNA fragments are then separated on 5 % SDS denaturing polyacrylamide gels and transferred onto nylon membranes. The blot is subsequently hybridized with a probe corresponding to the last 300 nt of the transcript under study to visualize only the short 3′-fragments containing the poly(A) tail. The length of the poly(A) tail at time point 0 serves as a positive control. Changes in poly(A) tail length can be monitored on SDS-PAGE by comparing the different time points after transcriptional shutoff with ActD. The polyadenylation status should be correlated to the decay rate of the uncut transcript. The small histone 4A transcript can be used as a loading control.

2 Materials

2.1 RNase Protection Assay (RPA)

2.1.1 TA Cloning of SIDER2-Specific PCR Fragments and Analysis of Positive Clones

1. Taq polymerase.
2. 10× Taq buffer.
3. dNTPs.
4. Genomic DNA from *Leishmania*.
5. Specific oligonucleotide primer pairs to amplify DNA.
6. PCR apparatus.
7. 0.2 mL PCR tubes.
8. Agarose gel electrophoresis equipment.
9. 1× TBE (90 mM Tris–borate, 2 mM EDTA).
10. Ethidium bromide (10 mg/mL) (ethidium bromide is mutagenic and toxic, so gloves should be worn at all times when handling this reagent).
11. 6× DNA loading buffer (Fermentas).
12. Gel extraction/PCR purification kit (Qiagen).
13. Isopropanol.
14. Microcentrifuge and microfuge tubes.
15. Nuclease-free reaction tubes.
16. TA-cloning kit with pCR2.1 vector (Invitrogen).
17. Thermo block (14 °C; 37 °C; 42 °C).
18. DH5a competent cells.
19. Luria-Bertani (LB) medium.
20. Shaking incubator (37 °C).
21. LB agar plates containing 50 μg/mL ampicillin.
22. X-gal (40 mg/mL).
23. 0.1 M IPTG.

24. Bacteria culture tubes.
25. Plasmid preparation kit (Qiagen).
26. *Eco*RI restriction endonuclease (NEB).
27. 10× *Eco*RI buffer (NEB).

2.1.2 Linearization of pCR2.1 Plasmid and In Vitro Transcription

1. *Hind*III restriction endonuclease (NEB).
2. 10× buffer 2 (NEB).
3. Thermo block (37 °C).
4. Agarose gel electrophoresis system.
5. 1× TBE (90 mM Tris–borate, 2 mM EDTA).
6. Ethidium bromide.
7. DNA loading buffer (Fermentas).
8. Gel extraction kit (Qiagen).
9. Isopropanol.
10. Microcentrifuge and microfuge tubes.
11. Nuclease-free 1.5 mL microfuge tubes.
12. MEGAscript kit (Ambion) including T7 enzyme buffer, NTPs (75 mM each), and Turbo RNase-free DNase I (2 U/μL; Ambion).
13. 10× DNase I buffer (Ambion).
14. 0.5 M EDTA.
15. NucAway™ spin columns (Ambion).
16. Alpha-^{32}P-[UTP] (10 μCi/μL).
17. Formaldehyde-agarose gel and RNA gel migration equipment: 100 % formamide, 37 % formaldehyde, and filtered MOPS 20×: 0.2 M MOPS, 80 mM sodium acetate, 10 mM EDTA, pH 7.0.
18. RNA loading buffer: 1× MOPS, 6.5 % formaldehyde, 50 % formamide, 0.25 % bromophenol blue, and 0.025 μg/μL ethidium bromide.
19. Spectrophotometer (260 nm) for RNA quantification (i.e., NanoDrop).

2.1.3 Isolation of Total RNA from Leishmania

1. Trizol® reagent (Invitrogen).
2. Autoclaved filter tips and nuclease-free reaction tubes.
3. Cooling table centrifuge.
4. Chloroform.
5. Isopropanol.
6. Diethylpyrocarbonate (DEPC)-treated and autoclaved (nuclease-free) water (Ambion).

7. 70 % Ethanol (prepared in DEPC-treated ultrapure water).
8. 10× DNase I buffer (Ambion).
9. RNase-free DNase I (2 U/μL; Ambion).
10. 3 M sodium acetate (pH 5.5).
11. 100 % Ethanol.
12. Spectrophotometer.

2.1.4 RNase Protection Using RPAIII kit (Ambion)

1. RPAIII kit (Ambion).
2. Autoclaved filter tips and nuclease-free reaction tubes.
3. Thermo block (95 °C; 45 °C).
4. 100 % Ethanol.
5. Gel electrophoresis system (glass plates, spacers, combs, metal clips, running chamber).
6. Sequagel solutions (National Diagnostics).
7. TEMED (Biorad).
8. 10 % Ammonium persulfate (APS; Biorad).
9. 0.5× TBE (45 mM Tris–borate, 1 mM EDTA).
10. Film cassette and films.

2.2 Primer Extension

All solutions must be prepared in diethylpyrocarbonate (DEPC)-treated ultrapure water (purifying deionized water to attain a sensitivity of 18 M Ω cm at 25 °C) to avoid RNase contamination. The reagents used must be of analytical grade of high purity.

2.2.1 Primer Extension Assay

1. Gene-specific oligonucleotide primers (18-40 nt) to amplify cDNA.
2. 10× polynucleotide buffer (NEB).
3. 10 U/μL polynucleotide kinase (NEB).
4. 10 μCi/μL [γ-^{32}P]ATP (3,000 Ci/mmol).
5. 25 mM $MgCl_2$.
6. 100 mM Dithiothreitol (DTT).
7. RNaseOUT (40 U/μL) (Invitrogen).
8. SuperScript™ III RT (200 U/μL) (Invitrogen).
9. 10 mM dNTP mix (Invitrogen).
10. 10× RT buffer (Invitrogen).
11. 2× Gel Loading Buffer II (Denaturing PAGE) (Ambion), 95 % Formamide, 18 mM EDTA, and 0.025 % SDS, xylene cyanol, and bromophenol blue.
12. Total *Leishmania* RNA.

13. Primer extension marker (lΦX174 DNA/HinfI dephosphorylated DNA marker from Promega, which is labeled with [γ-^{32}P] ATP and PNK (New England Biolabs) as per the manufacturer's instructions).
14. 5× Tris–borate–EDTA buffer (54 g Tris–Cl, 27.5 g boric acid, and 20 mL 0.5 M EDTA pH 8.0): Sterilize the solution by autoclave.

2.2.2 Sequencing Gel Reagents and Materials

1. Sequagel-Ureagel system (National Diagnostics) containing Ureagel Concentrate (237.5 g/L of acrylamide, 12.5 g/L of methylene bisacrylamide, and 7.5 M urea), Ureagel Diluent (7.5 M urea in deionized aqueous solution), and Ureagel Buffer (0.89 M Tris–borate–20 mM EDTA buffer pH 8.3 (10× TBE) and 7.5 M urea).
2. 0.8 mL 10 % ammonium persulfate (APS)/100 mL gel casting solution.
3. 40 μL TEMED/100 mL gel casting solution.
4. Whatman no. 3 filter paper.
5. Erlenmeyer flask.
6. Sigmacote (Sigma).

2.3 Deadenylation Assay

2.3.1 Inhibition of Active Transcription in Leishmania

1. Actinomycin (ActD) stock solution 5 mg/mL in dimethyl sulfoxide (DMSO): Dissolve ActD under the hood in the original glass tube by injecting DMSO through the rubber lid. After dissolution, transfer ActD carefully with syringe (solution is bright red) to a fresh reaction tube. Take extreme caution because ActD is highly toxic and DMSO is a skin-penetrating solvent. Aliquot solution in brown reaction tubes (ActD is very sensitive to light) and store at −20 °C.
2. Cell culture flasks.
3. SDM-79 medium supplemented with 10 % heat-inactivated FCS (Wisent) and 5 μg/mL hemin.
4. Spectrophotometer.
5. Autoclaved filter tips and nuclease-free reaction tubes.
6. Trizol® reagent (Invitrogen).
7. Liquid nitrogen and/or −80 °C freezer.

2.3.2 Isolation of Total RNA from Leishmania (Materials are as in Subheading 2.1.3)

2.3.3 RNase H Digest

1. Specific antisense DNA oligonucleotides (100 μM stock) for RNase H digest: We use antisense oligonucleotides complementary to a region 300 nt upstream of the poly(A) site for optimal resolution.
2. Oligo-dT primer (18 Ts; 100 μM stock).

3. RNaseOUT (Invitrogen).
4. RNase H (2 U/μL; Invitrogen).
5. 10× RNase H buffer (500 mM Tris–HCl, pH 7.4; 100 mM $MgCl_2$; 10 mM DTT; 800 mM KCl).
6. DEPC-treated and autoclaved (nuclease-free) water (Ambion).
7. Thermo block (37 °C).
8. 0.5 M EDTA (pH 8.0; sterile-filtered).
9. Glycogen.
10. 100 % Ethanol.
11. 2× RNA loading buffer (Ambion).

2.3.4 Polyacrylamide Gel Electrophoresis and Transfer

1. Gel system (glass plates, spacers, combs, metal clips, running chamber).
2. Sequagel solutions (National Diagnostics).
3. TEMED (Biorad).
4. 10 % ammonium persulfate (APS; Biorad).
5. 0.5× TBE (45 mM Tris–borate, 1 mM EDTA).
6. Thermo block (65 °C).
7. Filter paper (Whatman no. 3).
8. Wet-transfer system (sponges, sandwich holder, tank, cool blocks).
9. Nylon membrane (Hybond XL, Amersham).
10. 2× Saline-sodium citrate (SSC) buffer.
11. Stratalinker® (UV-cross-linking device).

2.3.5 Preparation of Gene-Specific Probes

1. Taq polymerase.
2. 10× Taq buffer.
3. dNTP mix.
4. Specific oligonucleotide primer pairs.
5. Genomic DNA from *Leishmania*.
6. 0.2 mL PCR tubes.
7. PCR apparatus.
8. Agarose gel electrophoresis system.
9. 1× TBE (90 mM Tris–borate, 2 mM EDTA).
10. Ethidium bromide.
11. 6× DNA loading buffer (Fermentas).
12. Gel extraction/PCR purification kit (Qiagen).
13. High/low-molecular mass ladder.

2.3.6 Northern Blot Hybridization

1. DEPC-treated water.
2. 50× Denhardt's solution.
3. 20× SSC.
4. 10 % SDS.
5. Salmon sperm DNA.
6. Formamide.
7. Hybridization oven and hybridization tubes.
8. Thermo block (37 °C and 95 °C).
9. 5× Oligonucleotide labeling buffer (1 M HEPES, pH 6.6; 250 mM Tris–HCl, pH 8.0; 25 mM $MgCl_2$; 30 U/mL random hexamers pol(N)6; 100 μM dATP; 100 μM dGTP; 100 μM dTTP; 0.36 % 2-mercaptoethanol).
10. Alpha-^{32}P-[dCTP] (10 μCi/μL).
11. Klenow enzyme (NEB).
12. Stop mix for probes (100 mM Tris–HCl (pH 8.0), 12.5 mM EDTA, 0.5 % SDS).
13. NucAway™ spin columns (Ambion).
14. Scintillation counter (optional) and scintillation solution.
15. Washing buffer 1 (2× SSC; 0.5 % SDS).
16. Washing buffer 2 (0.1× SSC; 0.5 % SDS).
17. Film cassette and film (Amersham).

3 Methods

3.1 RNase Protection Assay

3.1.1 TA Cloning of SIDER2-Specific PCR Fragments and Analysis of Positive Clones

1. Amplify DNA fragments by PCR from genomic DNA to generate SIDER2-specific antisense probes.
2. Purify PCR products either using gel extraction or PCR purification kits (Qiagen).
3. Quantify the purified PCR products on an agarose gel using a mass ladder.
4. Ligate the purified PCR products immediately into vector pCR2.1 (Invitrogen). Mix 1 μL plasmid, 1 μL 10× ligase buffer, and 1 μL T4 ligase with 7 μL PCR product. Incubate overnight at 14 °C.
5. Transform competent DH5a *E. coli* with the ligated pCR2.1 plasmid. Add 100 μL bacteria cells to the ligation mix and leave for 30 min on ice. Place cells for 2 min on a heat block (exactly 42 °C). Chill cells on ice for 2 min. Add 800 μL LB medium without antibiotics and shake at 37 °C for 60 min in a shaking incubator (250 rpm) to allow expression of ampicillin-resistance genes.

6. Spread 50 μL IPTG and 20 μL X-gal on pre-warmed LB agar plates containing ampicillin. Let dry for 20 min.
7. Spread 150 μL of bacteria suspension on the plate. Incubate overnight at 37 °C.
8. Pick five white colonies into 3 mL LB medium containing ampicillin and incubate overnight in a shaking incubator (250 rpm).
9. Spin down bacteria cells for 5 min in a microcentrifuge (10,000 rpm).
10. Isolate plasmid DNA with Qiagen Miniprep Spin Kit. Elute plasmids in 50 μL TE buffer.
11. Digest plasmid DNA with *Eco*RI. An *Eco*RI restriction site is present in the vector on either side of the inserted PCR product. Mix 19.5 μL water with 2.5 μL 10× *Eco*RI buffer, 1 μL *Eco*RI enzyme, and 2 μL plasmid DNA. Incubate for at least 1 h at 37 °C.
12. Stop the reaction by adding 5 μL 6× DNA loading buffer. Run 15 μL on a 1 % agarose gel and verify the gel for the presence of the insert.
13. Verify the orientation of inserted PCR products by sequencing.

3.1.2 Linearization of pCR2.1 Plasmid and In Vitro Transcription (IVT)

1. Use plasmid preparation in which SIDER2 had inserted in the reverse orientation. This allows in vitro transcription of antisense SIDER2 RNA using the T7 promoter sequence present in the pCR2.1 vector. Inclusion of vector-specific sequences in the in vitro-transcribed RNA simplifies discrimination of protected SIDER2 RNA versus undigested probe later on.
2. Linearize the plasmid downstream of the inserted SIDER2 fragment using a *Hin*dIII restriction site present in the multicloning site opposite to the T7 promoter to stop IVT at this point. Mix 30 μL plasmid DNA with 5 μL 10× buffer, 2 μL *Hin*dIII enzyme, and 13 μL water. Incubate for 3 h or overnight at 37 °C.
3. Stop the reaction by adding 10 μL 6× DNA loading buffer. Prepare a preparative 1 % agarose gel and separate digested from partially or undigested plasmid.
4. Cut the digested plasmid out of the gel and purify the DNA using the gel extraction kit from Qiagen.
5. Quantify purified plasmid on an agarose gel using a mass ladder.
6. Use the MEGAscript IVT kit (Ambion) (2 × 20 μL reactions) according to the manufacturer's instructions. In a nuclease-free microfuge tube, mix reagents in the following order at

room temperature: 0.5–1 μg of purified DNA template, 1 μL UTP solution, 2 μL CTP solution, 2 μL GTP solution, 2 μL ATP solution, and 2 μL 10× reaction buffer. Mix well and place sample on ice. Work now in the radioactive room. Add 3 μL labeled alpha-^{32}P-[UTP] (10 μCi/μL) and 2 μL enzyme mix (T7 RNA polymerase). Mix well and incubate at 37 °C for 1–2 h. Add 2 μL of Turbo DNase to each probe to remove the plasmid template DNA and mix well. Incubate at 37 °C for 30 min. Add 1 μL of 0.5 M EDTA to terminate the reaction.

7. Remove non-incorporated nucleotides with NucAway™ spin columns (Ambion) according to the manufacturer's instructions. Rehydrate the resin for 5 min with 650 μL of nuclease-free water, centrifuge for 2 min at 3,000 rpm, place the columns on fresh tubes, add sample on the top of the column, and centrifuge for 2 min at 3,000 rpm.
8. Scintillate 1 μL of the eluted radiolabeled IVT RNA in 5 mL scintillation solution. Store on ice until further use.
9. The quality of RNA synthesis is evaluated on gel. Add 5 μL of RNA loading buffer to 1 μL of eluted RNA and heat-denature at 65 °C for 10 min. Keep on ice before loading. Load it on 1× MOPS-1.5 % formaldehyde-2 % agarose gel. Migrate at 120 V in 1× MOPS. RNA quantification can be obtained using a spectrophotometer (usually 2–3 μg/μL).

3.1.3 Isolation of Total RNA from Leishmania

1. Spin 10 mL log-phase *Leishmania* culture by centrifugation at 3,000 rpm for 5 min.
2. Resuspend the pellet in 1 mL Trizol® and gently lyse the cells by slowly pipetting up and down three times. Transfer samples to nuclease-free reaction tubes, snap-freeze in liquid nitrogen, or transfer immediately to −80 °C. Store samples overnight.
3. Thaw samples on ice. Wipe working surfaces, pipettes, and racks with RNAzol (Ambion) to remove RNase contaminations! Work with gloves!
4. Add 200 μL chloroform to each sample and shake vigorously 15 times to extract proteins (do not vortex to avoid shearing of genomic DNA).
5. Spin samples for 10 min (10,000 rpm) at 4 °C. Prepare and label fresh reaction tubes.
6. Carefully transfer aqueous phase containing RNA into the fresh tubes (around 500 μL). Avoid touching the interphase or the organic solvent phase.
7. Add 1 volume of isopropanol to each sample and mix gently by inverting. Store RNAs on ice for at least 30 min to precipitate the RNA.
8. Pellet RNA for 15–20 min at 14,000 rpm and 4 °C.

9. Remove supernatant and wash pellets once with 400 μL of 70 % ethanol (prepared with DEPC-treated water).
10. Spin for 5 min at 10,000 rpm and 4 °C.
11. Remove supernatant carefully and dry pellets for 10 min at the bench. Avoid overdrying.
12. Resuspend RNA in 34 μL nuclease-free water. Incubate samples for 10 min at 37 °C and for 30 min on ice to complete dissolution.
13. Add 4 μL 10× DNase I buffer and 2 μL RNase-free DNase I (2 U/μL; Ambion) to each sample, mix, and incubate for 30 min at 37 °C to digest genomic DNA.
14. Add 160 μL nuclease-free water, 1/10 volume 3 M sodium acetate (20 μL), and 2.5 volumes 100 % ethanol (525 μL) to each sample and mix by inverting. Precipitate RNA at −20 °C for at least 1 h.
15. Repeat **steps 8–11**.
16. Resuspend RNA in 10 μL nuclease-free water. Incubate samples for 10 min at 37 °C and for 30 min on ice to complete dissolution.
17. Measure RNA concentrations with a spectrophotometer at 260 nm.

Safety instructions: Isolate the RNA in a chemical hood. Avoid contact with skin and clothing; Trizol contains phenol, which is corrosive. Always use gloves to protect from corrosive chemicals and to avoid RNase contamination.

3.1.4 RNase Protection Using RPA III kit (Ambion)

1. For each probe prepare N+2 tubes (one tube for positive control and one tube for negative control).
2. Mix 2.5 μL total RNA (50–100 μg) with 2.5 μL radiolabeled IVT antisense SIDER2 RNA ($1–2\times10^5$ cpm) and 10 μL hybridization buffer (RPA III kit; Ambion). For negative and positive controls use 2.5 μL yeast RNA (5 μg/μL; RPA III kit).
3. Vortex, quick spin, and heat samples at 95 °C for 4 min.
4. Immediately transfer samples to a thermo block at 45 °C and incubate samples overnight to allow annealing of complementary RNA sequences.
5. Prepare a 1:50 dilution of RNases T1/A in 150 μL digestion buffer (RPA III kit) for each sample except the negative control.
6. Add 150 μL diluted RNase T1/A to the samples and mix well. Add 150 μL digestion buffer without RNases to the negative control sample.
7. Incubate for 30–60 min at 37 °C. All single-stranded RNA will now be efficiently degraded.

8. To stop the reaction, add 225 μL RNase inactivation solution to each sample (RPA III kit).
9. Precipitate the RNA by adding 2 μL yeast RNA, 30 μg glycogen (to increase the pellet), and 100 μL 100 % ethanol. Store samples at −80 °C for at least 30 min.
10. Centrifuge samples (13,000 rpm) for 20 min at 4 °C.
11. Remove supernatant carefully (radioactive) and dry pellets at 37 °C for 5 min.
12. Resuspend pellets in 20 μL loading buffer II (RPA III kit).
13. To separate and visualize the protected fragments, prepare a 5 % urea-acrylamide gel (Sequagel) and store it for 30 min to ensure complete polymerization. For one gel mix 10 mL Sequagel concentrate with 35 mL Sequagel Diluent, 5 mL Sequagel Buffer, and 0.4 mL 10 % APS. Start polymerization with 20 μL TEMED.
14. Prepare 2 L of 0.5× TBE.
15. Assemble the acrylamide gel running system, fill tank with 0.5× TBE, remove comb, and complete a 30-min pre-run at 150 V.
16. Wash gel slots to remove excess urea. Load entire sample on the equilibrated acrylamide gel and separate them at 250 V until the light blue dye is at 2/3 of the gel (ca. 2 h).
17. Remove one glass plate and wrap gel including the second glass plate in Saran wrap. Place glass plate in film cassette and fix it. Expose film and incubate gel overnight at −80 °C.

3.2 Primer Extension

3.2.1 5′-End Labeling of Oligonucleotide Primers

1. The following reagents are mixed in a 1.5 mL microfuge tube for a total volume of 10 μL: 10 pmol gene-specific oligonucleotide primer (1 μL of 10 μM), 1 μL 10× polynucleotide buffer, 1 μL polynucleotide kinase (NEB), 1 μL 10 μCi/μL [γ-^{32}P] ATP, and 6 μL DEPC-treated water.
2. Centrifuge shortly.
3. Incubate the reaction at 37 °C for 1 h.
4. Adjust the volume to 100 μL by adding 90 μL DEPC-treated water (diluted primer) and keep at −20 °C until use.

3.2.2 Annealing Primer to RNA for Primer Extension Reaction

1. Add the reaction components in a microfuge tube (1.5 mL) in the following order: 10–50 μg of total RNA (X μL; *see* **Note 5**), 1 μL of end-labeled primer (diluted), and 1 μL of 10 mM dNTP mix and adjust the volume up to 10 μL with DEPC-treated water. Short spin and collect the pellet.
2. Incubate the 10 μL reaction mixture at 65 °C for 10 min, and then place on ice for 1 min.
3. Collect the pellet by a short spin.

4. Add the SuperScript™ III mixture in the following order: 2 μL of 10× RT buffer, 4 μL 25 mM $MgCl_2$, 2 μL of 100 mM DTT, 1 μL RNaseOUT (40 U/μL), and 1 μL of SuperScript™ III RT (200 U/μL).
5. Incubate the reaction at 42–50 °C (depending upon the primer) for 1 h.
6. Add 20 μL of 2× gel loading buffer II.
7. Heat the sample at 95–100 °C for 3–5 min.
8. Collect the pellet by short spin.
9. Load the sample on an 8 % SDS denaturing urea acrylamide gel.

3.2.3 Preparation of Urea Acrylamide Gel

1. Mix the components of Sequagel reagents according to the manufacturer's recommendations. For a 6 % sequencing gel (100 mL) the following reagents have to be added in an Erlenmeyer flask in the following order: 24 mL of Ureagel concentrate, 76 mL of Ureagel diluent, and 10 mL of Ureagel buffer.
2. Add 800 μL of 10 % APS and 40 μL of TEMED.
3. Apply Sigmacote (Sigma) to the glass plates (this will facilitate the removal of the gel from the glass plate).
4. Pour the acrylamide mix between two glass plates with spacer, which is well sealed to avoid leakage. Avoid air bubbles.
5. Insert the comb on the top of the gel and wait for 30 min for the gel to polymerize.
6. Once the gel is polymerized, assemble the glass plates with the gel in a vertical gel apparatus with 0.5 × TBE running buffer.
7. The wells must be flushed with a syringe using the running buffer to remove urea.

3.2.4 Gel Running and Band Detection

1. Pre-run the gel for 45 min at high voltage (1,500 V).
2. Switch off the power supply and flush the wells once again with the running buffer to remove urea.
3. Load the preheated samples into each well carefully and also the radiolabeled DNA marker in one of the wells to compare the size of the primer extension products.
4. Switch on the power supply and run the gel until the bromophenol blue is at 2 cm from the bottom of the gel.
5. Switch off the power supply and carefully remove the plates from the vertical apparatus.
6. Carefully split apart the plates with a spatula and make sure that the gel remains intact.

7. Keep a Whatman no. 3 filter paper on the top of the gel and press gently against the gel.
8. Carefully lift the Whatman paper with the gel.
9. Wrap the Whatman paper along with the gel by a Saran wrap.
10. Dry in a vacuum gel drying system, expose to a photographic film, and keep at −80 °C for different time periods depending upon the intensity of the signal (alternatively, the Whatman paper along with the gel can be exposed to a photographic film after covering with Saran wrap without drying).
11. Compare the size of the primer extension products with the primer extension ladder.
12. The exact positions of the cleavage sites can be identified by running in parallel a sequencing reaction.

3.3 Deadenylation Assay

3.3.1 Inhibition of Active Transcription in Leishmania Cells

1. Start a 30 mL *Leishmania* culture for six different time points (5 mL each).
2. Measure the OD_{600} on day 4 (exponential growth phase; OD should be maximal 0.5).
3. Transfer 5 mL culture into a new culture flask. Add 10 μL DMSO and keep cells for zero time point.
4. Add 50 μL ActD (stock 5 mg/mL) to the remaining *Leishmania* culture (25 mL) to obtain a final concentration of 10 μg/mL. Work without light and wrap cell culture flasks with aluminium foil.
5. Incubate cultures for the appropriate time points (include 5-min centrifugation time).
6. Harvest 5 mL cells per time point by centrifugation (5 min at 3,000 rpm).
7. Discard medium supernatant and lyse cell pellets in 1 mL Trizol® by slowly pipetting up and down three times. Transfer samples to nuclease-free reaction tubes, snap-freeze in liquid nitrogen, or transfer immediately to −80 °C. Store samples overnight.

3.3.2 Isolation of Total RNA from Leishmania (See method in Subheading 3.1.3)

3.3.3 RNase H Digest

1. Mix 40 μg of total RNA from each time point with 600 ng antisense oligonucleotide (~2 μL of 100 μM stock) in a fresh nuclease-free tube.
2. For the negative control, mix 40 μg total RNA (time point 0), 600 ng oligo(dT) primer (~2 μL of 100 μM stock), and 600 ng antisense primer.
3. For the positive control, mix 40 μg total RNA (time point 0) and 600 ng antisense oligonucleotide (do not add RNase H to this sample).

4. Heat samples for 10 min at 65 °C to denature the RNA and immediately chill on ice for 2 min.
5. Add 1 μL RNaseOUT, 2 μL 10× RNase H buffer, and 1 μL RNase H (2 U/μL) to each sample, except the positive control. Adjust samples to a final volume of 20 μL with DEPC water.
6. Incubate samples at 37 °C for 1 h in a thermo block.
7. Add 1 μL 0.5 M EDTA (sterile filtered) to each reaction to stop RNase H enzymatic activity.
8. Add 175 μL DEPC water and 1 μL glycogen to each sample. Precipitate RNA with 2.5 volumes 100 % ethanol (500 μL) at −20 °C for at least 1 h.
9. Pellet RNA by centrifugation at 14,000 rpm for 15 min at 4 °C.
10. Dry pellets for 5 min at 37 °C in a thermo block.
11. Resuspend pellets in 2× RNA loading buffer. Dissolve RNA at 37 °C for 5 min.
12. Denature samples at 65 °C for 10 min. Chill on ice until loading onto the gel.

3.3.4 Polyacrylamide Gel Electrophoresis and Transfer

1. Cast a 5 % acrylamide gel (Sequagel) and store it for 30 min to ensure complete polymerization. For one gel mix 10 mL Sequagel, concentrate with 35 mL Sequagel Diluent, 5 mL Sequagel Buffer, and 0.4 mL 10 % APS. Start polymerization with 20 μL TEMED.
2. Prepare 4 L of 0.5× TBE and store 2 L at 4 °C.
3. Assemble the acrylamide gel running system, fill tank with 0.5× TBE, remove comb, and complete a 30-min pre-run at 150 V.
4. Wash gel slots to remove excess urea. Load samples on the equilibrated acrylamide gel and separate them at 150 V until the light blue dye is at 2/3 of the gel (ca. 3 h).
5. Cut 6 Whatman filter papers and a nylon membrane and soak them together with two sponges for 15 min in precooled 0.5× TBE.
6. Build a gel sandwich with three filter papers on either side. Remove air bubbles carefully. Place the sandwich in the sandwich holder and tighten it carefully. Place the holder in the tank oriented so that negatively charged RNA will be transferred onto the nylon membrane. Fill the tank with cold 0.5× TBE buffer (2 L) and place it on a magnetic stirrer with a stir bar.
7. Transfer at 30 V with constant maximum stirring for 4 h and 4 °C.

8. Disassemble the wet-blot apparatus and rinse the membrane once with 2× SSC.
9. Air-dry the membrane for 15 min. UV-cross-link it twice with the Stratalinker automatic cross-link program.
10. Wrap membrane with Saran wrap and store it at 4 °C until hybridization.

3.3.5 Preparation of Gene-Specific Probes

1. To prepare gene-specific probes, amplify DNA fragments of interest from genomic DNA by PCR. To visualize only the 3′ fragment that was digested after RNase H treatment, design primers to amplify the 300 bp region immediately upstream of the poly(A) site.
2. Purify PCR products either using gel extraction or PCR purification kits (Qiagen).
3. Verify the correct sequence of the PCR product by sequencing.
4. Quantify the purified PCR product on agarose gel using a mass ladder.

3.3.6 Northern Blot Hybridization

1. Prepare 10 mL of pre-hybridization solution P per membrane. Mix 5 mL DEPC water with 1 mL 50× Denhardt's, 3 mL 20× SSC, and 1 mL 10 % SDS. Add ingredients in this order to avoid precipitation of SDS!
2. Warm solution P to 65 °C to completely dissolve SDS.
3. Heat salmon sperm DNA at 95 °C for 5 min. Chill on ice for at least 2 min.
4. Add 100 μL salmon sperm DNA to 10 mL solution P (unspecific blocking agent).
5. Place membrane in a hybridization tube, add 10 mL of solution P, and pre-hybridize for 2 h at 42 °C with constant rotation.
6. To radiolabel the probe, dilute the specific PCR fragment to a final concentration of 100 ng in 20 μL water. Work in the radioactivity room!
7. Heat probe at 95 °C for 5 min. Chill on ice for 2 min.
8. Add 6 μL 5× oligonucleotide labeling buffer, 3 μL alpha-dCTP32 (30 μCi), and 1 μL Klenow enzyme. Mix carefully, quick spin, and incubate at 37 °C for 60 min.
9. Stop reaction by adding 100 μL stop mix.
10. Remove unincorporated nucleotides by passing the probes over NucAway™ spin columns (Ambion).
11. Verify incorporation of radioactive nucleotides by counting 1 μL probe in 5 mL scintillation solution.

12. Prepare 10 mL hybridization solution H per membrane. Mix 5 mL formamide with 3 mL 20× SSC, 0.6 mL DEPC water, 1 mL 10 % SDS, and 0.2 mL 50× Denhardt's.
13. Heat salmon sperm DNA at 95 °C for 5 min. Chill on ice for at least 2 min.
14. Add 100 μL salmon sperm DNA to 10 mL solution H.
14. Exchange solution P for 10 mL solution H.
15. Denature labeled probes for 5 min at 95 °C.
16. Add the radiolabeled probe directly into solution H. Hybridize overnight at 42 °C with constant rotation.
17. Remove the radiolabeled probe and rinse membrane once with 10 mL wash buffer 1.
18. Wash membrane once with 25 mL wash buffer 1 at 25 °C for 15 min. Discard wash buffer.
19. Wash membrane 2× 30 min with 25 mL wash buffer 1 at 65 °C. Discard wash buffer.
20. Wash membrane once with 25 mL wash buffer 2 at 25 °C for 30 min. Discard wash buffer.
21. Wrap membrane with Saran wrap, fix membrane in a film cassette, and expose a film.

4 Notes

1. The RNA annealing gene-specific primer should be 100–200 nt downstream of the 5′-end of the RNA cleavage site. Primers annealed too far from the expected cleavage site on the RNA may lead to premature abortion of the primer extension reaction.
2. More than one gene-specific primer must be used to validate the size of extension products.
3. The annealing temperature of each gene-specific primer must be tested individually to find out the optimal conditions.
4. The reverse transcriptase reaction can be inhibited by RNA secondary structures or by modified bases [17]. Primer extension products generated by the falloff of RT enzyme due to RNA secondary structures appear like bands. Therefore, when interpreting the data, great caution must be taken to misleading results.
5. Since RNA integrity is important in order to obtain reproducible results for cDNA synthesis, the quality of RNA should be checked on agarose gel.

References

1. Parker R, Song H (2004) The enzymes and control of eukaryotic mRNA turnover. Nat Struct Mol Biol 11:121–127
2. Houseley J, LaCava J, Tollervey D (2006) RNA-quality control by the exosome. Nat Rev Mol Cell Biol 7:529–539
3. Coller J, Parker R (2004) Eukaryotic mRNA decapping. Annu Rev Biochem 73:861–890
4. Tourrière H, Chebli K, Tazi J (2002) mRNA degradation machines in eukaryotic cells. Biochimie 84:821–837
5. Liu H, Kiledjian M (2007) An erythroid-enriched endoribonuclease (ErEN) involved in alpha-globin mRNA turnover. Protein Pept Lett 14:131–136
6. Lee CH, Leeds P, Ross J (1998) Purification and characterization of a polysome-associated endoribonuclease that degrades c-myc mRNA in vitro. J Biol Chem 273:25261–25271
7. Stevens A, Wang Y et al (2002) Beta-Globin mRNA decay in erythroid cells: UG site-preferred endonucleolytic cleavage that is augmented by a premature termination codon. Proc Natl Acad Sci U S A 99:12741–12746
8. Bringaud F, Müller M et al (2007) Members of a large retroposon family are determinants of post-transcriptional gene expression in *Leishmania*. PLoS Pathog 3:1291–1307
9. Smith M, Bringaud F, Papadopoulou B (2009) Organization and evolution of two SIDER retroposon subfamilies and their impact on the *Leishmania* genome. BMC Genomics 10:240
10. Müller M, Padmanabhan PK et al (2010) Rapid decay of unstable Leishmania mRNAs bearing a conserved retroposon signature 3′-UTR motif is initiated by a site-specific endonucleolytic cleavage without prior deadenylation. Nucleic Acids Res 38:5867–5883
11. Clayton C, Shapira M (2007) Post-transcriptional regulation of gene expression in trypanosomes and leishmanias. Mol Biochem Parasitol 156:93–101
12. Haile S, Papadopoulou B (2007) Developmental regulation of gene expression in trypanosomatid parasitic protozoa. Curr Opin Microbiol 10:569–577
13. Haile S, Dupe A, Papadopoulou B (2008) Deadenylation-independent stage-specific mRNA degradation in Leishmania. Nucleic Acids Res 36:1634–1644
14. Schwede A, Ellis L et al (2008) A role for Caf1 in mRNA deadenylation and decay in trypanosomes and human cells. Nucleic Acids Res 36:3374–3388
15. Otsuka Y, Schoenberg DR (2008) Approaches for studying PMR1 endonuclease- mediated mRNA decay. Methods Enzymol 448: 241–263
16. Boorstein WR, Craig EA (1989) Primer extension analysis of RNA. Methods Enzymol 180:347–369
17. Yu E, Fabris D (2003) Direct probing of RNA structures and RNA-protein interactions in the HIV-1 packaging signal by chemical modification and electrospray ionization fourier transform mass spectrometry. J Mol Biol 330: 211–223
18. Sallés FJ, Richards WG, Strickland S (1999) Assaying the polyadenylation state of mRNAs. Methods 17:38–45

Chapter 8

Gene Suppression in Schistosomes Using RNAi

Akram A. Da'dara and Patrick J. Skelly

Abstract

Schistosomiasis is a neglected tropical disease responsible for the death of more than 300,000 people every year. The disease is caused by intravascular parasitic platyhelminths called schistosomes. Treatment and control of schistosomiasis rely on a single drug, praziquantel, and concern exists over the possible emergence of resistance to this drug. The recent completion of the genome sequences of the three main worm species that cause schistosomiasis in humans has raised hope for the development of new interventions to treat the disease. RNA interference (RNAi), a mechanism by which gene-specific double-stranded RNA (dsRNA) triggers degradation of homologous mRNA transcripts, has emerged as an important tool to evaluate and validate new potential drug targets. In addition, RNAi has been used to explore the basic biology of these debilitating parasites. RNAi can be achieved in all stages of the parasite's life cycle in which it has been tested. In this review, we describe methods for applying RNAi to suppress gene expression in the intra-mammalian life stages (adults and schistosomula) of *Schistosoma mansoni*. We describe procedures for isolating and culturing the parasites, preparing and delivering dsRNA targeting a specific gene, as well as a procedure to evaluate gene suppression by quantitative real-time PCR.

Key words Schistosome, Schistosomiasis, Trematode, Schistosomula, RNA interference, RNAi, siRNA, dsRNA, Electroporation, qRT-PCR

1 Introduction

Schistosomiasis is a neglected tropical disease caused by helminth parasites of the genus *Schistosoma* which affects more than 200 million people worldwide and is responsible for 300,000 deaths annually [1]. There are three main species that cause schistosomiasis in human: *Schistosoma mansoni*, *S. haematobium*, and *S. japonicum*. Schistosomes have complex life cycle with several distinct developmental stages [2]. Adult *S. mansoni* and *S. japonicum* parasites live largely in the mesenteric veins, whereas adults of *S. haematobium* live mostly in the veins of the vesical plexus around the urinary bladder. Adult female parasites produce hundreds of eggs each day, some of which are passed from the body into the environment. In freshwater, schistosome eggs hatch to release miracidia, which can infect specific species of snails. In the snails, the miracidia

Christopher Peacock (ed.), *Parasite Genomics Protocols*, Methods in Molecular Biology, vol. 1201,
DOI 10.1007/978-1-4939-1438-8_8, © Springer Science+Business Media New York 2015

transform into sporocysts which replicate asexually and then develop further to produce infectious larvae called cercariae. The cercariae leave the snail and swim freely in freshwater. Upon contact with human skin, the cercariae penetrate and transform into the schistosomula life stage. These invade a blood vessel, migrate through the bloodstream, and eventually develop into adult male and female parasites. Adults pair in the portal vasculature and the couples migrate to the preferred egg-laying sites noted above.

So far, there is no vaccine for schistosomiasis, and for the past three decades treatment and control of this disease have relied largely on chemotherapy using a single drug called praziquantel (PZQ) [3]. Concerns exist over the possible emergence and establishment of drug resistance to PZQ [4–6]. Thus, it remains a priority to identify and develop novel chemical and/or immunological therapeutic interventions for this disease. The recent completion of the genome sequence of *S. mansoni* [7], *S. japonicum* [8], and *S. haematobium* [9], as well as the large amount of schistosome transcriptome data collected, offer a fresh opportunity to identify new drug target molecules. However, until recently, the lack of genetic tools to evaluate and validate drug targets for schistosome parasites has hindered the development of new interventions. The development of RNA interference (RNAi) in schistosomes offers a valuable tool to overcome this limitation and permit new drug target evaluation [10, 11]. In addition, RNAi can be used to explore the basic molecular and cellular biology of these debilitating parasites.

RNAi is a mechanism by which exogenous gene-specific double-stranded RNA (dsRNA) triggers degradation of homologous mRNA transcripts which results in effective, sequence-specific, post-transcriptional gene silencing [12–15]. RNAi has been described in a diversity of organisms, including plants, fungi, arthropods, protozoans, and vertebrates [16–19]. RNAi technology has influenced strategies for the pharmacological treatment of many conditions including cancer, inflammatory diseases, and bacterial and viral infections [20–23]. Likewise for parasitic diseases, RNAi holds real potential. For instance, RNAi can serve as an effective tool to identify and test new antiparasitic drug targets. RNAi treatment that leads to parasite debility and/or death suggests that the specific gene product is a potential target for the development of antiparasite treatments. RNAi screening, therefore, should help to rapidly identify target molecules of interest and facilitate the development of new therapies. Recently, RNAi was successfully used to validate the enzyme thioredoxin-glutathione reductase as a drug target for schistosomes [24]. In addition, since its discovery in schistosomes [10, 11], RNAi has also been used to investigate several aspects of basic parasite biology [25–34]. In this review, we describe our preferred methodology for using RNAi to suppress gene expression in *S. mansoni*. We focus on the intra-mammalian life stages of the parasite (adults and schistosomula). We describe protocols for isolating and culturing the parasites, preparing and delivering dsRNAs, as well as evaluating gene suppression by quantitative real-time PCR.

2 Materials

2.1 Parasite Material

1. *Snails*: *Biomphalaria glabrata* snails, infected with *Schistosoma mansoni* (Puerto Rican strain), are obtained from the NIAID Schistosomiasis Resource Center at the Biomedical Research Institute, Rockville, MD, USA (http://www.schisto-resource.org/) and maintained in the laboratory.
2. *Cercariae*: Infectious schistosome larvae, cercariae, are obtained from infected snails as described in Subheading 3.2.1.
3. *Schistosomula*: Schistosomula are prepared from cercariae as described in Subheading 3.2.
4. *Adult parasites*: Adult male and female parasites are obtained by vascular perfusion of infected mice at 6–7 weeks after infection as previously described [35].

2.2 Parasite Culture Medium

1. DMEM/F12 (1:1) (Invitrogen).
2. HyClone Fetal Bovine Serum (FBS, Thermo Scientific).
3. Penicillin-Streptomycin solution (Invitrogen).
4. Triiodo-L-thyronine (Sigma).
5. Serotonin (Sigma).
6. Human insulin (Sigma).

2.3 Schistosomula Preparation and Purification Reagents

1. Percoll (Sigma).
2. RPMI 1640 (Invitrogen).
3. Microslides (25 × 75 × 1 mm) (VWR) and cover slips (25 mm^2) (Corning Inc.).
4. 15 ml and 50 ml BD Falcon Conical Centrifuge Tubes (BD Bioscience).
5. Trypan blue: 0.4 % solution (Sigma).
6. 100 μm cell strainers (BD Bioscience).
7. 25 μm nylon cell micro-sieves (Biodesign Inc., NY).
8. Tissue culture plates (Corning Inc.).

2.4 Long dsRNA Preparation

1. AccuPrime *Taq* DNA Polymerase High Fidelity (Invitrogen).
2. QIAquick gel extraction kit (Qiagen).
3. MEGAscript RNAi Kit (Ambion).

4. *Agarose gel electrophoresis*: Agarose (Sigma); 1× TBE buffer: 89 mM Tris base, 89 mM boric acid, 2 mM EDTA, pH 8.0; ethidium bromide: 10 mg/ml solution (*see* **Note 1**); 6× gel loading buffer: 0.25 % bromophenol blue, 0.25 % xylene cyanol, and 33 % glycerol; wide range DNA ladder (Sigma).

2.5 Electroporation

1. siRNA (obtained as described in Subheading 3.3) or long dsRNA (prepared as described in Subheading 3.4) targeting a specific gene of interest.
2. siRNA resuspension buffer: 100 mM potassium acetate, 30 mM HEPES, pH 7.5 (IDT).
3. 0.4 cm gene pulser electroporation cuvettes (Bio-Rad).
4. Electroporation buffer (Bio-Rad).
5. Gene Pulser Xcell Electroporator (Bio-Rad).
6. Tissue culture plates (Corning Inc.).

RNA isolation

1. TRIzol Reagent (Invitrogen).
2. RNase decontaminating wipes (RNase*Zap* Wipes, Ambion).
3. Disposable RNase-free pellet pestle and motor (VWR).
4. Chloroform (Sigma).
5. Isopropyl alcohol (Sigma).
6. Nuclease-free and/or DEPC-treated water (Ambion).
7. 75 % ethanol: Prepared in nuclease-free water.
8. TURBO DNA-free kit (Ambion).
9. Nanodrop or other UV spectrophotometer.

2.6 cDNA Synthesis

1. Oligo $(dT)_{12-18}$ Primer (0.5 μg/μl), RNaseOUT Recombinant Ribonuclease Inhibitor (40 U/μl), RNase H (2 U/μl), SuperScript III Reverse Transcriptase (200 U/μl), 0.1 M DTT, and 10× Reverse Transcriptase Buffer (10× RT Buffer): 200 mM Tris–HCl, 500 mM KCl, pH 8.4 (all from Invitrogen).
2. $MgCl_2$: 25 mM (Qiagen).
3. 10 mM dNTP mix (Invitrogen): Prepared by adding 50 μl of 100 mM each dATP, dCTP, dGTP, and dTTP to 300 μl of nuclease-free water.
4. RNase-Free 0.2 ml PCR Tubes (Ambion).
5. Thermal cycler (Bio-Rad).

2.7 Quantitative Real-Time-PCR (qRT-PCR)

1. Custom TaqMan Gene Expression Assay: Each assay is a 20× concentrated mix of forward primer (18 μM), reverse primer (18 μM), and 6-carboxyfluorescein (FAM)-labeled MGB reporter probe (5 μM) (Applied Biosystems, Foster City, CA).
2. TaqMan Universal PCR Master Mix (2×) (Applied Biosystems).

3. Nuclease-free water (Ambion).
4. MicroAmp Fast optical 96-well reaction plates (Applied Biosystems).
5. MicroAmp optical adhesive film (Applied Biosystems).
6. Real-Time PCR machine: StepOne Plus Real Time PCR System with StepOne Software v2.0 (Applied Biosystems).

3 Methods

3.1 Parasite Culture Medium

Schistosomula as well as adult parasites can be cultured in complete DMEM/F12 culture medium. Prepare complete culture medium by adding the following ingredients to DMEM/F12:

1. Add heat-inactivated fetal bovine serum (FBS) to a final concentration of 10 % (*see* **Note 2**).
2. Add penicillin to 200 U/ml and streptomycin to 200 μg/ml final concentration.
3. Add triiodo-L-thyronine to 0.2 μM (*see* **Note 3**).
4. Add serotonin to 1.0 μM.
5. Add human insulin to 8 μg/ml (*see* **Note 4**).
6. Mix all ingredients and sterilize the complete medium by filtration under vacuum.
7. Store at 4 °C.

3.2 Schistosomula Preparation

The following sections describe the in vitro production of schistosomula. The process involves first preparing cercariae (Subheading 3.2.1), from which pure schistosomula can be generated (Subheading 3.2.2). The major steps in this process are summarized in Fig. 1.

3.2.1 Cercariae Preparation

Cercariae are the infectious stage of the parasite which can directly penetrate the skin. Therefore, proper precautions must be taken to prevent contaminated water from coming into contact with skin. Researchers should receive proper training and wear protective clothing and gloves when handling infected snails and/or cercariae. Cercariae develop and emerge from snails usually ~40 days after initial infection. To obtain this life cycle stage:

1. Collect and clean infected snails.
2. Place the snails in a glass beaker containing clean water and expose to light for ~1–2 h to promote cercarial emergence (Fig. 1a) (*see* **Note 5**).
3. Filter cercariae through a mesh sieve to eliminate snail excrement and other debris. A large filter can be used to remove bigger particles and a smaller filter (e.g., a 100 μm cell strainer) to further clean the cercariae (Fig. 1b).

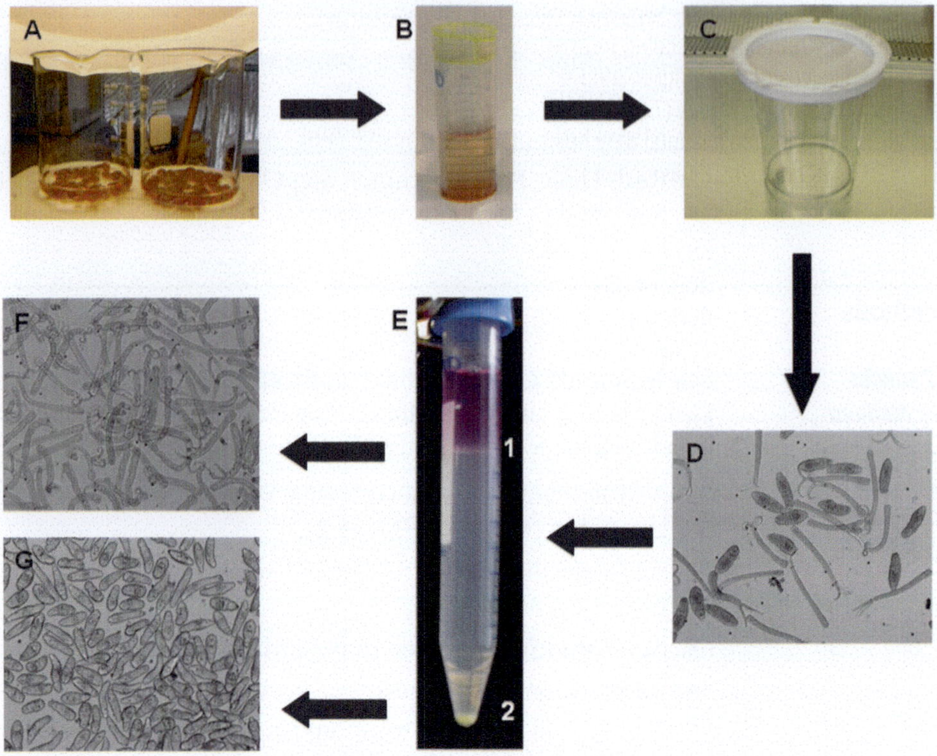

Fig. 1 Preparation and purification of schistosomula by Percoll gradient centrifugation. Cercariae are obtained by first exposing clean infected snails to light for 1–2 h. Two beakers containing many *Biomphalaria glabrata* snails at the *bottom* are shown (**a**). Water containing cercariae that emerge from the snails is filtered through 100 μm filters to remove debris (**b**). Cercariae are then washed with water containing antibiotics and concentrated on a 25 μm filter affixed over a collecting beaker as shown in (**c**). Cercarial bodies are next separated from cercarial tails by vortexing the parasites. (**d**) Shows a mixture of slender, forked cercarial tails and ovoid bodies that result from this procedure. Bodies (now called schistosomula) are separated from tails on a 30 % Percoll gradient. (**e**) Shows the Percoll gradient after centrifugation; tails remain on the top of the gradient as indicated by number 1, whereas bodies (schistosomula) are pelleted, as indicated by number 2. When examined microscopically, the material in band 1 is confirmed to be cercarial tails (**f**) while the material in band 2 is exclusively bodies (**g**). The procedure generates a pure population of schistosomula (**g**)

4. Wash with water and concentrate cercariae using a 25 μm nylon cell micro-sieves (Fig. 1c) as follows:
   (a) Cut an appropriately sized piece of 25 μm nylon filter and place it on the top of a clean, sterile beaker.
   (b) Secure the filter using plastic rings or a rubber band.
   (c) Wash the filter with water containing antibiotics (2× Pen/Strep: 200 U/ml penicillin and 200 μg/ml streptomycin).
   (d) Slowly pour the water containing cercariae onto the filter (Fig. 1c).
   (e) Wash cercariae with water containing antibiotics.
   (f) Wash with RPMI containing antibiotics.

5. Collect cercariae in 3 ml of RPMI medium.

   Cercariae possess forked tails which they use for swimming. These are discarded once the parasites have located and penetrated a suitable host. The following steps are designed to shear the tails from the cercariae:
6. Vortex cercariae in RPMI for 3 min.
7. Place on ice for 3 min.
8. Vortex two additional times, 3 min each.
9. Check under microscope for tail separation (Fig. 1d).

3.2.2 Purifying Schistosomula Using a Percoll Gradient

The following procedure is designed to separate the cercarial tails from the remaining parasite bodies (which are now referred to as schistosomula). The inherent stickiness of the tails can be problematic in culture, so their removal is recommended as follows:

1. Prepare 30 % Percoll (10 ml): In a 15 ml Falcon tube, mix 7.0 ml water containing antibiotics with 3.0 ml Percoll.
2. Gently load 3 ml of RPMI containing parasites onto the Percoll gradient.
3. Centrifuge at 300 × *g* for 15 min at 4 °C. Use a swinging bucket rotor. Cercarial tails remain at the top of this gradient (Fig. 1e band # 1 and f), and schistosomula pellet to the bottom of the tube (Fig. 1e band # 2 and g).
4. Carefully aspirate the overlay.
5. Resuspend the pelleted schistosomula in 50 ml RPMI containing antibiotics and centrifuge for 5 min at 400 × *g*. Use a swinging bucket rotor.
6. Aspirate the overlay and resuspend pelleted schistosomula in complete DMEM/F12 culture media at 1 ml per ~1,000 parasites.
7. To determine the number and the viability of schistosomula:
   (a) Take 10 μl of medium containing schistosomula.
   (b) Place schistosomula onto a glass slide.
   (c) Add 5 μl of trypan blue and mix gently.
   (d) Cover with a cover slip.
   (e) Observe under a microscope. Dead schistosomula will stain blue.
8. Incubate at 37 °C and change the medium every 3–4 days.

3.3 Preparation of siRNA

We use commercially synthesized gene-specific 27-mer dicer substrate siRNAs (IDT, Coralville, IA). For siRNA design, we use the online siRNA design tool found on the IDT website (http://www.idtdna.com). Generally, we target the 5′ half of the mRNA with siRNAs. Due to variable effectiveness of different siRNAs, it is advisable to synthesize and test several for each new gene target.

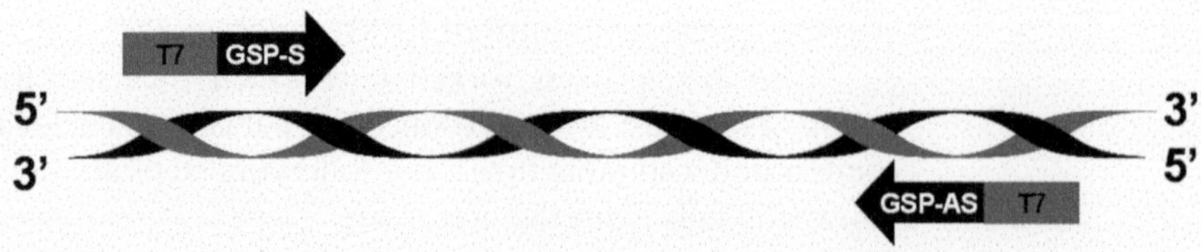

T7 promoter sequence: 5'-TAATACGACTCACTATA**G**GGAGA-3'

Fig. 2 Amplification of target gene by PCR for long dsRNA synthesis. The hypothetical target gene lies between the two primers indicated by *arrows*. GSP is gene-specific primer, sense (S) or antisense (AS). The sequence of the T7 primer is shown below. The larger, *bolded* "**G**" is the first base incorporated into RNA during transcription. *Underlined* is the minimum promoter sequence needed for efficient transcription

In most cases, complementary single-stranded RNAs are delivered and these must be annealed to generate double-stranded siRNAs prior to use. Under sterile conditions, prepare a 100 μM siRNA duplex as follows:

1. Spin down the tubes briefly to collect the contents.
2. Add an appropriate volume of siRNA resuspension buffer to yield a 100 μM solution.
3. Incubate at 94 °C for 2 min.
4. Cool slowly to room temperature.
5. Aliquot and store at −20 °C.

3.4 Preparation of Long dsRNA

Schistosome parasites can be targeted using long dsRNA or siRNA. The efficiency of gene silencing in schistosomes using long dsRNA is similar to that observed with siRNA [26]. However, preparation of long dsRNA can be expensive and labor intensive. It is noteworthy that long dsRNA can be delivered by electroporation or by soaking. Therefore, laboratories lacking an electroporator can generate long dsRNA and deliver it by soaking, as described below. There are several methods for preparing long dsRNA; here we describe one method which uses the MEGAscript RNAi Kit (Ambion). This system utilizes only T7 RNA polymerase to generate RNA. Prior to long dsRNA synthesis, a specific region of the target gene needs to be amplified by PCR (Subheading 3.4.1), and then the PCR product is used as a template to synthesize dsRNA (Subheading 3.4.2).

3.4.1 Amplification of the Target Region by PCR

1. Design a set of target-specific primers, perhaps with the help of primer design software such as PrimerQuest (IDT, www.idtdna.com). Include the T7 RNA polymerase promoter sequence at the 5′-end of both primers (Fig. 2). A PCR product of about 500–900 bp in length is suitable (*see* **Note 6**).

2. Using the primers and a suitable template (e.g., cDNA from adult parasites or a plasmid containing target cDNA) amplify the fragment of interest by conventional PCR, following the recipe below (*see* **Note 7**):

Component	Volume (μl)
10× PCR buffer I *(contains 20 mM $MgSO_4$ and 2 mM dNTPs)*	5
Forward primer (100 pmol/μl)	1
Reverse primer (100 pmol/μl)	1
cDNA	1
AccuPrime Taq DNA Polymerase (5 U/μl)	0.2
Nuclease-free H_2O	41.8

3. Cap the tubes, tap gently to mix, and centrifuge briefly to collect the contents.
4. Place the tubes in the thermal cycler and run the following program:
   (a) Initial denaturation: 94 °C for 2 min.
   (b) 35 cycles of:
   - Denaturation: 94 °C for 30 s.
   - Annealing: 50–60 °C for 30 s (*see* **Note 8**).
   - Extension: 68 °C for 1 min per kb of PCR product.

   (c) Final extension: 68 °C for 10 min.
5. Analyze PCR products by conventional agarose gel electrophoresis: Prepare 1 % agarose in 1× TBE buffer. Boil in a microwave. Add ethidium bromide to 0.5 μg/ml, prepare, pour, and run the gel (*see* **Note 1**).
6. Purify PCR products using QIAquick gel extraction kit (Qiagen). The protocol, essentially following the manufacturer's instructions, is as follows:
   (a) Visualize PCR products resolved following gel electrophoresis using a UV light box and carefully excise the DNA band of interest from the agarose gel using a razor blade.
   (b) Estimate the volume of the gel piece by weighing it (where each 1 mg is equivalent to 1 μl).
   (c) Add to the gel slice 3 volumes of QG buffer.
   (d) Incubate at 55 °C for 5–10 min (or until the agarose is completely melted).
   (e) Add 1 gel volume of isopropyl alcohol (2-propanol).

(f) Apply the sample to a QIAquick spin column placed in a 2 ml collection tube and centrifuge for 30 s at maximum speed.

(g) Discard the flow-through.

(h) Add 700 μl of wash buffer (PE, make sure that ethanol is added to PE buffer before use) and centrifuge for 30 s at maximum speed.

(i) Discard the flow-through.

(j) Wash again with 300 μl of PE as described in **step h**.

(k) Discard the flow-through and, to remove residual buffer, centrifuge the empty tubes for 1 min at maximum speed.

(l) Place the column into a new 1.5 ml tube.

(m) To elute the DNA, add 50 μl elution buffer (EB): 10 mM Tris–HCl, pH 8.5.

(n) Let the tubes stand at room temperature for 1–2 min and then centrifuge for 1 min at maximum speed.

(o) Discard the column and store the recovered DNA.

7. Determine the DNA concentration at OD_{260} using Nanodrop or another spectrophotometer containing a UV lamp (1 OD_{260} is equivalent to 50 μg/ml DNA). The purified product can now be used as a template for the synthesis of dsRNA as described next.

3.4.2 Long dsRNA Synthesis

Here we use the MEGAscript RNA kit, for long dsRNA synthesis. This kit uses T7 RNA polymerase to synthesize both sense and antisense RNAs. The protocol, essentially following the manufacturer's instructions, consists of four major steps: (A) generating complementary single-stranded RNAs (ssRNAs), (B) annealing the complementary ssRNAs to generate dsRNA, (C) removing plasmid DNA and residual ssRNA from the mixture by nuclease digestion, and (D) purifying the final dsRNA product.

(A) Generating complementary single-stranded RNAs

1. Mix the following components:

Component	Volume
PCR template	*x*μl (0.5–2.2 pmol, *see* **Note 9**)
10× T7 reaction buffer	2 μl
75 mM ATP	2 μl
75 mM CTP	2 μl
75 mM GTP	2 μl

(continued)

Component	Volume
75 mM UTP	2 μl
T7 Enzyme mix	2 μl
H_2O	*y*μl
Final volume	20 μl

2. Mix well and spin down briefly.
3. Incubate at 37 °C for 4 h.

(B) Annealing the RNA

1. Incubate the tube containing ssRNA products that was generated in step A at 75 °C for 5 min.
2. Leave the mixture on the bench to cool slowly to room temperature. Do not put the reaction on ice to cool.

(C) Nuclease digestion

This step is included to get rid of template DNA and single-stranded RNA.

1. Mix the following components:

Component	Volume (μl)
dsRNA (generated in steps A–B)	20
H_2O	21
10× digestion buffer	5
DNase I	2
RNase	2

2. Incubate at 37 °C for 1 h.

(D) Purification of dsRNA

Preheat the elution buffer to 95 °C.

1. Mix the following reagents in a clean tube:

Component	Volume (μl)
dsRNA	50
10× binding buffer	50
H_2O	150
100 % ethanol	250

2. Mix well and load onto the filter cartridge.
3. Centrifuge at maximum speed for 1 min.
4. Wash the cartridge two times with 500 μl wash buffer each time.
5. Centrifuge the empty tube at maximum speed for 1 min to remove residual wash buffer.
6. Transfer the filter to a new tube.
7. To elute the filter-bound dsRNA, add 50–100 μl of hot (95 °C) elution buffer and centrifuge at maximum speed for 1 min.
8. Add an additional 50–100 μl hot elution buffer and centrifuge again at maximum speed for 1 min, to recover any remaining bound dsRNA.
9. Pool both eluates and determine the final dsRNA concentration at OD_{260} using a nanodrop or another spectrophotometer containing a UV lamp. (1 OD_{260} is equivalent to 40 μg/ml dsRNA).
10. Analyze recovered dsRNA by 1.2 % agarose gel electrophoresis. dsRNA runs similar but slightly faster than its equivalent dsDNA and notably faster than either ssRNAs from which it was generated.

3.5 Electroporation of Schistosomula with siRNA or Long dsRNA

Schistosomula can be electroporated at different ages, from freshly prepared to several weeks old, as follows:

1. Transfer schistosomula to a 15 ml Falcon tube.
2. Count parasites as described above (Subheading 3.2.2).
3. Pellet schistosomula by centrifugation (300 × *g* for 5 min, at room temperature).
4. Resuspend in electroporation buffer at 1,000 parasites/50 μl buffer.
5. Transfer schistosomula (50 μl) into a 0.4 cm electroporation cuvette.
6. Add 2.5–5.0 μg siRNA (3–6 μM) or 5–10 μg of long dsRNA (0.25–0.5 μM; calculated for a dsRNA fragment of ~600 nucleotides) (*see* **Notes 10** and **11**).
7. Electroporate the parasites by applying a square wave with a single 20 ms pulse, at 125 V in 4 mm cuvette at room temperature.
8. Immediately after electroporation add 500 μl pre-warmed complete DMEM/F12 culture medium to each cuvette.
9. Transfer the schistosomula to a well of a 48-well culture plate and add an additional 200 μl of complete DMEM/F12 culture medium.

10. Incubate the parasites at 37 °C in a humidified incubator with 5 % CO_2.
11. The next day, and every 3 days thereafter, replace the medium with fresh culture medium.

3.6 Electroporation of Adult Parasites with siRNA or Long dsRNA

Electroporation can be performed on male, female, and mixed (male/female) adult parasites.

1. Transfer 6–12 parasites to an electroporation cuvette.
2. Wash with media lacking serum.
3. Add 100 μl of electroporation buffer to each cuvette.
4. Add 5–10 μg of siRNA (3–6 μM) or 10–20 μg of dsRNA (0.25–0.5 μM) (*see* **Notes 10** and **11**).
5. Electroporate the parasites by applying a square wave with a single 20 ms pulse, at 125 V in 4 mm cuvettes at room temperature.
6. Add 500 μl of prewarmed complete DMEM/F12 medium to each cuvette.
7. Transfer the parasites to a well of a 12-well plate.
8. Adjust the volume to 2 ml with extra medium.
9. After overnight incubation, and every other day thereafter, exchange the media with fresh complete DMEM/F12 medium.

3.7 Soaking Parasites with dsRNA

1. To ~1,000 schistosomula in 500 μl complete media, or 10–15 adult parasites in 1.5 ml medium, add 50 μg long dsRNA diluted in 50 μl medium without serum. After overnight incubation, remove ~80 % of the medium and replace with 300 μl of complete DMEM/F12 medium for schistosomula and 1 ml for adult parasites.
2. Replace the media with fresh culture media every 1–2 days for adult parasites and every 3 days in the case of schistosomula.

3.8 RNA Isolation Using TRIzol Reagent

Generally, we use TRIzol Reagent (Invitrogen) to purify total RNA from schistosomula and adult parasites as follows:

1. Transfer parasites from culture wells to 1.5 ml Eppendorf tubes.
2. Wash the parasites three times with nuclease-free PBS.
3. Add 1 ml of TRIzol Reagent to schistosomula and 50 μl of TRIzol Reagent to adult parasites.
4. Homogenize adult parasites on ice using a pellet pestle mortar for about 1 min.

5. Add an additional 950 μl TRIzol Reagent to each tube containing adult parasites to bring the volume to 1 ml, and incubate at room temperature (RT) for 10 min.
6. Add 200 μl of chloroform to each sample. Shake the tubes vigorously by hand for 15 s.
7. Incubate for 5 min at RT.
8. Centrifuge at 12,000 × *g* for 15 min at 4 °C.
9. Carefully collect about 500 μl of the upper (aqueous) layer in a new tube. This layer contains the RNA. Avoid drawing any of the interphase or the organic phase (red layer).
10. Add 500 μl of isopropanol to each tube. Vortex briefly and incubate for 10 min at RT.
11. Centrifuge at 12,000 × *g* at 4 °C for 10 min (*see* **Note 12**).
12. Carefully aspirate the supernatant.
13. Wash the pellet by adding 1 ml of 75 % ethanol.
14. Centrifuge at 12,000 × *g* for 5 min at 4 °C (*see* **Note 12**).
15. Carefully aspirate the ethanol.
16. Air-dry the pellet for 5–10 min at RT (*see* **Note 13**).
17. Add 50 μl RNase-free or DEPC-treated water to RNA pellets from adult parasites and 20 μl to RNA pellets generated from schistosomula. Keep on ice for 30 min to help dissolve the pellet completely (*see* **Note 14**).

To remove any traces of genomic DNA that may be present in the RNA preparation, use RNase-free DNase I digestion. The following protocol is for 50 μl RNA samples; for different volumes, adjust accordingly:

1. Add 5 μl 10× TURBO DNase Buffer to 50 μl RNA solution.
2. Add 1 μl TURBO DNase.
3. Mix well by tapping gently and incubate for 20 min at 37 °C.
4. Add 5 μl DNase I-inactivation reagent.
5. Mix well by gently tapping the tube and incubate for 3–5 min at RT. Mix the tubes occasionally.
6. Centrifuge at 10,000 × *g* for 2 min at RT and transfer the supernatant to a new tube. Make sure not to carry over any inactivation resins with the RNA.
7. Measure the RNA concentration using nanodrop, or any UV spectrophotometer, at OD 260 nm.
8. Immediately proceed to cDNA preparation (described next) or store RNA at −80 °C.

3.9 cDNA Synthesis

To prepare cDNA for quantitative real-time PCR, a minimum of 40 ng total RNA per cDNA reaction is needed.

1. Place RNA samples, SuperScript III, RNaseOUT, dNTP mix, 10× RT buffer, $MgCl_2$, and DTT on ice.
2. Combine the following in a nuclease-free tube (1 tube/RNA sample):

Component	Volume
RNA	*x* µl (≥40 ng RNA)
dNTP mix (10 mM mix)	1 µl
Oligo-dT (0.5 µg/µl)	1 µl
H_2O	*y* µl
Final volume	10 µl

3. Incubate all samples at 65 °C for 5 min, and then immediately cool on ice for at least 1 min.
4. Prepare the following reaction mixture for all RNA samples. To ensure that there is ample material, we routinely prepare enough master mix for all samples plus one spare:

Component	1× (µl)
10× RT buffer	2
$MgCl_2$ (25 mM)	4
DTT (0.1 M)	2
RNaseOUT (RNase inhibitor)	1

5. Add 9 µl of reaction mixture to each RNA/Oligo-dT mixture, mix gently, and centrifuge briefly.
6. Incubate at 42 °C for 2 min.
7. Add 1 µl (200 U) SuperScript III reverse transcriptase to each tube and mix.
8. Incubate at 42 °C for 60 min.
9. Inactivate the reaction mixtures by heating at 70 °C for 15 min. Chill on ice.
10. Collect mixtures by brief centrifugation.
11. To remove RNA, add 1 µl RNase H and incubate for 20 min at 37 °C.
12. Store cDNA at −20 °C.

Table 1
Sequences of endogenous control primers and probes used in qRT-PCR

Gene	Primer name	Sequence
SmTPI[a]	TPI-F	5′-CATACTTGGACATTCTGAGCGTAGA-3′
	TPI-R	5′-ACCTTCAGCAAGTGCATGTTGA-3′
	TPI-Probe	5′-FAM-CAATAAGTTCATCAGATTCAC-3′
α-Tubulin	Tub-F	5′-GGTTGACAACGAGGCCATTTATG-3′
	Tub-R	5′-TGTGTAGGTTGGACGCTCTATATCT-3′
	Tub-Probe	5′-FAM-ATATTTGTCGACGGAAT-3′

[a] *Schistosoma mansoni* triose phosphate isomerase

3.10 Quantitative Real-Time-PCR (qRT-PCR)

We design our 20× TaqMan Gene Expression Assays (forward primer (18 μM), reverse primer (18 μM), and 6-FAM-probe (5 μM)) using Applied Biosystems' Custom TaqMan Assay Tools (www.appliedbiosystems.com). As noted earlier, we generally use the 5′-end of the mRNA to design siRNAs, and we use the remainder of the mRNA for qRT-PCR primer design. Use the available *S. mansoni* genome sequence [7] to target the TaqMan probe to an exon/exon junction of the target gene. This ensures that any residual genomic DNA contaminating the preparation will not be amplified during qRT-PCR. We routinely use the α-tubulin gene as the endogenous control for comparisons within the same parasite life cycle stage and we use the housekeeping triose phosphate isomerase (SmTPI) gene for comparisons between the different life cycle stages of the parasite. Table 1 provides primer and probe sequences for these control genes. Routinely, knockdown at the RNA level is measured 2 days after RNAi treatment. Sustained gene suppression for up to 40 days has been observed in cultured parasites [27]. The following protocol is a step-by-step description of the qRT-PCR reaction:

1. Thaw on ice the TaqMan Gene Expression Assay reagent (20×) and the cDNAs.
2. Place the TaqMan master mix (2×) on ice.
3. Calculate the number of reactions that you need for each assay, and label a PCR plate. Be sure to include the following on each plate:
   - A gene-specific TaqMan Gene Expression Assay for each cDNA sample.
   - An endogenous control assay (α-tubulin) for each cDNA sample.
   - No template controls (NTCs, i.e., blank wells without cDNA) for each gene expression assay.

4. Prepare a master mix to measure the expression of the specific target gene and a master mix to measure the expression of the endogenous control gene. Make enough mix to be able to test all cDNA samples plus one spare, in triplicate. The following is an example of a master mix (10×) used to test three samples in triplicate: three wells for RNAi-treated cDNA, three wells for control treatment, and three wells for NTC:

Component	1 reaction (1×) (μl)	10 reactions (10×) (μl)
TaqMan PCR master mix (2×)	10	100
TaqMan Gene Expression Assay (20×)	1	10
Nuclease-free water	8	80

5. Pipet 19 μl of the mix into the fast optical plate well marked for each reaction.
6. Carefully pipet 1 μl of cDNA into each corresponding well.
7. Seal the plate with optical adhesive film, to avoid evaporation during amplification.
8. Load the plate into the real-time PCR machine, e.g., StepOne Plus PCR machine. Using StepOne software, choose "Advanced Setup" and select the following major parameters:
   (a) Under "Experiment Type," select Quantitation-Comparative Ct (cycle threshold) [delta-delta Ct (ΔΔCt)] [36].
   (b) Under "Select Reagents," select "TaqMan Reagents."
   (c) Under "Ramp Speed," select "Standard."
   (d) In the Plate Setup menu, define the different targets—gene-specific versus endogenous control. In the same menu, define the different samples in the plate (e.g., RNAi treatment, control treatment, and non-template control).
   (e) Select relative quantitation settings by designating a target as the endogenous control (e.g., α-tubulin) and a sample as the reference sample (e.g., an untreated sample or one treated with an irrelevant siRNA). The reference sample will be set as 100 % gene expression for the relative quantification of gene knockdown.
   (f) Run the qRT-PCR reaction using the universal cycling conditions (40 cycles of the following: 95 °C, 15 s and 60 °C, 1 min) of Applied Biosystems.
9. At the end of the run, StepOne software will display the amplification plots, Ct values, and the relative quantitation (RQ). This value is the fold change in gene expression in a sample relative to the reference sample. The RQ is calculated from

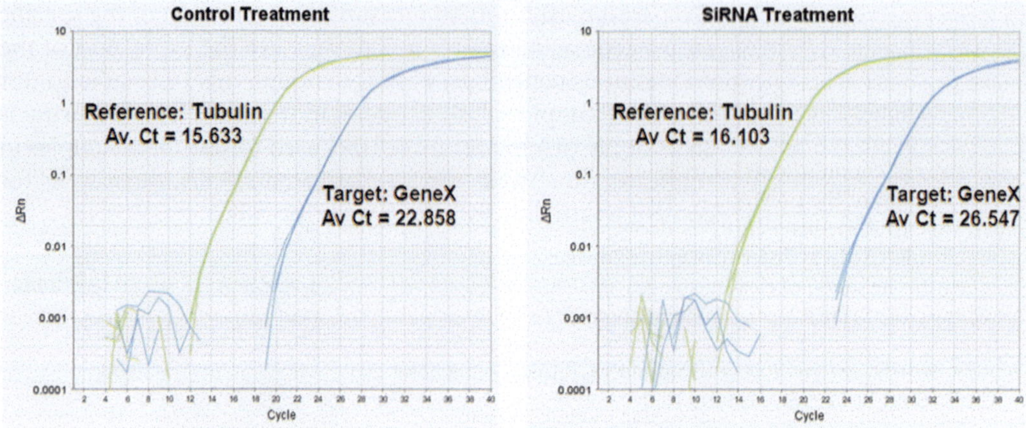

$\Delta Ct = Ct_{GeneX} - Ct_{Tub}$

$\Delta Ct = 7.225$

$\Delta Ct = Ct_{GeneX} - Ct_{Tub}$

$\Delta Ct = 10.444$

$\Delta\Delta Ct = \Delta Ct_{siRNA} - \Delta Ct_{Control.}$

$\Delta\Delta Ct = 3.219$

- Fold difference in GeneX expression in siRNA treatment relative to control = $2^{-\Delta\Delta Ct}$ = 0.107

- This suggests that the expression of GeneX was suppressed 9 fold relative to the control, i.e. The expression of GeneX was suppressed by ~ 90% as a result of siRNA treatment.

Fig. 3 qRT-PCR results and analysis of hypothetical target GeneX expression in control- and siRNA-treated adult schistosome parasites. The *left panel* (control) shows the amplification plots of the reference gene (α-tubulin) and the target gene (GeneX) in control RNA samples; whereas the *right panel* shows the amplification plots of these genes from samples treated with siRNA specific for GeneX. Samples were run in triplicate, and the average Ct (threshold cycle) values are given. The lower part of the figure shows the calculations of the delta (Δ)Ct values, $\Delta\Delta$Ct values, as well as the fold difference

the $\Delta\Delta$Ct values using the following equation: $RQ = 2^{-\Delta\Delta Ct}$. To further analyze the data, export the results into Microsoft Excel. This will allow you to analyze the $\Delta\Delta$Ct data, calculate the relative gene expression, plot the data, and perform statistical analysis [36]. An example of the results produced and the calculations undertaken is shown in Fig. 3.

4 Notes

1. Ethidium bromide is a known mutagen and should be handled as a hazardous chemical—wear gloves and follow proper safety procedures while handling.
2. FBS is heat-inactivated by being incubated in a water bath at 56 °C for 30 min as follows: Thaw serum completely and equilibrate to 37 °C in a water bath. Raise the temperature setting of the water bath to 56 °C. Once the temperature reaches

56 °C, incubate the serum for 30 min. Invert the bottle every 5–10 min. Cool the serum to room temperature, aliquot, and freeze at −20 °C.

3. Prepare triiodo-L-thyronine stock solution at 0.2 mM in 0.02 N NaOH, aliquot, and store at −20 °C.
4. Prepare 10 mg/ml human insulin solution in water. In order to enhance solubility add a few microliters of 0.2 N HCl. (The solubility of insulin is enhanced when the pH of the solution reaches 2–3.) Filter the clear solution through a 0.2 μm syringe filter. Store at 4 °C.
5. Do not use municipal water when working with cercariae since the chlorine concentration in the water can be detrimental to the parasites. We routinely use Poland Spring water. If you use tap water, then it must be conditioned first by passing it through a charcoal filter which reduces chlorine concentration to acceptable levels.
6. An irrelevant long dsRNA should be used as a negative control. This control should not have any significant similarity within the *S. mansoni* genome. To ensure this, blast the sequence against the schistosome genome at http://www.sanger.ac.uk/cgi-bin/blast/submitblast/s_mansoni. In our laboratory, we have used a sequence derived from a yeast expression vector that has no counterpart in the *S. mansoni* genome [26, 27]. Other laboratories have used the firefly luciferase gene sequence as a negative control. Control dsRNA should be prepared in an identical manner to parasite-specific long dsRNA.
7. We recommend the use of a high-fidelity Taq DNA polymerase such as Accuprime Taq polymerase high fidelity (*see* Subheading 2.4). In order to generate sufficient template for dsRNA synthesis, we often perform multiple, identical PCRs at the same time.
8. Annealing temperature varies depending on the sequences of the primers used as well as on the concentration of magnesium in the PCR. Make sure to check the melting temperature of your primers. Normally, programs used to predict/synthesize primers provide the melting temperature for each primer. However, be sure to exclude the T7 sequence in calculating the melting temperature of the primer.
9. As a template for dsRNA synthesis, use 0.5–2.0 picomole (pmol) of the PCR product. The following is a general formula to convert micrograms of double-stranded DNA to picomoles of DNA:

$$X\,\mu\text{g of DNA} \times \left[\frac{\text{pmol}}{660\text{ pg}}\right] \times \left[\frac{10^6\text{ pg}}{1\,\mu\text{g}}\right] \times \left[\frac{1}{N}\right] = \#\text{of pmol of DNA}$$

N is the number of nucleotides in the PCR product.

660 pg/pmol is the average molecular weight of a nucleotide pair.

For example:

- 0.5 μg of a 900 bp PCR product is equivalent to 0.84 pmol.
- 0.5 μg of a 600 bp PCR product is equivalent to 1.26 pmol.

10. Be sure to include negative control groups such as those treated with sequence scrambled siRNAs or irrelevant dsRNAs that are not predicted to impact any schistosome gene. Additionally, including a control group which are electroporated but are not treated with any RNA is optimal. Positive controls, if available, should also be included.

11. The following formula is used to calculate the molecular weight (M.wt.) of dsRNA:

$$\text{M.wt. of dsRNA} = 2\times\left[\left(A_n \times 329.2\right)+\left(G_n \times 345.2\right)+\left(C_n \times 305.2\right)+\left(U_n \times 306.2\right)+159\right]$$

A_n, G_n, C_n, and U_n are the number of each respective ribonucleotide in the single-stranded RNA chain.

The addition of the number 159 to the M.wt. is to adjust for the molecular weight of the 5′ triphosphate.

To calculate the approximate molecular weight of dsRNA, use the following formula:

$$\text{M.wt. of dsRNA} = 2\times\left[\left(N \times 321\right)+159\right]$$

N: number of ribonucleotides in the single-stranded RNA molecule.

321: average molecular weight of the ribonucleotides.

159: adjusting for the molecular weight of the 5′ triphosphates.

12. When pelleting RNA, align all the tubes in the same orientation in the microcentrifuge. RNA forms a gel-like pellet on the side and bottom of the tube, and the RNA pellet can be difficult to see. Therefore, aligning the tubes will help locate the RNA pellets after centrifugation.

13. It is important not to allow the RNA pellet to dry completely. This will greatly decrease its solubility and reduce the yield.

14. If the RNA pellet is difficult to dissolve, incubate the RNA suspension at 55–60 °C for 10–15 min.

Acknowledgments

This work was supported by the National Institutes of Health-National Institute of Allergy and Infectious Diseases (grant number AI-056273). Schistosome-infected snails were provided by the Biomedical Research Institute through the National Institutes of Health (NIAID contract number HHSN272201000009I).

References

1. Steinmann P, Keiser J, Bos R, Tanner M, Utzinger J (2006) Schistosomiasis and water resources development: systematic review, meta-analysis, and estimates of people at risk. Lancet Infect Dis 6:411–425
2. Ross AG, Bartley PB, Sleigh AC, Olds GR, Li Y, Williams GM et al (2002) Schistosomiasis. N Engl J Med 346:1212–1220
3. Abdul-Ghani R, Loutfy N, el-Sahn A, Hassan A (2009) Current chemotherapy arsenal for schistosomiasis *mansoni*: alternatives and challenges. Parasitol Res 104:955–965
4. Doenhoff MJ, Pica-Mattoccia L (2006) Praziquantel for the treatment of schistosomiasis: its use for control in areas with endemic disease and prospects for drug resistance. Expert Rev Anti Infect Ther 4:199–210
5. Melman SD, Steinauer ML, Cunningham C, Kubatko LS, Mwangi IN, Wynn NB et al (2009) Reduced susceptibility to praziquantel among naturally occurring Kenyan isolates of *Schistosoma mansoni*. PLoS Negl Trop Dis 3:e504
6. Doenhoff MJ, Cioli D, Utzinger J (2008) Praziquantel: mechanisms of action, resistance and new derivatives for schistosomiasis. Curr Opin Infect Dis 21:659–667
7. Berriman M, Haas BJ, LoVerde PT, Wilson RA, Dillon GP, Cerqueira GC et al (2009) The genome of the blood fluke *Schistosoma mansoni*. Nature 460:352–358
8. Schistosoma japonicum Genome Sequencing and Functional Analysis and Consortium (2009) The *Schistosoma japonicum* genome reveals features of host-parasite interplay. Nature 460: 345–351
9. Young ND, Jex AR, Li B, Liu S, Yang L, Xiong Z et al (2012) Whole-genome sequence of *Schistosoma haematobium*. Nat Genet 44: 221–225
10. Skelly PJ, Da'dara A, Harn D (2003) Suppression of cathepsin B expression in *Schistosoma mansoni* by RNA interference. Int J Parasitol 33: 363–369
11. Boyle JP, Wu XJ, Shoemaker CB, Yoshino TP (2003) Using RNA interference to manipulate endogenous gene expression in *Schistosoma mansoni* sporocysts. Mol Biochem Parasitol 128:205–215
12. Fire A, Xu S, Montgomery MK, Kostas SA, Driver SE, Mello CC (1998) Potent and specific genetic interference by double-stranded RNA in Caenorhabditis elegans [see comments]. Nature 391:806–811
13. Tabara H, Grishok A, Mello CC (1998) RNAi in C. elegans: soaking in the genome sequence. Science 282:430–431
14. Hunter CP (1999) Genetics: a touch of elegance with RNAi. Curr Biol 9:R440–R442
15. Kennerdell JR, Carthew RW (1998) Use of dsRNA-mediated genetic interference to demonstrate that frizzled and frizzled 2 act in the wingless pathway. Cell 95:1017–1026
16. Wianny F, Zernicka-Goetz M (2000) Specific interference with gene function by double-stranded RNA in early mouse development. Nat Cell Biol 2:70–75
17. Schoppmeier M, Damen WG (2001) Double-stranded RNA interference in the spider Cupiennius salei: the role of Distal-less is evolutionarily conserved in arthropod appendage formation. Dev Genes Evol 211:76–82
18. Morris JC, Wang Z, Drew ME, Englund PT (2002) Glycolysis modulates trypanosome glycoprotein expression as revealed by an RNAi library. EMBO J 21:4429–4438
19. Shi H, Djikeng A, Tschudi C, Ullu E (2004) Argonaute protein in the early divergent eukaryote Trypanosoma brucei: control of small interfering RNA accumulation and retroposon transcript abundance. Mol Cell Biol 24: 420–427
20. Petrocca F, Lieberman J (2011) Promise and challenge of RNA interference-based therapy for cancer. J Clin Oncol 29:747–754
21. Nemunaitis J, Rao DD, Liu SH, Brunicardi FC (2011) Personalized cancer approach: using RNA interference technology. World J Surg 35(8):1700–1714
22. Davidson BL, McCray PB Jr (2011) Current prospects for RNA interference-based therapies. Nat Rev Genet 12:329–340
23. Hong-Geller E, Micheva-Viteva SN (2010) Functional gene discovery using RNA interference-based genomic screens to combat pathogen infection. Curr Drug Discov Technol 7:86–94
24. Kuntz AN, Davioud-Charvet E, Sayed AA, Califf LL, Dessolin J, Arner ES et al (2007) Thioredoxin glutathione reductase from *Schistosoma mansoni*: an essential parasite enzyme and a key drug target. PLoS Med 4:e206
25. Bhardwaj R, Krautz-Peterson G, Da'dara A, Tzipori S, Skelly PJ (2011) Tegumental phosphodiesterase SmNPP-5 is a virulence factor for schistosomes. Infect Immun 79:4276–4284
26. Ndegwa D, Krautz-Peterson G, Skelly PJ (2007) Protocols for gene silencing in schistosomes. Exp Parasitol 117:284–291
27. Krautz-Peterson G, Radwanska M, Ndegwa D, Shoemaker CB, Skelly PJ (2007) Optimizing gene suppression in schistosomes using RNA interference. Mol Biochem Parasitol 153: 194–202

28. Stefanic S, Dvorak J, Horn M, Braschi S, Sojka D, Ruelas DS et al (2010) RNA interference in *Schistosoma mansoni* schistosomula: selectivity, sensitivity and operation for larger-scale screening. PLoS Negl Trop Dis 4:e850
29. Mourao MM, Dinguirard N, Franco GR, Yoshino TP (2009) Phenotypic screen of early-developing larvae of the blood fluke, *schistosoma mansoni*, using RNA interference. PLoS Negl Trop Dis 3:e502
30. Krautz-Peterson G, Simoes M, Faghiri Z, Ndegwa D, Oliveira G, Shoemaker CB et al (2010) Suppressing glucose transporter gene expression in schistosomes impairs parasite feeding and decreases survival in the mammalian host. PLoS Pathog 6:e1000932
31. Faghiri Z, Skelly PJ (2009) The role of tegumental aquaporin from the human parasitic worm, *Schistosoma mansoni*, in osmoregulation and drug uptake. FASEB J 23:2780–2789
32. Correnti JM, Brindley PJ, Pearce EJ (2005) Long-term suppression of cathepsin B levels by RNA interference retards schistosome growth. Mol Biochem Parasitol 143: 209–215
33. Krautz-Peterson G, Bhardwaj R, Faghiri Z, Tararam CA, Skelly PJ (2010) RNA interference in schistosomes: machinery and methodology. Parasitology 137:485–495
34. Rinaldi G, Morales ME, Alrefaei YN, Cancela M, Castillo E, Dalton JP et al (2009) RNA interference targeting leucine aminopeptidase blocks hatching of *Schistosoma mansoni* eggs. Mol Biochem Parasitol 167:118–126
35. Lewis F (2001) Schistosomiasis. Curr Protoc Immunol Chapter 19: Unit 19 11
36. Livak KJ, Schmittgen TD (2001) Analysis of relative gene expression data using real-time quantitative PCR and the 2(-Delta Delta C(T)) Method. Methods 25:402–408

Chapter 9

Construction of *Trypanosoma brucei* Illumina RNA-Seq Libraries Enriched for Transcript Ends

Nikolay G. Kolev, Elisabetta Ullu, and Christian Tschudi

Abstract

High-throughput RNA sequencing (RNA-Seq) has quickly occupied center stage in the repertoire of available tools for transcriptomics. Among many advantages, the single-nucleotide resolution of this powerful approach allows mapping on a genome-wide scale of splice junctions and polyadenylation sites, and thus, the precise definition of mature transcript boundaries. This greatly facilitated the transcriptome annotation of the human pathogen *Trypanosoma brucei*, a protozoan organism in which all mRNA molecules are matured by spliced leader (SL) *trans*-splicing from longer polycistronic precursors. The protocols described here for the generation of three types of libraries for Illumina RNA-Seq, 5′-SL enriched, 5′-triphosphate-end enriched, and 3′-poly(A) enriched, enabled the discovery of an unprecedented heterogeneity of pre-mRNA-processing sites, a large number of novel coding and noncoding transcripts from previously unannotated genes, and quantify the cellular abundance of RNA molecules. The method for producing 5′-triphosphate-end-enriched libraries was instrumental for obtaining evidence that transcription initiation by RNA polymerase II in trypanosomes is bidirectional and biosynthesis of mRNA precursors is primed not only at the beginning of unidirectional gene clusters, but also at specific internal sites.

Key words *Trypanosoma brucei*, RNA-Seq, 5′-SL enriched, 5′-Triphosphate-end enriched, 3′-Poly(A) enriched, Terminator exonuclease

1 Introduction

RNA-Seq was developed and first used for studying the transcriptome of yeast [1]. Surveys in many other organisms followed soon, and the number of applications for this method for analyzing RNA and its metabolism in the cell is limited only by the imagination of the investigator. The power of RNA-Seq [2, 3] can be explained in part by the versatility of this technology. Depending on the type of question to be addressed, different types of cDNA libraries can be prepared. Additionally, specific protocols for enrichment of transcript molecules with unique properties can be devised and tailored for providing single-nucleotide resolution answers on a genome-wide scale.

Christopher Peacock (ed.), *Parasite Genomics Protocols*, Methods in Molecular Biology, vol. 1201,
DOI 10.1007/978-1-4939-1438-8_9,

Libraries for RNA-Seq can be categorized into two major groups based on their capacity to retain information about the direction of the RNA sequence. The ones that retain the orientation of the transcript are usually generated by fragmenting long RNAs (most often by metal-facilitated limited hydrolysis) or purifying small RNAs (e.g., siRNAs, miRNAs, or piRNAs). Two different adapters are sequentially ligated to the RNA molecules with (repaired, if needed) 3′-hydroxyl and 5′-monophosphate ends. Alternatives include (a) the ligation of a 3′-adapter, conversion to cDNA, and subsequent ligation of a 5′-adapter; (b) ligation of a 5′-adapter and reverse transcription with a primer containing the 3′-adapter sequence and a randomized region; and (c) cDNA synthesis with a primer containing both 5′- and 3′-specific adapter sequences separated by an abasic spacer furan, circularization of the reverse transcription product, and cleavage at the abasic site [4].

Libraries that do not retain the directional information about the transcripts are typically produced from RNA that is first converted to double-stranded (ds) cDNA. The cDNA is then fragmented mechanically (e.g., nebulization) or enzymatically by limited DNase I digestion and adapters are added simultaneously to both (repaired) ends of the ds cDNA fragments. While most of the reads obtained from these libraries lack orientation information, sequences spanning *trans*-splice junctions and poly(A)-tail addition sites can provide directionality for the ends of the analyzed mRNA molecules [5]. One of the features of libraries produced by RNA fragmentation is the uniform coverage of the entire body of the transcripts by sequencing reads. Unfortunately, the ends of the RNA molecules (5′ and 3′ alike) are severely underrepresented in these libraries [2]. This information is usually retained in libraries obtained by fragmentation of cDNA [2].

An additional important factor to consider when choosing a strategy for generating libraries for RNA-Seq is the method for depleting the abundant rRNA from the total RNA sample [6]. While selection of polyadenylated RNA is the traditional and still most common procedure for removal of rRNA, this approach may not be appropriate when transcripts other than mRNA, or mRNAs with shorter or absent poly(A) tails (e.g., metazoan histone mRNAs), are also a desired subject of the analysis. The RiboMinus technology [7] is an alternative that relies on hybridization of biotinylated antisense rRNA oligonucleotides to their targets and subsequent removal of the complexes by binding to streptavidin attached to beads. This strategy is convenient for organisms in which the large rRNAs are not fragmented as a normal step in their biogenesis. For trypanosomes, with fragmented 28S rRNA [8], this presents a challenge. A third approach for removal of rRNA is the treatment with terminator exonuclease, an enzyme that requires a 5′-monophosphate for its action, and this functional group is present on large rRNAs in all species. The enzyme also degrades

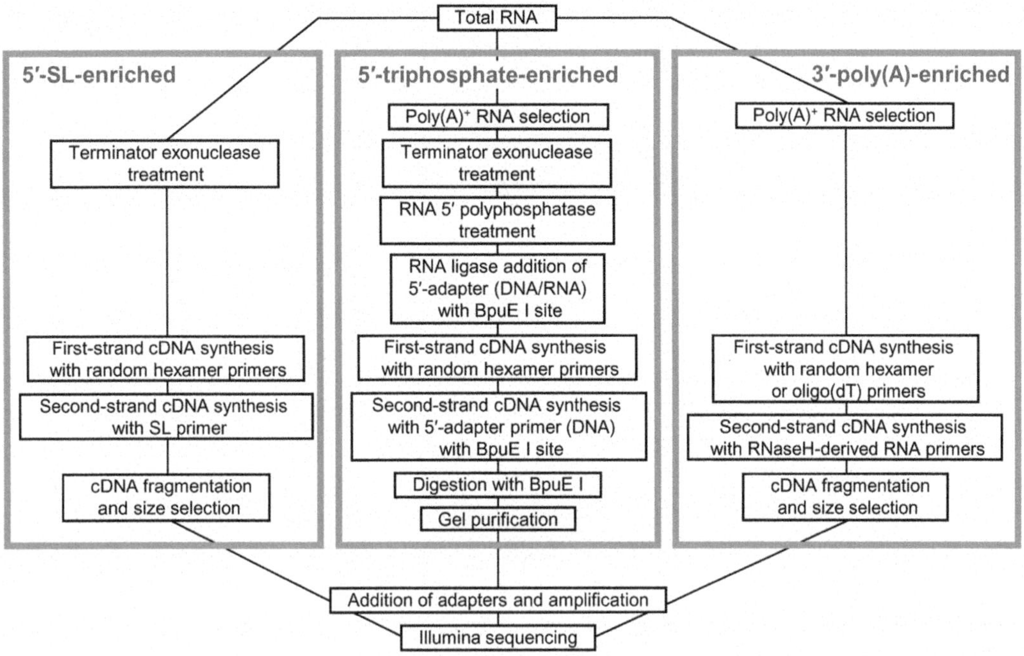

Fig. 1 Outline of the protocol steps for generating the three different types of RNA-Seq libraries

other RNAs bearing monophosphate at their 5′ end, and if these are needed for the transcriptome analysis, the RiboMinus method may be suitable.

Using different strategies for RNA-Seq analysis, we and others [5, 9, 10] detected an extensive heterogeneity of pre-mRNA processing sites in *T. brucei*. We describe here the protocols (Fig. 1) we used to generate the Illumina RNA-Seq libraries enriched for 5′-SL and 3′-poly(A) ends that led to our discovery of that phenomenon. Presented is also the procedure for synthesizing libraries enriched for 5′-triphosphate-end sequence reads that allowed us to map RNA Pol II transcription initiation sites on a genome-wide scale.

2 Materials

2.1 Reagents Required for the Preparation of all Described Libraries

1. Total *T. brucei* RNA prepared with TRIZOL (*see* **Note 1**).
2. RQ1 RNase-free DNase I and accompanying buffer (Promega).
3. Buffer-saturated phenol, pH 7.5.
4. Chloroform.
5. GlycoBlue (Ambion).
6. Random hexadeoxynucleotide primers (Promega).

7. SuperScript II reverse transcriptase and accompanying 5× first-strand buffer and 100 mM DTT (Invitrogen).
8. 10 mM dNTP solution.
9. Protector RNase inhibitor (Roche).
10. NucleoSpin Extract II columns and buffers (Macherey-Nagel).
11. End-It DNA end-repair kit (Epicentre).
12. Klenow fragment (3′–5′ exo^{-}) (New England BioLabs).
13. 1 mM dATP solution.
14. Illumina adapters.
15. LigaFast rapid DNA ligation system (Promega).
16. Illumina PCR primers.
17. Platinum Pfx DNA polymerase and accompanying 10× amplification buffer and enhancer solution (Invitrogen).
18. 100 bp DNA ladder.

2.2 Additional Reagents Required for the Preparation of 3′-Poly(A)-Enriched Libraries

1. Oligotex mRNA mini kit (Qiagen).
2. RNase H.
3. *E. coli* DNA polymerase I.
4. T4 DNA polymerase.

2.3 Additional Reagents Required for the Preparation of 5′-SL-Enriched Libraries

1. Terminator 5′-phosphate-dependent exonuclease and accompanying buffer (Epicentre).
2. 2.5 M sodium acetate, pH 5.0.
3. 1 N NaOH.
4. 1 N HCl.
5. 3 M sodium acetate, pH not adjusted.
6. SL primer, 5′-GCTATTATTAGAACAGTTTCTGTACTATATTG-3′.

2.4 Additional Reagents Required for the Preparation of 5′-Triphosphate-End-Enriched Libraries

1. Oligotex mRNA mini kit (Qiagen).
2. Terminator 5′-phosphate-dependent exonuclease and accompanying buffer (Epicentre).
3. RNA 5′-polyphosphatase (Epicentre).
4. 5′-Adapter with BpuE I site, 5′-GCACCATATAACCGCTTCCrUrUrGrArG-3′.
5. T4 RNA ligase (Ambion).
6. 1 N NaOH.
7. 1 N HCl.
8. 3 M sodium acetate, pH not adjusted.

9. BpuE I primer, 5′-GCACCATATAACCGCTTCCTTGAG-3′.
10. BpuE I (New England BioLabs).
11. pBR322 DNA-Msp I digest marker (New England BioLabs).

3 Methods

3.1 Preparation of 3′-Poly(A)-Enriched Libraries

3.1.1 Selection of Poly(A)⁺ RNA

1. To remove any contaminating genomic DNA, incubate 100 μg total RNA with 1 U RQ1 RNase-free DNase I using the manufacturer-provided buffer in a total volume of 100 μL for 15 min at 37 °C (*see* **Note 2**).
2. Extract the RNA solution with 100 μL of buffer-saturated phenol, pH 7.5, and then with 100 μL chloroform. Be careful not to carry over any chloroform to the next step.
3. Perform two rounds of enrichment for polyadenylated RNA with the Oligotex mRNA mini kit following the manufacturer's instructions, but bind the RNA to the resin on ice. Use microcentrifuge cooled to 15 °C for all required spins. The volume of elution buffer for the first round of selection is 2×100 μL and for the second selection is 25 μL.

3.1.2 Synthesis of First-Strand cDNA

1. Reverse transcribe approximately half of the purified polyadenylated RNA in a 20 μL reaction with SuperScript II reverse transcriptase (RT) according to the manufacturer's instructions. Use 500 ng random hexadeoxynucleotides per reaction (*see* **Note 3**) and include 20 U Protector RNase inhibitor. Heat inactivate RT at 70 °C for 15 min.

3.1.3 Second-Strand cDNA Synthesis

1. Cool the reaction on ice, and briefly spin down any condensation from the walls of the microcentrifuge tube.
2. Sequentially add the following components to assemble a reaction with final volume of 100 μL: water, 10 μL 10× NEBuffer 2 (supplied with enzymes), 3 μL 10 mM dNTPs, 2.5 U RNase H, and 50 U *E. coli* DNA polymerase I.
3. Incubate for 2 h at 16 °C.
4. Add 9 U of T4 DNA polymerase and incubate for 2 min at 25 °C.
5. Purify the cDNA on NucleoSpin Extract II column using 45 μL of elution buffer (EB).

3.1.4 Fragmentation of Double-Stranded cDNA

1. Prepare serial dilutions of RQ1 DNase I (e.g., 1×10^{-1} U, 2×10^{-2} U, 1×10^{-2} U, 2×10^{-3} U, and 1×10^{-3} U per μL) in enzyme storage buffer (10 mM Tris–HCl, pH 8.0, 10 mM $CaCl_2$, 1 mM $MgCl_2$, 50 % glycerol) and store at −20 °C.
2. Use any plasmid DNA as a substrate to determine the optimal conditions for limited DNase I digestion (Fig. 2). Assemble

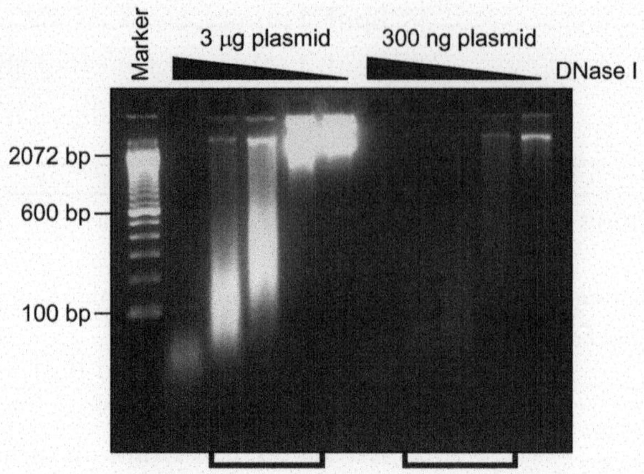

Fig. 2 Optimization of the conditions for DNase I fragmentation of the double-stranded cDNA library. Test for the amount of DNase I used with two sets of reactions containing 3 μg and 300 ng of plasmid DNA as a substrate and a series of DNase I amounts, 1×10^{-1} U, 2×10^{-2} U, 1×10^{-2} U, 2×10^{-3} U, and 1×10^{-3} U. Any plasmid can be used for this titration. The ethidium bromide-stained 1 % agarose gel shows the separated products of the DNase I digestion reaction. The three different amounts of enzyme suitable for library fragmentation are indicated with *brackets* below the gel

two sets of reactions with 3 μg plasmid and 300 ng plasmid by combining in 20 μL total volume 15 μL DNA solution, 2 μL 10× buffer (200 mM Tris–HCl, pH 7.5, 10 mM $CaCl_2$), 2 μL 100 mM $MnCl_2$ (*see* **Note 4**), and 1 μL RQ1 DNase I dilution.

3. After the addition of the enzyme, incubate reactions for exactly 10 min at room temperature.
4. Extract reactions with an equal volume of a 1:1 mixture of phenol:chloroform (*see* **Note 5**).
5. Analyze products by electrophoresis in a 1 % agarose gel (Fig. 2).
6. When a group of three dilutions of DNase I are chosen from the series as optimal, perform three separate reactions (*see* **Note 6**) with cDNA as substrate exactly as described above (**steps 2–4**) and then combine the phenol:chloroform-extracted samples.

3.1.5 Size Selection of cDNA Fragments

1. Separate fragmented cDNA by electrophoresis in a preparative 1 % agarose gel with the 100 bp DNA ladder as a marker. Make sure to leave an empty lane in the gel between the marker and the samples, and between different samples.
2. When the xylene cyanol dye from the loading buffer in the samples has migrated at least 1.8 cm inside the gel, carefully

excise approximately a 2 mm thick slice from the lane of interest with a scalpel blade (*see* **Note 7**). The size of the selected fragments should correspond to the 200 bp band of the marker.

3. Purify the DNA from the gel slice on a NucleoSpin Extract II column. Dissolve the agarose by incubating with the provided buffer at room temperature without heating [11].

   Elute the cDNA fragments in 35 μL EB.

3.1.6 Repair of the Ends of the cDNA Fragments

1. Assemble and incubate a 50 μL reaction with the End-It DNA end-repair kit according to the manufacturer's instructions.
2. Purify the DNA on NucleoSpin Extract II column. Elute in 33 μL EB.

3.1.7 Addition of dA Nucleotide to the 3′ End

1. Assemble a 50 μL reaction with the eluted cDNA, 5 μL 10× NEBuffer 2 (provided with the enzyme), 10 μL 1 mM dATP, and 3 μL (15 U) Klenow fragment (3′–5′ exo$^-$).
2. Incubate for 30 min at 37 °C.
3. Purify on a NucleoSpin Extract II column and elute in 23 μL EB.

3.1.8 Ligation of Illumina Adapters

1. Dilute the chosen Illumina forked adapters to 10 μM with 10 mM Tris–HCl, pH 7.5, and 10 mM NaCl.
2. Assemble a 50 μL reaction with the LigaFast rapid DNA ligation system by combining the eluted DNA with 25 μL 2× Ligation buffer, 1 μL 10 μM adapters, and 2 μL (6 U) T4 DNA ligase.
3. Incubate for 15 min at room temperature.
4. Purify on a NucleoSpin Extract II column and elute with 40 μL EB.

3.1.9 Amplification of the Library by PCR

1. Combine in a thin-wall PCR tube 10 μL (~1/4th) of the ligated DNA, 5 μL 10× Pfx amplification buffer, 2 μL of 25 μM first Illumina PCR Primer, 2 μL of 25 μM second Illumina PCR Primer, 2 μL 50 mM $MgSO_4$, 2 μL 10 mM dNTPs, and 0.8 μL (2 U) Platinum Pfx DNA polymerase [11].
2. Perform PCR with the following steps: (1) 5 min at 94 °C, (2) 15 s at 94 °C, (3) 30 s at 65 °C, (4) 30 s at 68 °C, (5) repeat **steps 2–4** 15 times, (6) 5 min at 68 °C, and (7) hold at 4 °C.
3. Separate 5 μL of the product by analytical electrophoresis in a 1 % agarose gel.
4. Optimize the amount of template or the number of cycles for the PCR, to ensure synthesis of maximal amount of product while remaining in the linear range for amplification (Fig. 3).

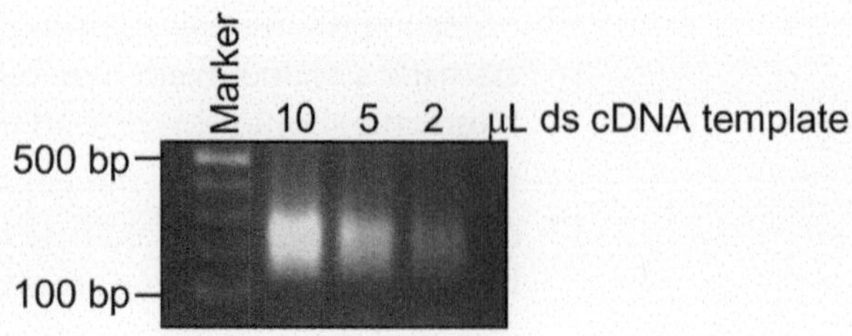

Fig. 3 Confirmation that the amplification of the final library product by PCR is in the linear range. A set of three reactions with the indicated volumes of purified adapter-ligated double-stranded cDNA as a template were separated by electrophoresis in 1 % agarose gel and stained with ethidium bromide. The amount of PCR product corresponds to the amount of used cDNA template

5. Purify the final product by preparative electrophoresis in a 1 % agarose gel and extraction from the gel with a NucleoSpin Extract II column. Elute in 100 μL and perform a second round of purification on a NucleoSpin Extract II column. Elute in 30 μL EB (*see* **Note 8**).

3.2 Preparation of 5′-SL-Enriched Libraries

3.2.1 Degradation of rRNA to Enrich for mRNA

1. Assemble a 20 μL reaction containing 10 μg or less of total RNA, 1 U terminator 5′-phosphate-dependent exonuclease, and 1× accompanying buffer according to the manufacturer's instructions.
2. Incubate for 1 h at 30 °C.
3. Add 1 U RQ1 RNase-free DNase I.
4. Incubate for 5 min at 37 °C.
5. Dilute to 100 μL with water.
6. Extract with 100 μL of buffer-saturated phenol, pH 7.5.
7. Extract with 100 μL of chloroform.
8. Precipitate the RNA with 10 μL 2.5 M sodium acetate, pH 5.0, 2 μL GlycoBlue, and 300 μL ethanol for at least 3 h at −20 °C. Spin for 15–20 min at 13,200 rpm in a refrigerated microcentrifuge, wash the pellet with 700 μL 70 % EtOH, air-dry, and dissolve in 10.5 μL water.

3.2.2 Synthesis of First-Strand cDNA (See Subheading 3.1.2)

Follow Subheading 3.1.2 as described above for 3′-poly(A)-enriched libraries.

3.2.3 Second-Strand cDNA Synthesis

1. To hydrolyze template RNA, add 7 μL 1 N NaOH to the reverse transcription reaction and incubate for 15 min at 65 °C.
2. Cool down and add 6 μL 1 M Tris–HCl, pH 8.0, 6.5 μL 1 N HCl, and 4 μL 3 M sodium acetate with mixing after each addition.
3. Precipitate the cDNA with 1 μL GlycoBlue and 120 μL ethanol at −20 °C, wash the precipitate with 70 % EtOH, and dissolve in 66 μL water.

4. To the DNA solution (in a thin-wall PCR tube) add 10 μL 10× Pfx amplification buffer, 10 μL PCR enhancer solution (provided with the enzyme), 5 μL 20 pmol/μL SL Primer, 4 μL 50 mM $MgSO_4$, 4 μL 10 mM dNTPs, and 1 μL (2.5 U) Platinum Pfx DNA polymerase.
5. Perform the reaction with the following thermocycler parameters: (1) 7 min at 94 °C, (2) 5 min at 40 °C, (3) 20 min at 68 °C, and (4) hold at 4 °C.
6. Purify the cDNA on a NucleoSpin Extract II column and elute in 45 μL EB.

3.2.4 (See Subheadings 3.1.4 through 3.1.9)

Follow Subheadings 3.1.4 through 3.1.9 as described above for 3′-poly(A)-enriched libraries.

3.3 Preparation of 5′-Triphosphate-End-Enriched Libraries

3.3.1 Selection of Poly(A)+ RNA

1. Follow Subheading 3.1.1 and use 500 μg total RNA as starting material. Elute in 35 μL.

3.3.2 Treatment with Terminator 5′-Phosphate-Dependent Exonuclease

1. Follow Subheading 3.2.1 and perform the reaction in a total volume of 40 μL and with 2 U of terminator enzyme. Dissolve the final RNA pellet in 34 μL water.

Treatment with RNA 5′-Polyphosphatase

1. Assemble and incubate a 40 μL reaction with the dissolved RNA and 40 U RNA 5′-polyphosphatase according to the manufacturer's instructions.
2. Extract with 40 μL of buffer-saturated phenol, pH 7.5.
3. Extract with 40 μL of chloroform.
4. Precipitate the RNA with 4 μL 2.5 M sodium acetate, pH 5.0, 2 μL GlycoBlue, and 120 μL ethanol for at least 3 h at −20 °C. Spin for 15–20 min at 13,200 rpm in a refrigerated microcentrifuge, wash the pellet with 700 μL 70 % EtOH, air-dry, and dissolve in 15 μL water.

3.3.3 Ligation of 5′-BpuE I-Adapter

1. Assemble a 20 μL reaction containing 200 pmol 5′-adapter, 1× RNA ligase buffer (supplied with enzyme), and 10 U T4 RNA ligase.
2. Incubate overnight at 4 °C.
3. Extract with 20 μL of buffer-saturated phenol, pH 7.5.
4. Extract with 20 μL of chloroform.
5. Precipitate the RNA with 2 μL 2.5 M sodium acetate, pH 5.0, 2 μL GlycoBlue, and 60 μL ethanol for at least 3 h at −20 °C. Spin for 15–20 min at 13,200 rpm in a refrigerated microcentrifuge, wash the pellet with 700 μL 70 % EtOH, air-dry, and dissolve in 10.5 μL water.

3.3.4 Synthesis of First-Strand cDNA (See Subheading 3.1.2)

Follow Subheading 3.1.2 as described above for 3′-poly(A)-enriched libraries.

3.3.5 Second-Strand cDNA Synthesis

1. Follow Subheading 3.2.3, but use the BpuE I Primer instead of the SL primer. Perform the reaction with the following thermocycler parameters: (1) 7 min at 94 °C, (2) 5 min at 46 °C, (3) 20 min at 68 °C, and (4) hold at 4 °C.
2. Purify the cDNA on a NucleoSpin Extract II column and elute in 25 μL EB.
3. Purify all DNA fragments larger than 100 bp on a 1.2 % agarose gel which is run for a short time, but enough to separate the 100 bp marker band from the rest of the marker fragments. Avoid prolonged exposure to UV light.

3.3.6 Digestion with BpuE I

1. Assemble a 100 μL reaction with the purified DNA and 25 U BpuE I in the presence of *S*-adenosylmethionine (provided with the enzyme) according to the manufacturer's instructions.
2. Purify the DNA on NucleoSpin Extract II column loading the column twice with the same sample to ensure good binding of the 40 bp fragments. Elute with 25 μL EB.

3.3.7 Size Selection of Digested DNA Fragments

1. Separate DNA on a 2 % agarose gel alongside the 100 bp ladder and pBR322 DNA-Msp I digest markers.
2. Excise several thin gel slices covering the region corresponding to ~40 bp.
3. Purify on a NucleoSpin Extract II column and elute with 35 μL EB.

3.3.8 (See Subheadings 3.1.6 through 3.1.9)

Follow Subheadings 3.1.6 through 3.1.9 as described above for 3′-poly(A)-enriched libraries. Choose the reaction with the amplified product of the expected size (130 bp) resulting from one of the samples from the different gel slices.

4 Notes

1. The integrity of the molecules in the total RNA sample must be verified prior to initiating library preparation. We recommend visualizing the large rRNAs after electrophoresis under denaturing conditions.
2. We strongly recommend the use of nonstick (or low-binding) microcentrifuge tubes for all steps of the protocols.
3. Alternatively, 5′-T_{15}VN-3′ oligodeoxynucleotide (V = A, G, or C; N = T, A, G, or C) can be used instead of the random hexadeoxynucleotide primers, but only for the 3′-end-enriched library.
4. DNase I generates double-stranded cuts in DNA in the presence of Mn^{2+} (in contrast to producing only single-stranded

cuts in the duplex when only Mg^{2+} is present). This presumably facilitates repairing the ends of the DNA fragments in the protocol steps that follow fragmentation.

5. Extraction of small volumes is facilitated by the use of 0.6 mL microcentrifuge tubes; alternatively, the sample can be diluted prior to extraction; however this will require subsequent precipitation with ethanol.
6. Three different amounts of enzyme in separate reactions will ensure that the substrate is not over- or underdigested.
7. Two additional gel slices can be cut (above and below the original excision) and kept frozen as backup fragmented cDNA material.
8. Quick spin (1–2 min) at high speed of the final sample and collection of the top 2/3 volume will ensure that there are no particulates that sometimes are produced in small amounts from the columns for DNA purification.

Acknowledgments

Work in our laboratory was funded in part by National Institute of Allergy and Infectious Diseases Grants AI28798 to EU and AI43594 and AI078333 to CT.

References

1. Nagalakshmi U, Wang Z, Waern K et al (2008) The transcriptional landscape of the yeast genome defined by RNA sequencing. Science 320:1344–1349
2. Wang Z, Gerstein M, Snyder M (2009) RNA-Seq: a revolutionary tool for transcriptomics. Nat Rev Genet 10:57–63
3. Waern K, Nagalakshmi U, Snyder M (2011) RNA sequencing. Methods Mol Biol 759: 125–132
4. Ingolia NT, Ghaemmaghami S, Newman JR et al (2009) Genome-wide analysis in vivo of translation with nucleotide resolution using ribosome profiling. Science 324: 218–223
5. Kolev NG, Franklin JB, Carmi S et al (2010) The transcriptome of the human pathogen *Trypanosoma brucei* at single-nucleotide resolution. PLoS Pathog 6(9):e1001090. doi:10.1371/journal.ppat.1001090
6. Raz T, Kapranov P, Lipson D et al (2011) Protocol dependence of sequencing-based gene expression measurements. PLoS One 6:e19287. doi:10.1371/journal.pone.0019287
7. Cui P, Lin Q, Ding F et al (2010) A comparison between ribo-minus RNA-sequencing and polyA-selected RNA-sequencing. Genomics 96:259–265
8. White TC, Rudenko G, Borst P (1986) Three small RNAs within the 10 kb trypanosome rRNA transcription unit are analogous to domain VII of other eukaryotic 28S rRNAs. Nucleic Acids Res 14:9471–9489
9. Siegel TN, Hekstra DR, Wang X et al (2010) Genome-wide analysis of mRNA abundance in two life-cycle stages of *Trypanosoma brucei* and identification of splicing and polyadenylation sites. Nucleic Acids Res 38:4946–4957
10. Nilsson D, Gunasekera K, Mani J et al (2010) Spliced leader trapping reveals widespread alternative splicing patterns in the highly dynamic transcriptome of *Trypanosoma brucei*. PLoS Pathog 6(8):e1001037. doi:10.1371/journal.ppat.1001037
11. Quail MA, Kozarewa I, Smith F et al (2008) A large genome center's improvements to the Illumina sequencing system. Nat Methods 5: 1005–1010

Chapter 10

Techniques to Study Epigenetic Control and the Epigenome in Parasites

Sheila C. Nardelli, Li-Min Ting, and Kami Kim

Abstract

Epigenetics is the study of heritable changes in gene expression that occur independent of the DNA sequence. Due to their intimacy with DNA, histones have a central role in chromatin structure and epigenetic regulation. Their tails are subject to posttranslational modifications (PTMs) that together with chromatin-remodeling proteins control the access of different proteins to DNA and allow a precise response to different environmental conditions. The first part of this chapter is dedicated to histone enrichment methods that allow the study of histones using techniques such as immunoblot or mass spectrometry for the mapping of the histone PTM network. Next we describe chromatin immunoprecipitation-based techniques (ChIP) for study of the epigenome. ChIP followed by microarray or next-generation sequencing enables the precise genomic localization of protein-DNA interactions. These techniques for genome-wide profiling of chromatin provide powerful and efficient tools to study the epigenome.

Key words Histone, Chromatin immunoprecipitation, Microarray, Sequencing

1 Introduction

Chromatin is a very organized and dynamic structure, formed by the association of DNA with histone proteins. Two copies of each histone (H2A, H2B, H3, H4), or their variants, assemble in an octamer to form the nucleosome, the basic unit of chromatin. In addition to the four canonical histones, eukaryotes also have an additional histone, histone H1, which facilitates a higher degree of compaction of chromatin. Each histone has a globular domain and flexible N- and C-terminal domains that project outside the nucleosome and are targets for many posttranslational modifications (PTMs). These modifications act interdependently, generating several combinations that can affect gene expression [1–3].

Histones are small, basic proteins that are highly conserved in eukaryotes. *Toxoplasma gondii*, the etiologic agent of toxoplasmosis, has the four canonical histones, all very similar to human histones. Some divergences are observed, especially at the N-terminal

Christopher Peacock (ed.), *Parasite Genomics Protocols*, Methods in Molecular Biology, vol. 1201, DOI 10.1007/978-1-4939-1438-8_10, © Springer Science+Business Media New York 2015

tails of H2A and H2B [4–6]. Although a protein similar to *Kinetoplastidae* H1 can be found in the genome database (TGME49_315570), it is unclear whether this protein actually plays the role of an H1 in *Toxoplasma*. In addition, *T. gondii* has an extensive repertoire of chromatin-remodeling proteins that regulate histone modifications, as well as putative transcription factors and other enzymes that regulate gene transcription and other DNA-related processes [5–8].

In the past decades, histones have been studied in detail, due to their central role in epigenetic regulation. There are two methods used for isolation of histones. Acidic extraction was first described at the end of nineteenth century, by Albrecht Kossel who named the nuclear proteins soluble under acid conditions "histone" [9]. Since then, most histone studies have been based on acid extraction protocols (either HCl or H_2SO_4). This technique allows histone enrichment with minimal contamination by DNA, RNA, and non-histone proteins [10–12]. Alternatively, histone purification with high salt concentrations is recommended [13]. Both techniques are effective, but the advantage in the salt purification is that it enables the differential separation of histones. Using increasing salt concentrations, the histone H1 should be the first to be isolated, followed by H2A–H2B, and finally H3–H4 at the highest concentrations. Exact concentrations can vary, depending on the organism [13]. Here, we describe an acid extraction protocol that was first described for *Trypanosoma* species [14], but, due to similar characteristics between histone proteins, also is useful for efficient isolation of *Toxoplasma* histones. Many laboratories use this purification technique followed by one- or two-dimensional SDS-PAGE gels, TAU-PAGE gels (Triton-acetic acid-urea), HPLC, or mass spectrometry analysis to identify common or new PTMs and characterize histone variants.

Recently, many research groups have characterized parasite epigenetic machinery in order to understand gene expression and the importance of epigenetics in parasites. Analysis of histones from different parasites reveals similarities as well as divergences in primary sequence, PTMs, or histone-modifying enzymes. One major technique used for studying epigenetics mechanisms is chromatin immunoprecipitation (ChIP) [15, 16]. ChIP technique involves three basic steps: first, the extracted chromatin is fragmented into small pieces. Second, chromatin fragments are enriched for specific regions bound by the factor of interest using immunoprecipitation with specific antibody. Third, the precipitated DNA is subjected to microarray hybridization (ChIP-chip) [17, 18] or high-throughput sequencing (ChIP-seq) [19–22] or PCR [23, 24]. See Figure 1 for a schematic of the methodology.

There are two types ChIP: cross-linking ChIP (X-ChIP) or native ChIP (N-ChIP). For X-ChIP protocols, the DNA and adjacent protein complexes are mildly cross-linked, typically with small amounts of formaldehyde [15, 25]. The cross-linked DNA is sheared by sonication into small pieces (about 200–1,000 bp). In contrast, in N-ChIP DNA, the proteins are not chemically fixed, and instead of sonication, the fragmentation is obtained by Micrococcal nuclease (MNase) digestion [26, 27]. MNase digests between nucleosomes providing fragments of about 147 bp. When DNA is sequenced after immunoprecipitation, it can provide a precise nucleosome map. Although the MNase digestion is usually rapid, many investigators believe that during this step the nucleosome position and PTMs can change, since the cells are under environmental stress without the fixation step. In addition, N-ChIP normally is not efficient for non-histone proteins that are usually less tightly attached to DNA [19].

In ChIP-chip the DNA obtained from ChIP is amplified, labeled with a fluorophore, and hybridized to an array [17, 28]. Then, the fluorescence signal is determined using a scanner. These last two steps are the most controversial of this technology, since they can generate nonspecific binding and noise that can be difficult to distinguish from specific low-level hybridization. DNA microarray technology has, since its first description in 2000, undergone incredible advances and remains a valuable tool since the design of the array may be customized, depending on the requirements of the experiment. The composition of the array can encompass the whole genome and be composed of probes made from PCR products, oligonucleotides, and EST (or cDNA) arrays. Customized arrays can feature specific regions of the genome such as promoters or be enriched with genes from pathways of interest [29]. Commercial platforms offering high-density arrays successfully used for parasites include Affymetrix, Agilent, and Nimblegen.

Next-generation sequencing, developed in subsequent years, provides single base-pair resolution and higher quality data [30, 31]. The first described was pyrosequencing 454 by LifeSciences (later acquired by Roche) [32], the second was sequencing by synthesis ("Solexa") by Illumina [33], and finally sequencing by oligo ligation (SOLiD) by Applied Biosystems. The main difference between the methods is the size of the read sequences and number of reads. Pyrosequencing 454 technologies can sequence 200–400 bp with about one million reads, while Solexa and SOLiD show read lengths of about 35–100 bp with up to 200–300 million reads [34]. The capabilities have increased with each new generation of machine from each platform. More recently, two new systems were described: Helicos Heliscope (www.helicosbio.com) and Pacific Biosciences SMRT (www.pacificbiosciences.com) that enable single molecule sequencing and longer reads, respectively. As the cost of the technology has decreased, smaller systems are being developed that can be used in individual laboratories rather than within a core facility.

There are a few points to consider before choosing the technology to be used. Next-generation sequencing provides high resolution, coverage, and specificity; however, it is still too expensive to be used routinely, and assays take longer to perform and analyze than arrays. Prices have been decreasing rapidly, and multiplexing of samples within a single lane has also contributed to decreased costs. On the other hand, microarrays provide faster results and because the technology is based upon specific hybridization (and selection of nonrepetitive and nonredundant sequences for the array), contamination of parasite DNA or RNA by host sequences is not a major concern when performing experiments with *T. gondii* or other intracellular parasites. For a comparison of results obtained by X-ChIP-chip or N-ChIP-seq see Figure 2.

Similar advantages and challenges are observed in the analysis of transcriptomes. RNA profiling using microarrays involves RNA isolation (total or subspecies RNAs), conversion to cDNA, labeling with fluorescent dyes, and finally hybridization to an array. Microarray approaches depend on knowledge of the genome size, availability of genome sequence, and annotation. They can be limited by probe cross-reactivity, high background, and signal saturation. But depending on the array design, chips may provide rapid and high-quality data at lower cost. The methodology and depth of sequencing of RNA (RNA-seq) are being constantly improved and the advantage of RNA-seq is direct sequence information independent of prior genome sequence with single base resolution. Because millions of sequences are generated quantitation of relative abundance of transcripts is possible. RNA-seq is particularly useful for detection of polymorphisms, minor RNA species, and RNA editing [35].

ChIP-chip and ChIP-Seq have been employed in a wide variety of parasites to study histone posttranslational modifications, histone variants, and chromatin remodelers [36–41]. In parasites, these techniques are quite similar to those used for mammalian cells, but have required specific adaptions for parasite species such as *Trypanosome* and *Plasmodium*. Library construction in *Plasmodium* can be problematic. Sequencing can generate a series of artifacts in genomes with high or low GC content, especially during the PCR amplification. Choice of a high-fidelity polymerase that is capable of error-free extension during the library amplification step may also improve results. *Plasmodium* species have AT-rich genomes, and some groups have developed alternative library preparation protocols that exclude the PCR step or couple the T7 promoter to the adapters [42, 43].

Nowadays, high-throughput techniques are main strategies for the study of histones and DNA-binding or chromatin-associated factors. Here we describe the protocol standardized in our laboratory for *T. gondii*.

2 Materials (*See* Figs. 1 and 2)

2.1 Equipment

- Cell scraper.
- Centrifuge (benchtop) and microcentrifuge.
- 150 cm^3 Dishes or T175 flasks (tissue culture).
- Gel electrophoresis apparatus.
- Incubators and heat blocks.
- Tubes (15 and 1.5 ml).
- Micropipettor (and tips).
- PCR apparatus.
- Rotator.
- Sonicator.
- Spectrophotometer.

3 Methods

3.1 Parasite Preparation

In order to study gene expression, we recommend using intracellular parasites, because the gene expression profile will change after host cell lysis. *Toxoplasma gondii* is maintained in confluent cultures of human foreskin fibroblasts (HFF) in Dulbecco's Modified Eagle Medium (DMEM-Cat. no. 11965-092, Gibco) supplemented with 10 % fetal bovine serum, 2 mM glutamine (Cat. no. 25030-081, Gibco), 100 μg/ml penicillin, and 100 μg/ml streptomycin (PenStrep Cat. no. 15140-122, Gibco).

1. Infect confluent 150 cm^3 plates with $3–5 \times 10^7$ parasites. Allow parasites to grow for 40 h (*see* **Note 1**).
2. Wash the plates with cold PBS twice.
3. Harvest cells containing *Toxoplasma* with cell scrapers and centrifuge at 800 rcf for 10 min at 4 °C.
4. Resuspend the cell pellet in PBS (approximately 3 ml per 150 cm^3 plate) and disrupt HFF cells by passing sequentially through 20-23-25 gauge needles on a syringe, at least once each (until cells are lysed and the suspension passes easily through).
5. After lysis, purify the parasites from host cells using 3 μm Nuclepore membrane (Nuclepore Track-Etch-Membrane-Whatman). Due to the large amount of host debris, do not overload the membrane or many parasites will be lost.
6. Count parasites and centrifuge at 800 rcf for 10 min at 4 °C. Polystyrene tubes are preferable, so parasites do not stick to sides of the tubes.
7. The pellet can be stored at −80 °C or in liquid nitrogen until needed.

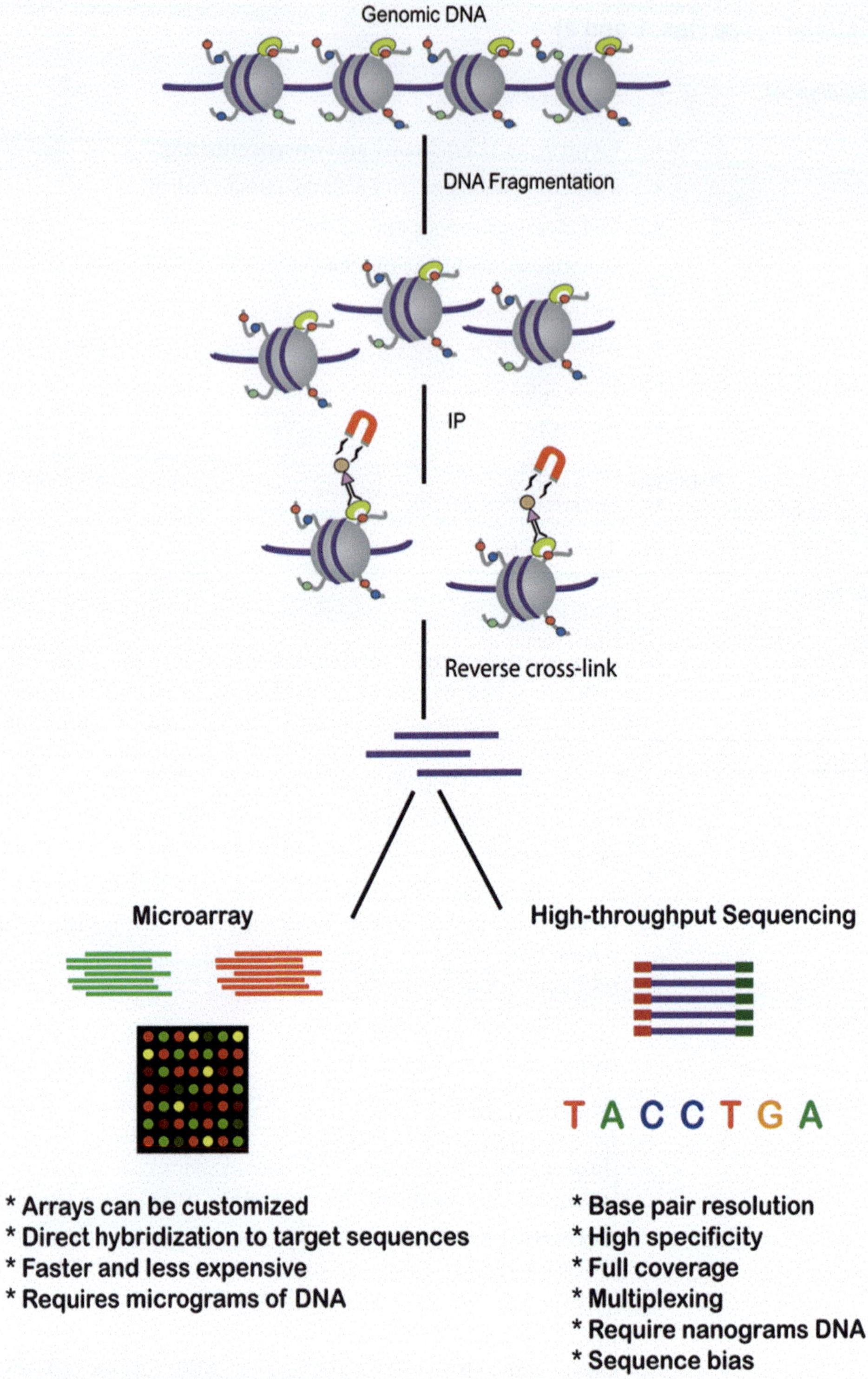

Fig. 1 Schematic view of a ChIP experiment: X-ChIP experiments start with formaldehyde cross-linking in order to preserve protein-DNA interactions, followed by DNA fragmentation by sonication (or MNase treatment). N-ChIP omits the cross-linking step and DNA is fragmented with MNase treatment. The complex is immunoprecipitated (IP)

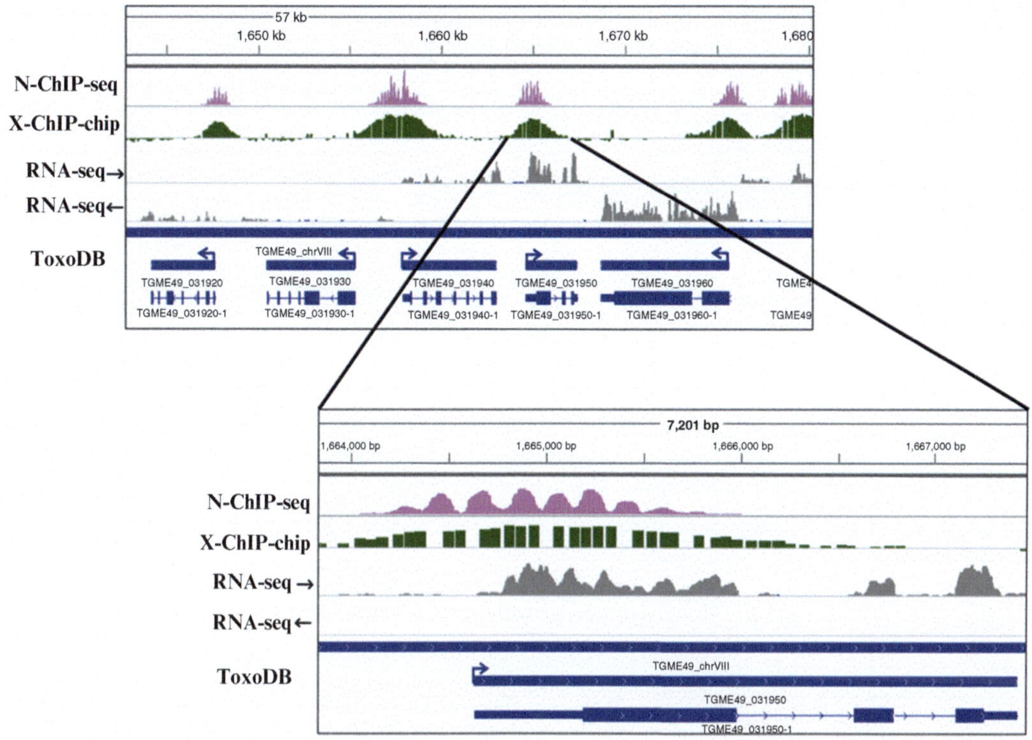

Fig. 2 Comparison between X-ChIP-chip and N-ChIP-seq in *T. gondii*: For this comparison, we used a commercial antibody specific for H3K4me3 (Millipore), a well-described posttranslational modification that is located in promoter regions of active genes of *T. gondii*. In *pink*, the results for N-ChIP followed by next-generation sequencing on the Illumina platform. Each peak corresponds to approximately 200 bp or one nucleosome. *Green* represents the results for X-ChIP followed by tiled genomic array hybridization (Nimblegen platform). The data are compared to RNA-seq results (*gray*, top is positive strand and lower is negative strand) and to the genome annotation (www.ToxoDB.org; in *blue* with *top* showing gene and *bottom* showing predicted introns and exons; note the agreement with RNA-seq data). The direction of the predicted gene is indicated with *blue arrows*. Data were visualized using the Integrated Genome Viewer (IGV, available at http://www.broadinstitute.org/igv/)

3.2 Histone Prep Solutions

All solutions described in this protocol must be prepared with ultrapure water and analytical grade reagents. The solutions can be prepared in advance, filtered with 0.2 mM filter and stored at −20 °C. Protease inhibitors or other inhibitors of interest should be added immediately before the solution is needed.

1. Solution A: 0.25 M Sucrose; 1 mM EDTA; 3 mM $CaCl_2$; 0.01 M Tris–HCl pH 7.4; 0.5 % saponin.

Fig. 1 (continued) by adding antibodies of interest, which will be subsequently affinity-purified with protein A (or G) coupled to magnetic beads. After repeated washings, the antigen-antibody-DNA complex is eluted. The cross-link is reversed in X-ChIP and the proteins are removed by proteinase K degradation. Finally, the resulting DNA is purified and subjected to microarray hybridization or sequencing, following standard procedures for each one (e.g., amplification, library preparation). The advantages of arrays vs. sequencing are listed

2. Solution B: 0.25 M Sucrose; 1 mM EDTA; 3 mM $CaCl_2$; 0.01 M Tris–HCl pH 7.4.
3. Solution C: 1 % Triton X-100; 0.15 M NaCl; 0.025 M EDTA; 0.01 M Tris–HCl pH 8.

3.2.1 Histone Preparation Method

1. Resuspend frozen or freshly collected parasites (*see* **Note 2**) in 1 ml Buffer A. Mix by vortexing and centrifuge at 4,000 rcf for 10 min (4 °C) (*see* **Note 3**).
2. Resuspend pellet in 1 ml Buffer B and centrifuge at 4,000 rcf for 10 min (4 °C).
3. Resuspend pellet in 1 ml Buffer C and centrifuge at 12,000 rcf for 20 min (4 °C).
4. Remove supernatant and wash the pellet three times in 100 mM Tris pH 8 (*see* **Note 4**).
5. Carefully remove supernatant and resuspend the resulting pellet in 1 ml of 0.4 N HCl (*see* **Note 5**).
6. Incubate the solution under rotation for 2 h at 4 °C.
7. Recover the acid-soluble proteins in the supernatant by centrifuging at 10,000 rcf for 15 min at 4 °C.
8. Add 8 volumes of acetone to the supernatant and store overnight at −20 °C.
9. On the next day, centrifuge for 20 min at 3,500 rcf at 4 °C (*see* **Note 6**).
10. Remove the supernatant carefully and wash three times with acetone.
11. The resulting pellet can be resuspended in water (for further analysis by MS) or specific sample buffers (SDS-PAGE or TAU-PAGE) (*see* **Note 7**).

3.3 N-ChIP Solutions

Solutions must be freshly prepared using ultrapure water and analytical grade reagents and inhibitors should be added immediately before use.

1. Cell suspension buffer: 300 mM sucrose; 15 mM Tris–HCl pH 7.5; 5 mM $MgCl_2$; 15 mM NaCl; 60 mM KCl; 0.1 mM EDTA; 0.1 mM PMSF; protease inhibitors.
2. MNase digestion buffer: 0.32 M sucrose; 50 mM Tris–HCl pH 7.5; 4 mM $MgCl_2$; 1 mM $CaCl_2$; 0.1 mM PMSF and protease inhibitors.
3. Lysis buffer: 1.0 mM Tris–HCl pH 7.5; 0.2 mM EDTA; 0.3 mM PMSF.
4. ChIP dilution buffer: 0.01 % SDS; 1.1 % Triton X-100; 1.2 mM EDTA; 16.7 mM Tris–HCl; 167 mM NaCl and protease inhibitors; pH 8.1.

5. Low-salt wash buffer: 0.1 % SDS; 1 % Triton X-100; 2 mM EDTA; 20 mM Tris–HCl; 150 mM NaCl; pH 8.1.
6. High-salt wash buffer: 0.1 % SDS; 1 % Triton X-100; 2 mM EDTA; 20 mM Tris–HCl; 500 mM NaCl; pH 8.1.
7. LiCl wash buffer: 0.25 M LiCl; 1 % NP40; 1 % deoxycholate; 1 mM EDTA; 10 mM Tris–HCl; pH 8.1.
8. TE buffer: 10 mM Tris–HCl; 1 mM EDTA; pH 8.
9. Elution buffer: 1 % SDS; 0.1 M $NaHCO_3$.

3.3.1 N-ChIP Method

1. Resuspend frozen or freshly collected *Toxoplasma* pellet in 500 μl of ice-cold cell suspension buffer (*see* **Note 8**).
2. Add 500 μl of cell lysis buffer containing 2× NP40 (0.4 %; 1× is 0.2 %). Homogenize by pipetting and incubate on ice for 5 min (parasite lysis).
3. Centrifuge at 4,000 rcf for 10 min at 4 °C and save supernatant (cytosol fraction) in new tube. Wash the pellet again with 1 ml of ice-cold cell suspension buffer (without NP40).
4. Centrifuge at 4,000 rcf for 10 min at 4 °C.
5. Resuspend pellet (nucleus-enriched) with 1 ml of ice-cold MNase digestion buffer and spin at 4,000 rcf for 5 min at 4 °C.
6. Resuspend nuclei pellet in final volume of 100 μl and performed the MNase digestion at 37 °C for 5 min. The reaction should be stopped by addition of 5 mM EDTA (*see* **Note 9**). The sample must remain on ice for 5 min in order to inactivate the reaction completely.
7. Centrifuge at 8,000 rcf for 5 min at 4 °C and save the supernatant to a new tube (S1 fraction).
8. Resuspend pellet in 1 ml lysis buffer and dialyze overnight with lysis buffer (we use Slide-A-Lyzer Dialysis Cassette 3.5 K MWCO, 3 ml, Cat. no. 66330, Pierce).
9. Centrifuge at 500 rcf for 10 min at 4 °C and save the supernatant in a new tube (S2 fraction). Finally, the pellet should be resuspended in 200 μl lysis buffer (P fraction) (*see* **Note 10**).
10. Combine fractions S1 and S2 (*see* **Note 11**) and dilute the sample ten times in ChIP dilution buffer. Remove 200 μl as input DNA.
11. Add 80 μl of protein A coupled to magnetic beads to preclear the solution. Incubate for 30 min at 4 °C under gentle agitation (we use Dynabeads Protein A Cat. no. 100.02D, Invitrogen).
12. Separate beads using a magnetic rack and place the supernatant in a new sterile tube.
13. Add the antibody of interest. The immunoprecipitation should occur overnight at 4 °C with gentle agitation (*see* **Note 12**).

14. When the immunoprecipitation is complete, add 60 μl of protein A coupled to magnetic beads for 2 h at 4 °C under gentle agitation to collect the antibody-protein-DNA complex.
15. Pellet protein A-magnetic beads by using magnetic rack and save supernatant in a new tube (we use MagnaRack for microcentrifuge tubes Cat. no. CS15000, Invitrogen).
16. Wash pellet three times with 1 ml of low-salt buffer. Tube should be inverted 20–30 times between each wash and then pelleted using the magnetic rack.
17. Wash three times with high-salt buffer and LiCl wash buffer and finally six times with TE buffer, inverting the tube between each wash.
18. Elute the protein-DNA complex twice with 250 μl each of elution buffer for 15 min at room temperature. Combine eluates.
19. Dilute input DNA (200 μl) in 10 mM Tris–HCl pH 8 to a final volume of 500 μl.
20. Add 10 μl of 0.5 M EDTA and 20 μl of 1 M Tris–HCl pH 6.5 and 1 μl of 20 mg/ml proteinase K to the combined eluates and input DNA and incubate for 1 h at 45 °C.
21. Recover DNA (we use MinElute PCR Purification Kit, Cat. no. 28006, Qiagen), according to the manufacturer's instructions. Elute from the column with 20 μl of EB buffer (provided by the kit).

3.4 X-ChIP

3.4.1 Parasite Preparation

Due to cross-linking, parasite preparation is different than described above. HFF (or other host cells) are infected with *Toxoplasma* RH strains in the manner reported in Subheading 2.1. However, the cross-linking step is performed with parasites still inside vacuoles. After 40 h of infection and two washes using cold PBS, follow the steps below:

1. Add 37 % formaldehyde to a final concentration of 1 %. The formaldehyde is added directly to plates (150 cm^3) containing 10 ml of cold PBS. Incubate at room temperature for 10 min under gentle agitation (*see* **Note 13**).
2. Stop the cross-linking by adding 125 mM of glycine directly to the plates. Incubate at room temperature for 5 min with gentle rotation. Remove the supernatant and wash the cells twice with ice-cold PBS (*see* **Note 14**).
3. Cells are scraped into ice-cold PBS and collected by centrifugation at 800 rcf for 10 min.
4. Resuspend cells in ice-cold PBS, disrupt them passing them sequentially through 20-23-25 gauge needles attached to a syringe, and centrifuge the cells (*see* **Note 15**). Parasites can be counted in a hemocytometer, transferred to a 1.5 ml

microcentrifuge tube (we use 5×10^8 parasites/per experiment), and centrifuged at 800 rcf per 10 min at 4 °C and place samples on ice (*see* **Note 16**).

5. Parasites can be stored after snap freezing samples in liquid nitrogen.

3.4.2 X-ChIP Solutions

1. Lysis buffer: 50 mM HEPES; 150 mM NaCl; 1 % NP40; 0.1 % SDS; 0.1 % sodium deoxycholate; 1 mM EDTA, pH 8; and protease inhibitors.
2. Wash buffer: 50 mM HEPES, pH 7.5; 150 mM NaCl; 1 mM EDTA pH 8; and protease inhibitors.
3. Elution buffer: 1 % SDS; 50 mM Tris–HCl pH 8; and 10 mM EDTA.

3.4.3 X-ChIP Methodology

1. Resuspend parasite pellet (around 5×10^8 parasites) in 1 ml of lysis buffer and incubate on ice for 10 min.
2. Using a microtip, sonicate chromatin to an average length of 0.8–1 kb. We use Branson Sonifier Disruptor (Model W140), power of 5, and a 50 % duty cycle; we sonicate cells 20 times, 10-s pulses, keeping samples on ice for at least 1 min between pulses (*see* **Note 17**).
3. Centrifuge samples at maximum speed in a microcentrifuge for 10 min at 4 °C. Transfer supernatant to another tube and centrifuge again for 15 min at maximum speed. Adjust the volume to 1.1 ml and save 0.1 ml as sample input.
4. Add the antibody (as described in item 3.3.1, section 13) and incubate overnight at 4 °C under rotation.
5. Add protein A coupled to magnetic beads previously washed with lysis buffer and incubate for 2 h at 4 °C under rotation (*see* **Note 18**).
6. Separate the beads using a magnetic field and wash column four times with wash buffer.
7. Add 100 μl elution buffer and collect the eluate. Repeat the elution once.
8. Incubate the combined eluates (immunoprecipitates) and input overnight at 65 °C in order to reverse cross-link (minimum of 6 h is necessary).
9. Purify DNA using MinElute PCR Purification Kit (Cat. no. 28006, Qiagen), according to the manufacturer's instructions. Elute from the column with 20 μl of EB buffer (provided by the kit).
10. For microarray analysis (ChIP-chip), DNA should be amplified. We use Genome Plex Complete-Whole Genome Amplification Kit (Cat. no. WGA2, Sigma-Aldrich). Typically the yield of DNA prior to amplification will be too low to

measure concentrations accurately even with a NanoDrop spectrophotometer. Start with 10 μl ChIP DNA and input as template. If necessary amplify the rest in a second reaction. The amount of DNA varies but for microarray experiments normally 1–2 μg DNA is necessary. Analyze the samples running an agarose gel.

4 Notes

1. The protocols presented here are standardized for RH strain (type I). For type II or III, the number of parasites/cell infection should be determined, but yields of parasite material will generally be lower.
2. For histone purification, a specific amount of parasites are not required; however, using small numbers of parasites increases the loss of material, so for best results we recommend using $3–5 \times 10^8$ parasites.
3. Our laboratory uses the purified histones for both basic detection protocols such as Western blot or Coomassie staining, but also for more sensitive techniques such as mass spectrometry to determine the histone PTMs in this parasite. For PTM analysis we use 0.1 mM PMSF, 5 mM sodium butyrate, 1 mM sodium fluoride, protease inhibitor cocktail (Complete Mini, EDTA-free Cat. no. 11836170001, Roche), and specific phosphatase inhibitors (Half Protease and Phosphatase Inhibitor Cocktail Cat. no. 78440, Thermo Scientific).
4. At this point the chromatin pellet may be viscous. Take care that it does not stick to the tip of your pipet.
5. Under acidic conditions, the DNA becomes insoluble and basic proteins (such as histones) become soluble. At this point the use of a pestle is recommended to dissociate histones from high-order chromatin.
6. The resulting pellet is fragile and should be handled carefully. Acetone is an organic solvent that must be stored for proper disposal.
7. For details about the technique of TAU-PAGE, *see* Shechter et al. [10].
8. The protocol was standardized to an average of $2.5–3 \times 10^8$ parasites/tube.
9. Reaction time for Micrococcal nuclease digestion is critical. We use an enzyme from USB (Cat. no. 70196Y), but we strongly recommend testing your enzyme first varying concentration and time of digestion. The expected fragments are approximately 150–200 bp corresponding to one nucleosome. The digestion conditions should be optimized to so that the majority of fragments correspond to single nucleosomes.

Preparations with partially digested fragments 400-600 nt, corresponding to 2 or 3 nucleosomes, may be used as long as the majority of fragments are 150-200 nt.

10. At this stage, we recommend that you measure the amount of DNA in each fraction (S1, S2, and P). The ratio 260/280 should be around 1.8, 1.5, and 1.3. These fractions correspond to different degrees of chromatin compaction. S1 corresponds to free nucleosomes, while the S2 corresponds to more condensed regions and finally, P, to high-compaction regions (heterochromatin).
11. As mentioned above, each fraction corresponds to different levels of compaction. In some laboratories, depending on the amount of DNA obtained, the fraction P is also used.
12. The antibody specificity is the critical step for this technique. Although antibody concentrations change, for an initial experiment 5–10 μg commercial antibody or about 15 μl of serum should be sufficient.
13. The cross-link is used in order to hold protein complexes attached to DNA. 1 % formaldehyde easily penetrates different cell types and usually is sufficient to maintain antigen associated to DNA. Furthermore the cross-linking is a time-critical procedure; excessive cross-links can generate artificial epitopes, or alter epitopes, so they are no longer recognized by antibody.
14. Formaldehyde is an organic reagent that must be stored and properly disposed.
15. After fixation, the cells are very difficult to disrupt. We recommend passaging at least six times through 25 gauge needles (prior passage through 20, 22, and 23 g needles will also probably be needed), but more passages may be necessary.
16. If the purpose of the experiment is hybridization to a microarray, parasite purification using a 3 μm Nuclepore membrane is not necessary, although the amount of sample used should take into account the proportion that is parasite material rather than relying solely on DNA concentration. In case of sequencing, parasite purification is highly recommended to minimize host cell contamination. In this case, start with a larger number of parasites, since parasites are more difficult to purify after cross-linking.
17. The sonication parameters should be standardized for each equipment.
18. Alternatively, protein G can be used, depending on the immunoglobulin type (Dynabeads Protein G, Cat. no. 100.03D, Invitrogen).

    There are protocols available with modifications from those described here that contain excellent suggestions and troubleshooting advice. We particularly recommend those from Abcam website (www.abcam.com) and Cold Spring Harbor Protocols [44].

Acknowledgement

This work was supported by NIH grants RC4AI092801 (KK), R01AI087625 (KK), and 5T32AI070117-04 (SCN). We thank members of the Kim laboratory for review and helpful suggestions for this chapter.

References

1. Strahl BD, Allis CD (2000) The language of covalent histone modifications. Nature 403:41–45
2. Kouzarides T (2007) Chromatin modifications and their function. Cell 128:693–705
3. Goldberg AD, Allis CD, Bernstein E (2007) Epigenetics: a landscape takes shape. Cell 128:635–638
4. Sullivan WJ Jr, Naguleswaran A, Angel SO (2006) Histones and histone modifications in protozoan parasites. Cell Microbiol 8: 1850–1861
5. Bougdour A, Braun L, Cannella D et al (2010) Chromatin modifications: implications in the regulation of gene expression in Toxoplasma gondii. Cell Microbiol 12:413–423
6. Saksouk N, Bhatti MM, Kieffer S et al (2005) Histone-modifying complexes regulate gene expression pertinent to the differentiation of the protozoan parasite Toxoplasma gondii. Mol Cell Biol 25:10301–10314
7. Hakimi MA, Deitsch KW (2007) Epigenetics in Apicomplexa: control of gene expression during cell cycle progression, differentiation and antigenic variation. Curr Opin Microbiol 10:357–362
8. Croken MM, Nardelli SC, Kim K (2012) Chromatin modifications, epigenetics, and how protozoan parasites regulate their lives. Trends Parasitol 28(5):202–213
9. Olins DE, Olins AL (2003) Chromatin history: our view from the bridge, *Nature reviews.* Mol Cell Biol 4:809–814
10. Shechter D, Dormann HL, Allis CD et al (2007) Extraction, purification and analysis of histones. Nat Protoc 2:1445–1457
11. Murray K (1966) The acid extraction of histones from calf thymus deoxyribonucleoprotein. J Mol Biol 15:409–419
12. Stedman E (1950) Cell specificity of histones. Nature 166:780–781
13. von Holt C, Brandt WF, Greyling HJ et al (1989) Isolation and characterization of histones. Methods Enzymol 170:431–523
14. Toro GC, Galanti N (1990) Trypanosoma cruzi histones. Further characterization and comparison with higher eukaryotes. Biochem Int 21:481–490
15. Solomon MJ, Larsen PL, Varshavsky A (1988) Mapping protein-DNA interactions in vivo with formaldehyde: evidence that histone H4 is retained on a highly transcribed gene. Cell 53:937–947
16. Gilmour DS, Lis JT (1984) Detecting protein-DNA interactions in vivo: distribution of RNA polymerase on specific bacterial genes. Proc Natl Acad Sci U S A 81:4275–4279
17. Ren B, Robert F, Wyrick JJ et al (2000) Genome-wide location and function of DNA binding proteins. Science 290:2306–2309
18. Lee TI, Rinaldi NJ, Robert F et al (2002) Transcriptional regulatory networks in Saccharomyces cerevisiae. Science 298:799–804
19. Park PJ (2009) ChIP-seq: advantages and challenges of a maturing technology, *Nature reviews.* Genetics 10:669–680
20. Goren A, Ozsolak F, Shoresh N et al (2010) Chromatin profiling by directly sequencing small quantities of immunoprecipitated DNA. Nat Methods 7:47–49
21. Johnson DS, Mortazavi A, Myers RM et al (2007) Genome-wide mapping of in vivo protein-DNA interactions. Science 316:1497–1502
22. Barski A, Cuddapah S, Cui K et al (2007) High-resolution profiling of histone methylations in the human genome. Cell 129: 823–837
23. Hecht A, Strahl-Bolsinger S, Grunstein M (1996) Spreading of transcriptional repressor SIR3 from telomeric heterochromatin. Nature 383:92–96
24. Rundlett SE, Carmen AA, Suka N et al (1998) Transcriptional repression by UME6 involves deacetylation of lysine 5 of histone H4 by RPD3. Nature 392:831–835
25. Orlando V (2000) Mapping chromosomal proteins in vivo by formaldehyde-crosslinked-

chromatin immunoprecipitation. Trends Biochem Sci 25:99–104

26. O'Neill LP, Turner BM (2003) Immunoprecipitation of native chromatin: NChIP. Methods 31:76–82
27. Schones DE, Cui K, Cuddapah S et al (2008) Dynamic regulation of nucleosome positioning in the human genome. Cell 132:887–898
28. Iyer VR, Horak CE, Scafe CS et al (2001) Genomic binding sites of the yeast cell-cycle transcription factors SBF and MBF. Nature 409:533–538
29. Buck MJ, Lieb JD (2004) ChIP-chip: considerations for the design, analysis, and application of genome-wide chromatin immunoprecipitation experiments. Genomics 83: 349–360
30. Shendure J, Ji H (2008) Next-generation DNA sequencing. Nat Biotechnol 26: 1135–1145
31. Mardis ER (2008) Next-generation DNA sequencing methods. Annu Rev Genomics Hum Genet 9:387–402
32. Margulies M, Egholm M, Altman WE et al (2005) Genome sequencing in microfabricated high-density picolitre reactors. Nature 437:376–380
33. Bentley DR (2006) Whole-genome re-sequencing. Curr Opin Genet Dev 16: 545–552
34. Metzker ML (2010) Sequencing technologies—the next generation, *Nature reviews.* Genetics 11:31–46
35. Wang Z, Gerstein M, Snyder M (2009) RNA-Seq: a revolutionary tool for transcriptomics, *Nature reviews.* Genetics 10:57–63
36. Salcedo-Amaya AM, van Driel MA, Alako BT et al (2009) Dynamic histone H3 epigenome marking during the intraerythrocytic cycle of Plasmodium falciparum. Proc Natl Acad Sci U S A 106:9655–9660
37. Flueck C, Bartfai R, Niederwieser I et al (2010) A major role for the Plasmodium falciparum ApiAP2 protein PfSIP2 in chromosome end biology. PLoS Pathog 6:e1000784
38. Brooks CF, Francia ME, Gissot M et al (2011) Toxoplasma gondii sequesters centromeres to a specific nuclear region throughout the cell cycle. Proc Natl Acad Sci U S A 108: 3767–3772
39. Lopez-Rubio JJ, Mancio-Silva L, Scherf A (2009) Genome-wide analysis of heterochromatin associates clonally variant gene regulation with perinuclear repressive centers in malaria parasites. Cell Host Microbe 5:179–190
40. Gissot M, Kelly KA, Ajioka JW et al (2007) Epigenomic modifications predict active promoters and gene structure in Toxoplasma gondii. PLoS Pathog 3:e77
41. Siegel TN, Hekstra DR, Kemp LE et al (2009) Four histone variants mark the boundaries of polycistronic transcription units in Trypanosoma brucei. Genes Dev 23: 1063–1076
42. Bartfai R, Hoeijmakers WA, Salcedo-Amaya AM et al (2010) H2A.Z demarcates intergenic regions of the plasmodium falciparum epigenome that are dynamically marked by H3K9ac and H3K4me3. PLoS Pathog 6:e1001223
43. Kozarewa I, Ning Z, Quail MA et al (2009) Amplification-free Illumina sequencing-library preparation facilitates improved mapping and assembly of (G+C)-biased genomes. Nat Methods 6:291–295
44. Carey MF, Peterson CL, Smale ST (2009) Chromatin immunoprecipitation (ChIP). Cold Spring Harbor protocols, pdb prot5279.

Chapter 11

The Genome-Wide Identification of Promoter Regions in *Toxoplasma gondii*

Junya Yamagish and Yutaka Suzuki

Abstract

Parasites change their transcriptional systems in different developmental stages and in response to environmental changes. To investigate the molecular mechanisms that underlie transcriptional regulation, it is essential to identify the exact positions of the transcriptional start sites (TSSs) and characterize the upstream promoter regions. However, it has been essentially impossible to obtain comprehensive information using conventional methods. Here, we introduce our TSS-seq method, which combines full-length technology, oligo-capping, and rapidly developing next-generation sequencing technology. TSS-seq has enabled identification of TSS positions and upstream promoter activities as digital TSS tag counts within a reasonable cost and time frame. In this chapter, we describe in detail the TSS-seq method for the identification and characterization of the promoters in *Toxoplasma gondii*.

1 Introduction

It has been reported that many transcription factor-binding sites and other important sequence motifs are embedded in overlapping or immediately upstream regions of transcriptional start sites (TSSs). Therefore, it is essential to identify and characterize the exact locations and frequencies of the transcriptional initiations (promoter activities) as a function of each cellular state to understand the molecular mechanisms and dynamics of transcriptional regulation. In conventional methods, such as primer extension and 5′ RACE, TSSs are detected by the identification of the 5′-end of the intact mRNA. However, the application of these methods is not practical for genome-wide analyses. Moreover, these conventional techniques are indirect methods that are dependent on the quality of the initial RNA material. When cDNA synthesis has not been carried out completely, it is difficult to distinguish between genuine TSSs and artifacts derived from truncated cDNAs.

By contrast, the oligo-capping method focuses on the cap structure of the mRNA, which is a direct signature of the TSS [1]. In this method, the cap structure is replaced with a synthesized

Christopher Peacock (ed.), *Parasite Genomics Protocols*, Methods in Molecular Biology, vol. 1201,
DOI 10.1007/978-1-4939-1438-8_11,

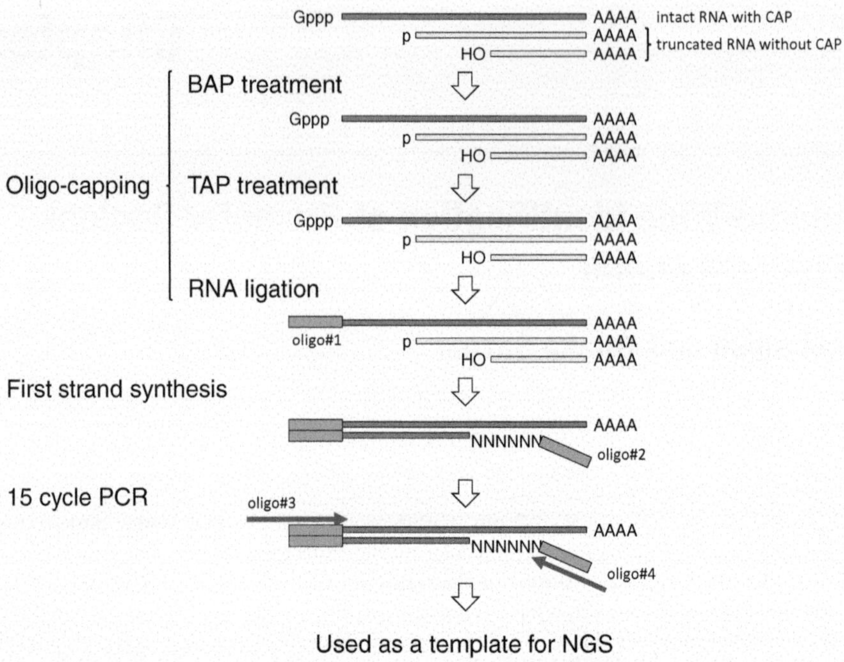

Fig. 1 Schematic diagram of the TSS-seq method. *Gray boxes* represent linker oligo RNAs, which are used for primer binding sites for Illumina GAII sequencing. Gppp: cap structure. AAA: polyA

oligo RNA linker (Fig. 1). Briefly, the replacement of the cap structure is carried out by three enzymatic reactions: (1) the elimination of the phosphate at the 5′-end of non-capped RNA, such as fragmented mRNA, rRNA, or tRNA, for the exclusion of these RNA species from subsequent reactions (note that the cap structure itself remains intact in this reaction); (2) the specific canalization of the cap structure by tobacco acid pyrophosphatase (TAP) to remove the cap structure and place a phosphate at the position where the cap structure was originally located; and (3) the ligation of the synthesized oligo RNA linker to the exposed 5′-end phosphate of the mRNA. The capped mRNA is then used for cDNA synthesis using either an oligo dT or a random primer such that the 5′-end of the RNA linker sequence is copied to the cDNA. After the cDNAs are marked with the known sequence, "full-length" cDNAs can be selectively amplified by RT-PCR using the 5′-end sequence as the PCR primer. By contrast, truncated cDNAs derived from incomplete elongation, which do not have the 5′-end sequence, cannot be amplified.

Genome-wide TSS analysis is further enabled by integrating oligo-capping with next-generation sequencing technology [2]. In this technique, called the TSS-seq method, the DNA sequence that is required for next-generation sequencing is embedded in the RNA linker sequence. Therefore, massively parallel sequencing of cDNA sequences immediately downstream of the cap site of the mRNA can be obtained using the obtained random primer-primed

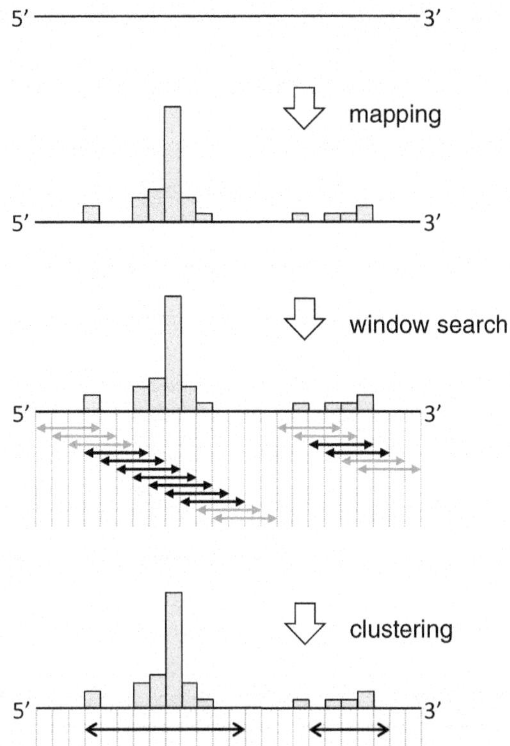

Fig. 2 Schematic diagram of promoter identification. The *line* and *bar* represent the genome and mapped SRS, respectively. The *black double-headed arrow* and *gray double-headed arrow* represent valid and invalid windows, respectively. The window size is 4 nt, and the threshold is 4 SRS per window. Overlapping valid windows are clustered to identify the promoter region

cDNA library (TSS-seq library) as a template in a next-generation sequencing platform. Typically, 36-base single-end sequencing is performed to read the sequence as a TSS tag, which corresponds to one molecular copy of the TSS. The generated short-read sequences are mapped to the reference genome sequence using a general mapping program, such as Bowtie [3, 4] or ELAND (Illumina, San Diego, CA), to locate the genomic coordinate of the TSS tag. Each of the mapped TSS tags is further clustered and associated using gene models based on the overlap of their genomic coordinates. Obtained positional information regarding the TSS tags and their frequencies (i.e., digital tag counts) are integrated to assign TSSs or promoters (Fig. 2). A similar method, called the CAGE (cap analysis of gene expression) technique, also utilizes the cap site of mRNA for the large-scale identification of TSSs. In CAGE, selection of the cap structure is enabled by chemical reactions. For further details, *see* ref. 5.

Both TSS-seq and CAGE were originally developed for representative model organisms, such as humans and mice. Recently, these methods have also been applied to non-model organisms as a growing number of the genomic sequences of non-model organisms become available. When coupled with RNA-seq analysis, these transcriptome-based approaches have often been highly useful for refining structural and functional annotations of genes in newly sequenced genomes. Unlike microarray analysis, the aforementioned sequence-based approaches do not require the a priori design of probes or primers, which has frequently hampered streamlined gene annotations. In the case of apicomplexan parasites, several series of intensive transcriptome analyses, such as full-length cDNA sequencing, microarray analysis, and RNA-seq analysis, have been conducted in *Plasmodium* species [6–10]. However, the accumulation of information on other parasites remains relatively poor. We recently applied TSS-seq for the identification and characterization of promoters in *T. gondii* [11]. In this chapter, we describe a detailed protocol for TSS-seq in *T. gondii*. We believe that a similar application of TSS-seq can produce comprehensive data in various other parasites and will further elucidate their diverse transcriptional programs, which are dependent on their life cycles, host preferences, and, above all, their etiologies, using a comparative genomics approach.

2 Materials

2.1 RNA Preparation

1. Phosphate-buffered saline (PBS) (137 mM NaCl, 2.7 mM KCl, 10 mM Na_2HPO_4, 1.76 mM KH_2PO_4).
2. TRI reagent (Sigma, St. Louis, MO).
3. RNeasy Maxi Kit (Qiagen, Valencia, CA).

2.2 Oligo-Capping

1. Bacterial alkaline phosphatase (BAP) (Takara, Shiga, Japan).
2. Phenol/chloroform (1:1 water-saturated phenol:chloroform).
3. Ethanol (EtOH).
4. Tobacco acid pyrophosphatase (TAP) (Wako, Osaka, Japan).
5. RNasin (Promega, Madison, WI).
6. T4 RNA ligase (TaKaRa).
7. $MgCl_2$.
8. PEG8000.
9. Adenosine triphosphate (ATP).
10. RNA and DNA oligos (*see* Table 1).
11. DNaseI (TaKaRa).
12. Tris–HCl (pH 7.0) (Sigma, T1819-100ML).
13. Dithiothreitol (DTT).
14. uMACs mRNA Isolation Kit (Miltenyi Biotec, Auburn, CA).
15. SuperScript II (Invitrogen).

Table 1
Synthesized oligo RNA and DNA

Name	Seqence (5'-3')
oligo#1	ACCGAGAUCUACACUCUUUCCCUACACGACGCUCUUCCGAUCUGG
oligo#2	CCTGCTGAACCGCTCTTCCGATCTNNNNNNC
oligo#3	AATGATACGGCGACCACCGAGATCTACACTCTTTCCCTACACGACGCTCTTCCGATCT
oligo#4	CAAGCAGAAGACGGCATACGAGATCGGTCTCGGCATTCCTGCTGAACCGCTCTTCCGATCT

16. dNTPs.
17. Ethylenediaminetetraacetic acid (EDTA).
18. NaOH.
19. Magnesium acetate (MgOAc).
20. GeneAmp XL PCR kit (Applied Biosystems).
21. Agilent RNA 6000 Nano Kit (Agilent Technologies, Santa Clara, CA).
22. Agilent DNA 7500 kit (Agilent Technologies).
23. Agilent High Sensitivity DNA kit (Agilent Technologies).

3 Methods

3.1 RNA Preparation

1. Suspend 10^9 purified parasites in 1 mL of PBS (*see* **Note 1**).
2. Add 20 mL of TRI reagent and then homogenize.
3. Purify the total RNA according to the instructions of the TRI reagent manufacturer.
4. Dissolve in 15 mL of RTL buffer included in the RNeasy Maxi kit.
5. Purify the total RNA according to the RNeasy Maxi kit instructions.
6. Check the RNA quality and quantity using an Agilent RNA 6000 Nano Kit (*see* **Note 2**).

3.2 Oligo-Capping

3.2.1 BAP Treatment

1. Mix purified total RNA (100 μg), 5× BAP buffer (40 μL), RNasin (5.4 μL), and BAP (50 U), and then bring to 200 μL with dH_2O.
2. Incubate at 37 °C for 1 h.
3. Extract with phenol/chloroform.
4. Precipitate with EtOH.
5. Suspend in 36.65 μL of dH_2O.

3.2.2 TAP Treatment

1. Add 5× TAP buffer (10 μL), RNasin (1.35 μL), and TAP (2 μL).
2. Incubate at 37 °C for 1 h.
3. Extract with phenol/chloroform.
4. Precipitate with EtOH.
5. Suspend in 21.7 μL of dH_2O.

3.2.3 Oligo Ligation

1. Add oligo#1 (12 μL), 10× ligation buffer (30 μL), 25 mM $MgCl_2$ (60 μL), 24 mM ATP (6.3 μL), RNasin (7.5 μL), T4 RNA ligase (50 U), and 50 % PEG8000 (150 μL).
2. Incubate at 20 °C for 3 h.
3. Add 300 μL of dH_2O.
4. Extract with phenol/chloroform.
5. Precipitate with EtOH.
6. Suspend in 54.3 μL of dH_2O.

3.2.4 DNaseI Treatment

1. Add 25 mM $MgCl_2$ (32 μL), 1 M Tris–HCl (4 μL), 0.1 M DTT (5 μL), RNasin (2.7 μL), and DNaseI (2 μL).
2. Incubate at 37 °C for 1 h.
3. Extract with phenol/chloroform.
4. Precipitate with EtOH.
5. Suspend in 100 μL of dH_2O.

3.2.5 Poly(A) Selection

1. Isolate poly(A)$^+$ RNA using the uMACs mRNA Isolation Kit according to the manufacturer's instructions.
2. Precipitate with EtOH.
3. Suspend in 21 μL of dH_2O.

3.2.6 Reverse Transcription

1. Add 5× first-strand buffer (10 μL), 5 mM dNTPs (8 μL), 0.1 M DTT (6 μL), oligo#2 (2.5 μL), RNasin (1 μL), and SuperScript II (2 μL).
2. Incubate at 12 °C for 1 h, and then keep at 42 °C overnight.
3. Add dH_2O (50 μL), 0.5 M EDTA (2 μL), and 0.1 N NaOH (15 μL).
4. Incubate at 65 °C for 40 min.
5. Add 20 μL of 1 M Tris–HCl.
6. Precipitate with EtOH.
7. Suspend in 10 μL of dH_2O.

3.2.7 Amplification and Size Selection

1. Mix first-stranded cDNA (2 μL), dH_2O (50.4 μL), 3.3× buffer (30.0 μL), dNTPs (8 μL), 25 mM MgOAc (4.4 μL), oligo#3 (1.6 μL), oligo#4 (1.6 μL), and DNA polymerase (2 μL).

2. Run the following thermal cycle: 94 °C for 1 min; 94 °C for 1 min, 58 °C for 1 min, and 72 °C for 2 min × 15 cycles; and 72 °C for 10 min.
3. Check PCR products using an Agilent DNA 7500 kit (*see* **Note 3**).
4. Fractionate the products using a 6 % PAGE gel; collect fragments between 250 and 300 bp, and then elute in 10 μL of dH_2O.
5. Check the purified fragments using an Agilent High Sensitivity DNA kit (*see* **Note 4**).

3.3 Sequencing Using an Illumina GAII System

Follow the manufacturer's instructions.

3.4 Mapping of Short Read Sequences (SRS) Using Bowtie

1. Download and install Bowtie (*see* **Note 5**).
2. Convert qseq-formatted output data (*see* **Note 6**) from Illumina NGS to FASTA or FASTQ format using a custom script or open script, such as qseq2fastq (*see* **Note 7**).
3. Construct indexes for alignment by running bowtie-build from the command line as in the following example (*see* **Note 8**):

   $ bowtie-build TgondiiME49Genomic_ToxoDB-7.2.fasta index_name
   where index_name specifies the name of your index.
4. Run Bowtie from the command line according to the following example:

   $ bowtie -q TgondiiME49Genomic_ToxoDB-7.2 reads.fq –v 2 –k 2 –5 2

   where the parameters –q, –v, –k, and –5 specify the input format of the SRS, the number of acceptable mismatches, the maximum number of reports, and the number of bases to be trimmed from the left end, respectively (*see* **Note 9**). For other parameters, *see* the Bowtie manual [3].

3.5 Identification of Promoter Regions

An open script for the following procedure is not available; therefore, you must write your own code to implement the following procedure. A schematic diagram of the procedure is also shown (Fig. 2).

1. Specify the TSS locations based on the 5′-end of mapped SRS according to the Bowtie results.
2. Count the redundancy of TSSs in each nucleotide in both strands (*see* **Note 10**).
3. Add up the number of TSSs within a 20-nt window (*see* **Note 11**).

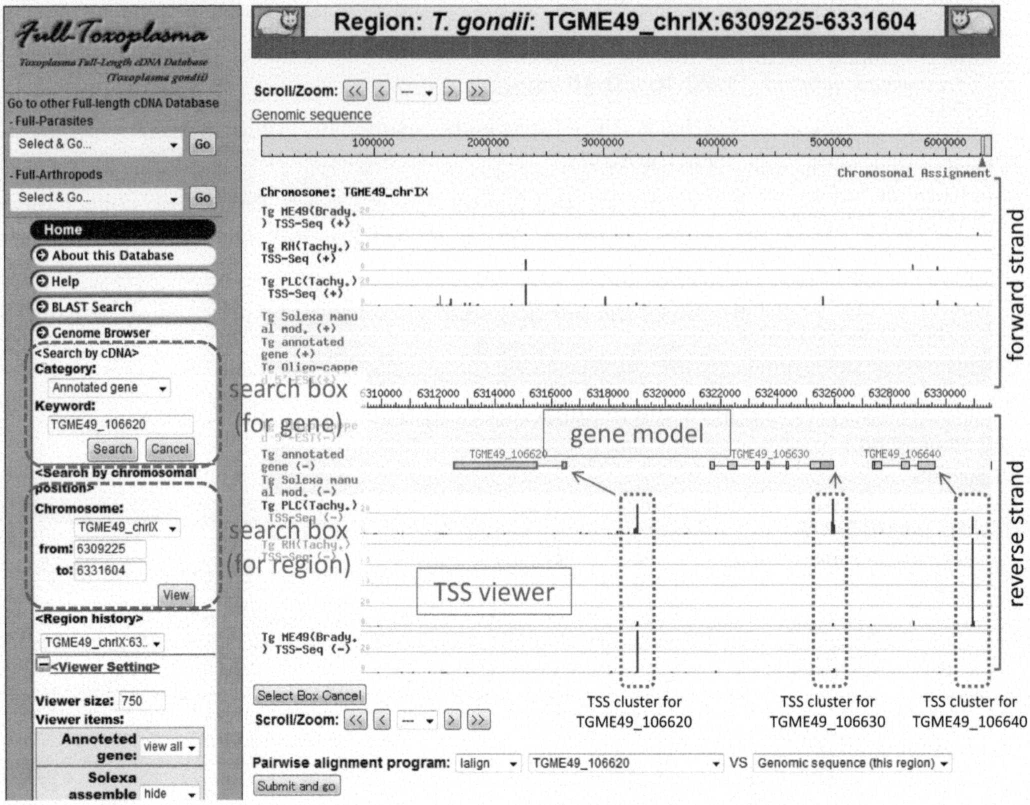

Fig. 3 Screenshot of the Full-Toxoplasma genome browser implemented using the TSS viewer

4. Repeat **step 3** over the entire genome while sifting the window by 1 nt.
5. Select a set of valid windows that have a greater number of TSSs than an arbitrary threshold (*see* **Note 12**).
6. Cluster the valid windows if they are overlapped.

3.6 Database

For the visualization and public use of our data, we have established a website called Full-Toxoplasma (Fig. 3) (http://fullmal.hgc.jp/index_tg_ajax.html) [12]. Mapping images of SRS obtained from TSS-seq analyses for the tachyzoite and bradyzoite stages of strain ME49 as well as the tachyzoite stage of strain RH are available in this database.

4 Notes

1. Both in vivo- and in vitro-cultured parasites can be used. High integrity of the RNA (RIN value of >9) is preferable. Contamination of the host RNA is allowable because the obtained tags are mapped on the parasite genome, and SRS from

contaminants, such as host cells, can be separated out in the mapping step.

2. Two sharp peaks corresponding to ribosomal RNA are observed for successfully purified total RNA (Fig. 4).
3. A smooth peak is observed at approximately 1,000–2,000 bp for a successfully amplified PCR product (Fig. 4).
4. A single peak is observed at approximately 200–300 bp in samples successfully recovered by PAGE purification (Fig. 4).
5. Bowtie is an ultrafast, memory-efficient short-read aligner [3] and is available from http://bowtie-bio.sourceforge.net/index.shtml. The downloadable files contain executable binaries for Linux, Windows, and Mac OS X, and there is no need for further installation. For more detail, *see* "MANUAL" in the extracted directory.
6. The qseq format output from Illumina sequencers is currently not allowed by Bowtie.
7. The qseq2fastq is available from http://sourceforge.net/projects/qseq2fastq/.
8. The script bowtie-build builds a Bowtie index from a set of DNA sequences. It produces six files with a common index name and different suffixes. The index name can be specified in the mapping step using Bowtie as reference. For further detail, *see* the Bowtie manual.
9. The parameters shown are for TSS analysis in *T. gondii*. Two mismatches are allowed because genomic sequences of the analyzed strain and the reference strain are not always identical in parasites. The k parameter is useful for detecting multiple mapped SRS. The first two bases of the TSS tag is always GG, which can be used to directly confirm the RNA oligo-mRNA ligation site, and thus should be masked in the following mapping step. Partial output from Bowtie is shown as an illustration (Table 2). For further detail, *see* the Bowtie manual.
10. Statistics for a mapped SRS are shown in Table 3 as an example. The results follow a power law-like distribution [11].
11. The window size can be arbitrarily defined. We tried a series of sizes, and a cluster derived from 20-nt windows was best suited for our cluster recognition.
12. The Poisson distribution is available to distinguish a significant accumulation of mapped SRS from the background signal, where the gamma parameter is defined as total SRS number × window size ÷ (2 × genome size). If multiple tests are used, then the level of significance can be corrected by the FDR

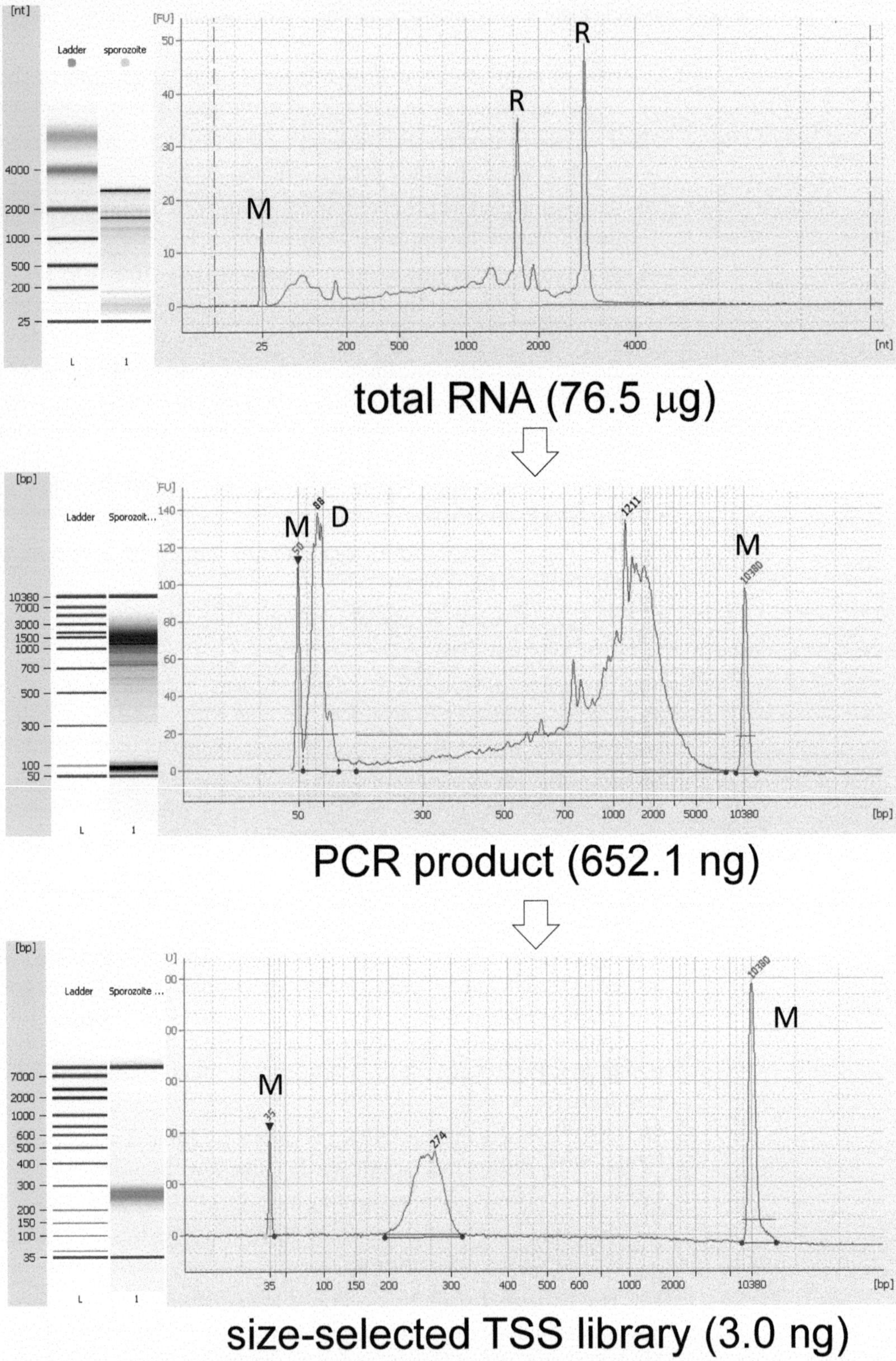

Fig. 4 Virtual gel images and output graphs for each sample preparation. The examples shown here correspond to the *T. gondii* sporozoite sample. (*Top*) Snapshot from Agilent RNA Nano analysis in the total RNA purification step. (*Center*) Snapshot from Agilent DNA 7500 analysis in PCR amplification step. (*Bottom*) Snapshot from Agilent DNA high-sensitivity analysis in the PAGE purification step. M, markers; R, ribosomal RNAs; D, putative primer dimer or nonspecific products

Table 2
Example of Bowtie output

ID	Ori	Chromosome	Position	SRS	Quality		Mismatch
HWI-EAS74:90:8:1:1266:949:0:1	–	gb|TGME49_chrVIIb	222603	AACACGGACGCCAAAGGGTC AAAAAAACGGTTCT	IIIIIIIIIIIIIIIIIIIIIIIIIIIIIIIIII	0	
HWI-EAS74:90:8:1:1296:930:0:1	–	gb|TGME49_chrXII	5585320	TAGAAAAAAATGAAAATACCT ATACAACAAACTC	IIIIIIIIIIIIIIIIIIIIIIIIIIIIIIIIII	0	
HWI-EAS74:90:8:1:1576:945:0:1	+	gb|TGME49_chrX	817229	CTTTGTAGACTTCTCCCTTC AGCGGCTTTCCAGC	IIIIIIIIIIIIIIIIIIIIIIIIIIIIIIIIII	0	
HWI-EAS74:90:8:1:1655:930:0:1	+	gb|TGME49_chrXII	5487545	ATTTTTCTTGTTTCGTTTTC TCGCAATTTTTGCG	IIIIIIIIIIIIIIIIIIIIIIIIIIIIIIIIII	0	0:C>A,1:A>T
HWI-EAS74:90:8:1:2259:943:0:1	+	gb|TGME49_chrV	2842174	ATACCTTCTCGGCCTTTTGG CTATGATCAAGTGT	IIIIIIIIIIIIIIIIIIIIIIIIIIIIIIIIII	0	
HWI-EAS74:90:8:1:2259:943:0:1	–	gb|TGME49_chrXI	3715483	ACACTTGATCATAGCCAAAA GGCCGAGAAGGTAT	IIIIIIIIIIIIIIIIIIIIIIIIIIIIIIIIII	0	
HWI-EAS74:90:8:1:1686:933:0:1	+	gb|TGME49_chrXII	1181011	AGCTGTTAAACAGACAATTT CTCGGCCGTAGAAA	IIIIIIIIIIIIIIIIIIIIIIIIIIIIIIIIII	0	
HWI-EAS74:90:8:1:2005:947:0:1	–	gb|TGME49_chrX	4123891	TCCCTCGCCGTCCGGCACT CTTCTCTTGTTGATG	IIIIIIIIIIIIIIIIIIIIIIIIIIIIIIIIII	0	

Table 3
Distribution of SRS

# of mapped SRS	# of site
1	31624
2	17702
3	13728
4	11239
5	8735
6	6710
7	4762
8	3594
9	2711
10	2134
…	…
58313	1
85604	1
88905	1
102573	1
133132	1
223828	1
390641	1

method. Q-VALUE, which is an R package available from http://genomics.princeton.edu/storeylab/qvalue/, is useful for the FDR calculation [13]. In addition, a simple index, such as the ratio of the population of mapped SRSs to the total number of SRSs, is also available as a threshold in case of comparative analysis.

Acknowledgements

This work was supported by a Grant from the Asia-Africa S&T Strategic Cooperation Promotion Program by the Special Coordination Funds for Promoting Science & Technology from the MEXT Japan.

References

1. Suzuki Y, Taira H, Tsunoda T et al (2001) Diverse transcriptional initiation revealed by fine, large-scale mapping of mRNA start sites. EMBO Rep 2(5):388–393
2. Tsuchihara K, Suzuki Y, Wakaguri H et al (2009) Massive transcriptional start site analysis of human genes in hypoxia cells. Nucleic Acids Res 37(7):2249–2263
3. Langmead B, Trapnell C, Pop M et al (2009) Ultrafast and memory-efficient alignment of short DNA sequences to the human genome. Genome Biol 10(3):R25
4. Li H, Durbin R (2009) Fast and accurate short read alignment with Burrows-Wheeler transform. Bioinformatics 25(14):1754–1760
5. Kawaji H, Kasukawa T, Fukuda S et al (2006) CAGE Basic/Analysis Databases: the CAGE resource for comprehensive promoter analysis. Nucleic Acids Res 34(Database issue): D632–D636
6. Watanabe J, Sasaki M, Suzuki Y et al (2002) Analysis of transcriptomes of human malaria parasite *Plasmodium falciparum* using full-length enriched library: identification of novel genes and diverse transcription start sites of messenger RNAs. Gene 291(1–2):105–113
7. Wakaguri H, Suzuki Y, Sasaki M et al (2009) Inconsistencies of genome annotations in apicomplexan parasites revealed by 5′-end-one-pass and full-length sequences of oligo-capped cDNAs. BMC Genomics 10:312
8. Hayward RE, Derisi JL, Alfadhli S et al (2000) Shotgun DNA microarrays and stage-specific gene expression in *Plasmodium falciparum* malaria. Mol Microbiol 35(1):6–14
9. Daily JP, Scanfeld D, Pochet N et al (2007) Distinct physiological states of *Plasmodium falciparum* in malaria-infected patients. Nature 450(7172):1091–1095
10. Otto TD, Wilinski D, Assefa S et al (2010) New insights into the blood-stage transcriptome of *Plasmodium falciparum* using RNA-Seq. Mol Microbiol 76(1):12–24
11. Yamagishi J, Wakaguri H, Ueno A et al (2010) High-resolution characterization of *Toxoplasma gondii* transcriptome with a massive parallel sequencing method. DNA Res 17(4): 233–243
12. Tuda J, Mongan AE, Tolba MEM et al (2011) Full-parasites: database of full-length cDNAs of apicomplexa parasites, 2010 update. Nucleic Acids Res 39(Database issue): D625–D631
13. Storey JD, Tibshirani R (2003) Statistical significance for genomewide studies. Proc Natl Acad Sci U S A 100(16):9440–9445

Chapter 12

RNA-Seq Approaches for Determining mRNA Abundance in *Leishmania*

Andrew Haydock, Monica Terrao, Aarthi Sekar, Gowthaman Ramasamy, Loren Baugh, and Peter J. Myler

Abstract

High-throughput sequencing of cDNA copies of mRNA (RNA-seq) provides a digital read-out of mRNA levels over several orders of magnitude, as well as mapping the transcripts to the nucleotide level. Here we describe an RNA-seq approach that exploits the 39-nucleotide mini-exon or spliced leader (SL) sequence found at the 5′ end of all *Leishmania* (and other trypanosomatid) mRNAs.

Key words RNA-seq, Transcriptome, mRNA, Differential gene expression

1 Introduction

In 2005, the publication of the TriTryp genomes [1–4] heralded a paradigm shift in the ability of researchers to investigate trypanosomatid gene expression. Coupled with the emerging microarray-based technology, it became possible to interrogate the mRNA levels for all (or at least, most) genes in whatever lifecycle stage or growth condition was accessible experimentally. These advances soon led to a number of publications that examined the genome-wide changes in gene expression between insect and mammalian stages of several *Leishmania* species [5–15]. While not without controversy, these studies generally concluded that, unlike many other organisms, only a small percentage (<10 %) of *Leishmania* genes showed significant changes in mRNA levels in these different lifecycle stages. However, microarray-based analysis of gene expression had several limitations, principally a lack of sensitivity and inability to distinguish between closely related genes, as well as a prohibitively expensive initial cost outlay for less well-studied organisms such as *Leishmania*. In addition, microarray analyses failed to precisely define the boundaries of the mRNAs being interrogated, and were thus unable to reveal

Christopher Peacock (ed.), *Parasite Genomics Protocols*, Methods in Molecular Biology, vol. 1201,
DOI 10.1007/978-1-4939-1438-8_12,

whether there were changes in the 5′ and 3′ untranslated regions (UTRs) during parasite development. Fortunately, the recent emergence of next-generation sequencing (NGS) technologies provided a solution: high-throughput sequencing of cDNA copies of mRNA (RNA-seq). This approach not only allows mapping of the transcripts to the nucleotide level, but it also provides a robust digital read-out of mRNA levels over a dynamic range of several orders of magnitude [16].

While there are several different NGS technologies currently available (and more being developed), most RNA-seq applications use the Illumina or SOLiD platforms, since they provide the massively parallel throughput (tens to hundreds of millions of reads per lane) necessary to obtain sufficient coverage of lower abundance mRNAs. Similarly, there are several approaches that can be used for cDNA library generation, depending on the specific question(s) being asked. However, since most investigators are interested (at least initially) in determining the steady-state levels of mRNAs in the sample(s) of interest, it is usually desirable to use a method that avoids making cDNA from the noncoding (ribosomal, transfer, small nuclear, and small nucleolar) RNAs that make up the majority of cellular RNA. While this can be readily achieved by purification of polyadenylated mRNA using oligo(dT) magnetic beads, this approach has the disadvantage of being subject to variable recovery (especially for small samples). An alternative approach is removal of the rRNA by hybridization with biotinylated probes, which has the advantage of maintaining other ncRNAs in the sample. However, the rRNA probes are commercially available only for common organisms (e.g., human). There are also several approaches for the reverse transcription step, which generates the cDNA that is subsequently amplified and sequenced. Oligo(dT) priming has the advantage of being selective for polyA mRNA, but the disadvantage of under-representing the 5′ end of most mRNAs. Random priming has the opposite bias (i.e., against the 3′ end of the mRNA), as well as not being suitable with polyA selection. There is also priming bias because of secondary structure at some regions of the RNA. Hydrolysis of the mRNA into 200–300 nucleotide fragments, followed by ligation of RNA adapters (which provide the primer sequence for cDNA synthesis), is probably the most common method used, but still suffers from some sequence bias.

Fortunately, the peculiarity of mRNA processing in trypanosomatids provides a unique opportunity to simplify the protocol for many (but not all) applications of RNA-seq in *Leishmania*. In these organisms, all (nuclear-encoded) mRNAs contain a common sequence at their 5′ end: the 39-nucleotide mini-exon or spliced leader (SL) sequence that is added posttranscriptionally by *trans-splicing* [17]. This sequence thus provides a convenient primer for second strand synthesis, ensuring that all cDNAs contain the 5′ end of mRNA. This confers the advantage of being able to use

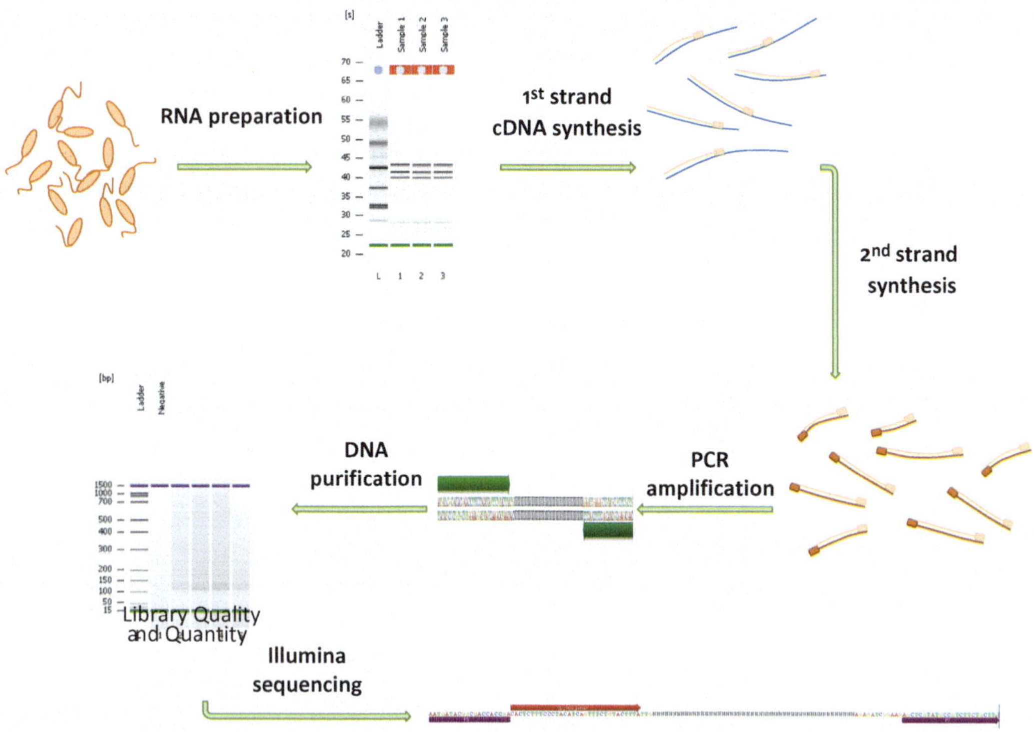

Fig. 1 Schematic representation of SL RNA-seq methodology, showing each of the steps described below. The last panel shows the final PCR product (*top* strand only), with the Illumina adaptor sequences (P5 and P7) indicated in purple and the SL SEQ_primer in *red*

random priming of unpurified RNA for first strand synthesis without being overwhelmed by ncRNA sequences. It also has the advantage of precisely defining the SL site(s) for each mRNA in a relatively quantitative manner, without having to resort to the extreme coverage needed by conventional methods. This approach has been used for *Trypanosoma brucei*, where it was called "Spliced Leader trapping" [18]. Here, we describe a similar approach (*see* Fig. 1) that we have used to sequence more than 50 libraries constructed from five different *Leishmania* species.

In *L. major*, we have compared gene expression in procyclic and metacyclic promastigotes (insect stage) and amastigotes (mammalian stage). We have also examined changes in gene expression during in vitro promastigote-to-amastigote differentiation in *L. donovani* (on the same samples as previously used for microarray), as well as comparing differences in mRNA levels between axenic and macrophage-derived amastigotes. For both of these species, we have also combined these results with those obtained from RNA-seq libraries prepared using oligo(dT)-primed first strand cDNA to accurately define the 5′ and 3′ UTRs of >95 % of all mRNAs. Similar experiments have also been carried out using *L. braziliensis*, although without the 3′ libraries. These results are being prepared for publication. In *L. amazonensis*, we used SL RNA-seq to examine changes in mRNA levels following iron-starvation of promastigotes, and have shown that this triggered many of the same changes seen during differentiation to amastigotes [19]. For *L. tarentolae*, we have used SL RNA-seq (as well as RNA-seq of small RNAs) to examine changes in transcript abundance associated with reduction of the hypermodified DNA base J [20]. Somewhat surprisingly, we were able to use the SL RNA-seq data to show that loss of J resulted in extensive transcriptional read-through at chromosome-internal sites (iJ) normally associated with transcription termination. These data have also enabled us to show that a relatively small number of genes show increased or decreased mRNA levels under these conditions. As expected, most of these genes occur close to iJ regions.

In conclusion, we have found that SL RNA-seq provides a rapid, quantitative and cost-effective method determining changes in mRNA levels in *Leishmania* parasites, and hope that the methods described herein will enable others implement this approach for their own purposes. As indicated below (*see* **Note 2**), the methods described here result in libraries that need to be sequenced individually, but we have recently modified the protocol to allow multiplexing multiple samples per lane.

2 Materials

2.1 RNA Preparation

1. RNase-free water, pipette tips, and centrifuge tubes, RNase*Zap*® (Applied Biosystems #AM9780).
2. TRIzol® Reagent (Invitrogen #15596-026).
3. Chloroform, isopropanol, and 75 % ethanol (RNase-free).
4. High-speed refrigerated centrifuge and fixed-angle rotor.
5. Qubit® Fluorometer (Applied Biosystems).
6. Quant-iT™ RNA Assay kit (Applied Biosystems #Q33140).
7. Bioanalyzer 2100 and RNA 6000 Nano Kit (Agilent #5067-1511).

2.2 First Strand cDNA Synthesis

1. RNase-free water, pipette tips, and centrifuge tubes.
2. DNase I (New England Biolabs #M0303L).
3. First strand primer (Random5 Random-CT, TCCGATCTCT NNNNNNN, *see* **Note 1**), HPLC-purified.
4. Deoxynucleotide Solution Set, containing 100 mM each of dATP, dCTP, dGTP, dTTP (New England Biolabs #N0446S).
5. SuperScript® III Reverse Transcriptase, including 5×SSIII RT buffer [250 mM Tris–HCl (pH 8.3), 375 mM KCl, 15 mM $MgCl_2$] and 100 mM dithiothreitol (DTT) (Invitrogen™ #18080-044).

6. RNaseH (Invitrogen™ #18021-014).
7. QIAquick® PCR Purification Kit, including PB, PE, and EB buffers (Qiagen #28106).
8. Microfuge and 1.5 ml centrifuge tubes.

2.3 Second Strand Synthesis

1. RNase-free water, pipette tips, and centrifuge tubes.
2. Second strand primer (SL_2nd_primer, TCAGTTTCTGTA, *see* **Note 1**), HPLC-purified.
3. Klenow Fragment (3′→5′ exo-), including 10× NEBuffer 2 (New England Biolabs #M0212L).
4. QIAquick® PCR Purification Kit, including PB, PE, and EB buffers (Qiagen #28106).
5. Microfuge and 1.5 ml centrifuge tubes.

2.4 PCR Amplification and DNA Purification

1. RNase-free water, pipette tips, and centrifuge tubes.
2. Forward primer (R-prime-CT, CAAGCAGAAGACGGCATACGAGCTCTTCCGATCTCT), HPLC-purified.
3. Reverse primer, (SL_PCR_primer, AATGATACGGCGACCACCGACACTCTTTCCCTACATCAGTTTCTGTACTTTA, *see* **Note 2**), HPLC-purified.
4. 96-Well PCR machine, with 0.2 ml thin-wall strip tubes (RNase-free).
5. Expand High Fidelity^{Plus} PCR System (Roche #3300242001).
6. QIAquick® PCR Purification Kit, including PB, PE, and EB buffers (Qiagen #28106).
7. Microfuge and 1.5 ml centrifuge tubes.
8. Quant-iT™ dsDNA BR Assay Kit (Invitrogen #Q32853).
9. Quant-iT™ dsDNA HS Assay Kit (Invitrogen #Q32854.
10. Qubit® 2.0 Fluorometer (Applied Biosystems).
11. Bioanalyzer 2100 and DNA 1000 Kit (Agilent #5067-1504).

2.5 Illumina Sequencing

1. TruSeq SR Cluster Kit v3 (cBot—HS) (Illumina # GD-401-3001).
2. 0.1 N NaOH.
3. TruSeq SBS (Sequencing-by-Synthesis) kit (Illumina # FC-401-3002).
4. Custom sequencing primer (SL_SEQ_primer, CACTCTTTCCCTACATCAGTTTCTGTACTTTA).
5. TruSeq Dual Index Sequencing Primer Box SR (Illumina # FC-121-1003).

2.6 Data Analysis

1. FastQC v0.10.1: http://www.bioinformatics.babraham.ac.uk/projects/fastqc/
2. Bowtie v0.12.8: http://bowtie-bio.sourceforge.net/index.shtml
3. Samtools v0.1.18: http://samtools.sourceforge.net/
4. Bioconductor v2.1: http://www.bioconductor.org/
5. edgeR v3.0.7: http://www.bioconductor.org/packages/release/bioc/html/edgeR.html

3 Methods

3.1 RNA Preparation

1. Pellet ~5×10^8 cells (*see* **Note 3**) at 3,000 × *g* for 10 min at 4 °C in a 50 ml conical tube and discard supernatant.
2. Add 1 ml of TRIzol® Reagent to each tube, pipette up and down to disrupt the pellet, before letting it sit at room temperature for 5 min.
3. Add 200 μl chloroform and shake vigorously by hand for 15 s, before letting the sample sit at room temperature for another 5 min.
4. Centrifuge for 15 min, 12,000 × *g* at 4 °C, and transfer 400 μl of the aqueous phase to a new tube.
5. Add 500 μl of isopropanol and mix by inversion, before incubating at room temperature for 10 min.
6. Centrifuge for 10 min, 12,000 × *g* at 4 °C, and discard supernatant.
7. Wash pellet with 1 ml ice-cold 75 % EtOH (prepared with RNase-free water). Vortex briefly to disrupt the pellet.
8. Centrifuge 7,500 × *g* for 5 min at 4 °C and remove all supernatant.
9. Let pellet air-dry at room temperature for 10 min (longer if necessary).
10. Resuspend in 40 μl RNase-free water and determine concentration using a Qubit® Fluorometer.
11. The RNA should be checked for degradation and/or DNA contamination by running on the Agilent Bioanalyzer 2100.
12. Store the RNA at −80 °C, avoiding freeze/thawing.

3.2 First Strand cDNA Synthesis

1. Dilute 1 μg of RNA (*see* **Note 4**) to 4.8 μl with RNase-free water and add 3.2 μl 5 × SSIII RT buffer and 1.0 μl DNase I.
2. Incubate at 37 °C for 10 min, then heat-kill the enzyme by incubating at 75 °C for 10 min.

3. Dilute first strand primer to 10 μM in RNase-free water and add 2 μl of this solution to the DNase-treated RNA. Heat at 65 °C for 5 min and snap-cool to 4 °C.
4. Add 0.8 μl 5 × SSIII RT buffer, 1.0 μl 0.1 M DTT, 1.6 μl 25 mM dNTPs, 1.0 μl SuperScript® III Reverse Transcriptase, and 6.4 μl RNase-free water.
5. Incubate at 40 °C for 90 min and 70 °C for 15 min, before cooling to 4 °C.
6. Add 1 μl of RNaseH, mix, and incubate at 37 °C for 20 min and 70 °C for 10 min before cooling to 4 °C.
7. Transfer sample to a 1.7 ml tube, add 500 μl PB buffer and mix.
8. Transfer to a QIAquick® PCR Purification spin column and spin for 1 min at 9,300 × *g* (10,000 rpm).
9. Discard flow-through and add 750 μl PE buffer to the spin column.
10. Spin for 1 min at ~9,300 × *g*, discard flow-through and spin for a further 3 min at 21,000 × *g* (15,000 rpm), or maximum speed.
11. Transfer column to elution tube, add 50 μl EB buffer and incubate at room temperature for 1 min.
12. Centrifuge at 21,000 × *g* or maximum speed for 1 min and collect the flow-through.

3.3 Second Strand Synthesis

1. Dilute second strand primer to 10 μM in RNase-free water and add 10 μl to 25 μl of the first strand cDNA above.
2. Incubate at 95 °C for 2 min and snap-cool on ice.
3. At room temperature, mix 10 μl 10× NEBuffer 2, 5 μl 10 mM dNTPs, 47 μl RNase-free water, and 3 μl Klenow Fragment (3′ → 5′ exo-).
4. Add to the cDNA/primer solution and incubate at 37 °C for 30 min, before cooling to 4 °C (*see* **Note 5**).
5. Transfer sample to a 1.7 ml tube, add 500 μl PB buffer, and mix.
6. Transfer to a QIAquick® PCR Purification spin column and spin for 1 min at 9,300 × *g* (10,000 rpm in a microfuge).
7. Discard flow-through and add 750 μl PE buffer to the spin column.
8. Spin for 1 min at ~9,300 × *g*, discard flow-through, and spin for a further 3 min at 21,000 × *g* (15,000 rpm), or maximum speed.
9. Transfer column to elution tube, add 50 μl EB buffer, and incubate at room temperature for 1 min.
10. Centrifuge at 21,000 × *g* for 1 min and collect the flow-through.

3.4 PCR Amplification and DNA Purification

1. Prepare the PCR reaction at room temperature by mixing:
   (a) 10 μl purified second strand cDNA above
   (b) 34 μl RNase-free water
   (c) 20 μl 5× Expand High Fidelity^{Plus} Reaction Buffer (without $MgCl_2$)
   (d) 10 μl 25 mM $MgCl_2$
   (e) 5 μl 10 mM dNTP mix
   (f) 10 μl 10 μM PCR Forward Primer (R-prime-CT)
   (g) 10 μl 10 μM PCR Reverse Primer (SL_PCR_primer)
   (h) 1 μl Expand High Fidelity^{Plus} Enzyme Blend
2. PCR amplification is carried out by heating to 94 °C for 2 min; followed by two cycles of 94 °C for 10 s, 40 °C for 2 min, 72 °C for 1 min; 12–20 cycles (*see* **Note 6**) of 94 °C for 10 s, 60 °C for 30 s, 72 °C for 1 min; and finally 72 °C for 5 min, before cooling to 4 °C.
3. Transfer the sample to a 1.7 ml tube, mix with 500 μl PB buffer, and transfer to a QIAquick® PCR Purification spin column.
4. Spin for 1 min at 9,300 × *g* (10,000 rpm) and discard the flow-through.
5. Add 750 μl 35 % guanidine hydrochloride solution to denature remaining primer dimers.
6. Spin for 1 min at 9,300 × *g* (10,000 rpm), discard the flow-through, and add 750 μl PE buffer.
7. Spin for 1 min at ~9,300 × *g*, discard flow-through, and spin for a further 3 min at 21,000 × *g* (15,000 rpm).
8. Transfer column to elution tube, add 50 μl EB buffer, and incubate at room temperature for 1 min.
9. Centrifuge at 21,000 × *g* for 1 min and collect the flow-through.
10. Quantify sample concentration using a Qubit® Fluorometer.
11. Verify the expected size range (100–500 bp) by running an aliquot on a FlashGel™ or Agilent Bioanalyzer DNA 1000 chip.

3.5 Illumina Sequencing (See Note 7)

1. Prepare a Single-Read (SR) cBot reagent plate for use by thawing, vortexing, and piercing the foil over each tube in row 10.
2. Denature the cDNA library template by mixing 10 μl of 2 nM template with 10 μl of 0.1 N NaOH for 5 min at room temperature.

3. Dilute the library to 20pM by adding 980 μl of HT1 (Hybridization Buffer), then load 120 μl into a tube in the eight-tube strip.
4. Dilute the custom sequencing primer (SL_SEQ_primer) to 0.5 μM using HT1 and load 120 μl into the corresponding tube(s) of the eight-tube strip.
5. Perform a pre-run wash, then load reagent plate, flow cell, manifold, and tube strips.
6. Perform pre-run checks and start the run. Monitor bridge amplification, linearization, blocking of free 3′-OH ends to prevent nonspecific binding, and sequencing primer hybridization steps.
7. After the run (~4 h), unload run components and confirm reagent delivery.
8. Prepare TruSeq SBS and Sequencing Primer reagents by thawing and inverting tubes.
9. Input run parameters using HiSeq Control Software.
10. Load SBS and Indexing reagents (*see* **Note 8**). Confirm proper flow and prime SBS reagents. Load flow cell that was previously clustered on the cBot.
11. Start sequencing run.
12. When run is complete (2–10 days), unload and weigh reagents.

3.6 Quality Check and Read Filtering

1. Fastq files are uncompressed using appropriate tool (e.g., gunzip) and the average read quality calculated using FastQC tool kit. Reads with a quality score of less than 20 are removed from subsequent analysis.
2. The average quality for each cycle/base position is calculated and plotted graphically using a customized R script. If there are striking discrepancies in read qualities of adjacent bases, the possible cause is investigated by consultation with the Sequencing facility.
3. The average GC content for each cycle/base is calculated and compared to that of *Leishmania* (~60 %). A non-random GC content (TTG) is expected at the 5′ end of the reads due to presence of the last three nucleotides of the SL sequence. Non-random distribution at other position would indicate a heavily skewed or contaminated library. Presence of significant Ns at any position would indicate problems with read quality.
4. The thousand most frequent reads are identified and aligned to reference genome (*see* below). The frequency and position of these alignments are manually reviewed to determine if there are a large number of reads from small number of genes (e.g., rRNAs, tubulin), indicating bias in library preparation.

3.7 Alignment to Reference Genome

1. A fasta file containing the reference genome sequence (usually obtained from TritrypDB) is indexed with "bowite-build" command of the Bowtie suite.
2. Fastq files are aligned against the indexed reference genome using Bowtie with following parameters: -maqerr 70, -seedmms 2, -seedlen 20, -trim5 3 (to trim the TTG SL sequence from the 5′ end), -nomaqround, -S(for SAM formatted output), -M 1 (for random assignment of nonuniquely aligning reads).
3. SAM output is converted into a binary format (BAM), sorted, and indexed using Samtools.
4. Custom scripts (written in BioPerl) are used to map the location of each SL site (i.e., the 5′ end of the read) and to determine the number of reads mapped to each SL site.
5. A custom script is used to associate each SL site with the nearest gene (*see* **Note 9**) and SL site read count used to determine the major and minor SL sites for each gene.
6. The gene-level data generated above is exported to a tab-delimited file with one line for each gene, containing the number of SL sites, the number of reads at each of the 3 most abundant sites, and the total number of reads for each gene. Separate files are made for SL sites on the sense and antisense strands.

3.8 Differential Expression Analysis (*See* Note 10)

For experiments with at least two biological replicates.

1. Raw count data (not RPKM) are loaded into the Bioconductor edgeR package via the readDGE() function.
2. The DGEListobject$samples$lib.size <- colSums(DGEList object$counts) function is used to recalculate library sizes after filtering out genes with fewer than 3 counts in any libraries.
3. Normalization factors are calculated using the calcNormFactors() function (*see* **Note 11**).
4. A sample-wide common Biological Coefficient of Variation (BCV) is calculated using the estimateGLMCommonDisp() function (*see* **Note 12**).
5. The estimateGLMTagwiseDisp() function is used to calculate gene-wise dispersion from common BCV (*see* **Note 13**).
6. The glmFit() function is used to fit the negative binomial GLM for each gene and the glmLRT() used to perform a differential expression test using likelihood ratio test (*see* **Note 14**).
   For experiments with only single biological replicate:
7. Gene-level read counts for each library are rescaled by median normalization (*see* **Note 15**).

8. The $\log_2$ ratio of median-normalized read counts between the libraries is calculated for each gene to determine the fold change in mRNA expression level between samples. In case of time-series experiments, the starting sample (time-zero) is usually used as a common reference.

4 Notes

1. The choice of first and second strand cDNA primers will depend on the details of the cDNA library being made. The first strand primer described here contains a random hexamer sequence at its 3′ end, while the second strand primer contains nucleotides 22–33 (TCAGTTTCTGTA) of the 39-nt Splice Leader (SL) or mini-exon sequence present at the 5′ end of all *Leishmania* mRNAs. Since this sequence is conserved in most trypanosomatids, the primer could also be used in other species.
2. The libraries described here were initially designed to be sequenced on the Illumina Genome Analyzer IIx, but may also be sequenced on the HiSeq® 2000 using settings for a custom primer in Read 1. However, since they lack sequence matching the Index primer they are unsuitable for multiplexing. We are currently developing a protocol that will allow multiplexing of at least 4 samples per lane.
3. The protocol is designed for use with cultured promastigotes, but can also be used for axenic, macrophage- or lesion-derived amastigotes. It is also suitable for preparation of total RNA from infected amastigotes or lesion material without isolation of amastigotes. In the latter cases, the amount of RNA used for cDNA synthesis may need to be increased, depending on the ratio of parasite to host mRNA, as determined in **step 11**.
4. **Steps 1–2** may be omitted if there is no contamination with genomic DNA, as determined in **step 11**, above. Reagent volumes would need to be adjusted accordingly.
5. The protocol can be stopped at this point and the cDNA stored at 4 °C overnight or –20 °C for longer periods.
6. In order to minimize the number of amplification cycles used, it is advisable to perform the PCR twice. The first time, remove a 10 μl aliquot from each sample after 10, 12, 14, 16, and 18 cycles and examine on a FlashGel™ to determine the minimum number of cycles needed to obtain sufficient DNA (~50–100 ng/20 μl). The second time, amplification is only continued for this number of cycles before processing the sample.
7. The protocol described here is based on clonal cluster generation and sequencing primer hybridization using the Illumina cBot system and sequencing on the Illumina HiSeq® 2000 at

the High Throughput Genomic Center facility, Department of Genome Sciences, University of Washington, Seattle WA, USA. Different protocols may be needed for other sequencing platforms.

8. Single-indexing sequencing involves Read 1 sequencing using the custom sequencing primer, Index sequencing using the Index 1 (i7) sequencing primer, and Read 2 sequencing steps. While the SL RNA-seq libraries described here contain no indexing sequence, other samples on the Illumina Flow Cell may be indexed.
9. Since the 3′ UTRs for most genes are not yet defined, all SL sites between the 3′ end of the CDS and the nearest boundary of the next gene upstream are associated with each particular gene. The major SL site is defined as that with the most reads and minor SL sites are ranked in decreasing order of read abundance.
10. The choice of approach used to identify differentially expressed genes will be influenced heavily by the experimental design. Here we describe the use of edgeR Bioconductor package [21] for cases where there are two or more biological replicates in a least one of the sample groups. edgeR can analyze two or more sample groups (e.g., drug-treated and un-treated) for one (treatment effect) or multiple compounding factors (batch effects, lifecycle stage, etc.) at the same time. While it is not possible to perform rigorous statistical analyses on experiments that lack replicates, we also describe the use of median normalization to correct for differences in library size, allowing descriptive analysis of single-replicate preliminary experiments.
11. calcNormFactors normalizes for RNA composition by finding a set of scaling factors for the library sizes that minimize the log-fold changes between the samples for most genes. The default method for computing these scale factors uses a Trimmed Mean of *M*-values (TMM) between each pair of samples [22]. The product of the original library size and the scaling factor is called the "effective library size," which replaces the original library size in all downstream analyses.
12. A lower BCV indicates higher consistency within biological replicates.
13. This estimates the unknown variation that exists between genes within biological replicates.
14. The method used to calculate False Discovery Rate (FDR) can be changed using p.adjust() function. The toptags() function can be used to retrieve the results of edgeR analysis in a tabular form (containing GeneID, $\log_2$-Fold Change, *P*-Values, and FDR, among others) for use in subsequent analyses using different software.

15. The raw read counts for gene is divided by the median read count of all genes in each library and multiplied by 100. To minimize artifacts due to low read counts, genes with raw read counts below 10 % of the median value are assigned a normalized read count of 10.

References

1. Ivens AC, Peacock CS, Worthey EA et al (2005) The genome of the kinetoplastid parasite, *Leishmania major* Science 309:436–442
2. Berriman M, Ghedin E, Hertz-Fowler C et al (2005) The genome of the African trypanosome, *Trypanosoma brucei* Science 309:416–422
3. El-Sayed NM, Myler PJ, Bartholomeu DC et al (2005) The genome sequence of *Trypanosoma cruzi*, etiologic agent of Chagas disease. Science 309:409–415
4. El-Sayed NM, Myler PJ, Blandin G et al (2005) Comparative genomics of trypanosomatid parasitic protozoa. Science 309:404–409
5. Saxena A, Lahav T, Holland N et al (2007) Analysis of the *Leishmania donovani* transcriptome reveals an ordered progression of transient and permanent changes in gene expression during differentiation. Mol Biochem Parasitol 152:53–65
6. Lahav T, Sivam D, Volpin H et al (2011) Multiple levels of gene regulation mediate differentiation of the intracellular pathogen *Leishmania*. FASEB J 25:515–525
7. Duncan R (2004) DNA microarray analysis of protozoan parasite gene expression: outcomes correlate with mechanisms of regulation. Trends Parasitol 20:211–215
8. Duncan RC, Salotra P, Goyal N et al (2004) The application of gene expression microarray technology to kinetoplastid research. Curr Mol Med 4:611–621
9. Guimond C, Trudel N, Brochu C et al (2003) Modulation of gene expression in *Leishmania* drug resistant mutants as determined by targeted DNA microarrays. Nucleic Acids Res 31:5886–5896
10. McNicoll F, Drummelsmith J, Muller M et al (2006) A combined proteomic and transcriptomic approach to the study of stage differentiation in *Leishmania infantum*. Proteomics 6:3567–3581
11. Rochette A, Raymond F, Ubeda JM et al (2008) Genome-wide gene expression profiling analysis of *Leishmania major* and *Leishmania infantum* developmental stages reveals substantial differences between the two species. BMC Genomics 9:255
12. Almeida R, Gilmartin BJ, McCann SH et al (2004) Expression profiling of the *Leishmania* life cycle: cDNA arrays identify developmentally regulated genes present but not annotated in the genome. Mol Biochem Parasitol 136:87–100
13. Holzer TR, McMaster WR, Forney JD (2006) Expression profiling by whole-genome interspecies microarray hybridization reveals differential gene expression in procyclic promastigotes, lesion-derived amastigotes, and axenic amastigotes in *Leishmania mexicana Mol.* Biochem Parasitol 146:198–218
14. Leifso K, Cohen-Freue G, Dogra N et al (2007) Genomic and proteomic expression analysis of *Leishmania* promastigote and amastigote life stages: The *Leishmania* genome is constitutively expressed. Mol Biochem Parasitol 152:35–46
15. Cohen-Freue G, Holzer TR, Forney JD, McMaster WR (2007) Global gene expression in *Leishmania*. Int J Parasitol 37:1077–1086
16. Wang Z, Gerstein M, Snyder M (2009) RNA-Seq: a revolutionary tool for transcriptomics. Nat Rev Genet 10:57–63
17. Clayton CE (2002) Life without transcriptional control? From fly to man and back again. EMBO J 21:1881–1888
18. Nilsson D, Gunasekera K, Mani J et al (2010) Spliced leader trapping reveals widespread alternative splicing patterns in the highly dynamic transcriptome of *Trypanosoma brucei*. PLoS Pathog 6:e1001037
19. Mittra B, Cortez M, Haydock A et al (2013) Iron uptake controls the generation of *Leishmania* infective forms through regulation of ROS levels. J Exp Med 210:401–416
20. van Luenen H, Farris C, Jan S et al (2012) Glucosylated hydroxymethyluracil, DNA base J, prevents transcriptional readthrough in *Leishmania*. Cell 150:909–921
21. Robinson MD, McCarthy DJ, Smyth GK (2010) edgeR: a Bioconductor package for differential expression analysis of digital gene expression data. Bioinformatics 26:139–140
22. Robinson MD, Oshlack A (2010) A scaling normalization method for differential expression analysis of RNA-seq data. Genome Biol 11:R25

Chapter 13

Protein Microarrays for Parasite Antigen Discovery

Patrick Driguez, Denise L. Doolan, Douglas M. Molina, Alex Loukas, Angela Trieu, Phil L. Felgner, and Donald P. McManus

Abstract

The host serological profile to a parasitic infection, such as schistosomiasis, can be used to define potential vaccine and diagnostic targets. Determining the host antibody response using traditional approaches is hindered by the large number of putative antigens in any parasite proteome. Parasite protein microarrays offer the potential for a high-throughput host antibody screen to simplify this task. In order to construct the array, parasite proteins are selected from available genomic sequence and protein databases using bioinformatic tools. Selected open reading frames are PCR amplified, incorporated into a vector for cell-free protein expression, and printed robotically onto glass slides. The protein microarrays can be probed with antisera from infected/immune animals or humans and the antibody reactivity measured with fluorophore labeled antibodies on a confocal laser microarray scanner to identify potential targets for diagnosis or therapeutic or prophylactic intervention.

Key words Schistosomiasis, Protein microarray, Parasite, Antibody/serum screening, Vaccine and diagnostic discovery

1 Introduction

The serological profile of a parasitic disease, such as schistosomiasis, is the result of the interaction between the host's immune system and exposed parasite antigens. The recognition by and affinity of host antibodies for specific components of the parasite proteome indicates which antigens are accessible to the host immune response. When correlated with disease immunity or severity, these data can provide important information for vaccine and diagnostic target selection. Conventionally, antibody specificity and reactivity against native or recombinant antigens are measured using techniques such as ELISA or two-dimensional protein gels but these methods are difficult to adapt for high-throughput screens [1, 2]. Clearly, a protein microarray comprising hundreds to thousands of antigens which can be probed with antisera and individual antigen

Christopher Peacock (ed.), *Parasite Genomics Protocols*, Methods in Molecular Biology, vol. 1201,
DOI 10.1007/978-1-4939-1438-8_13, © Springer Science+Business Media New York 2015

reactivity measured with a laser scanner is a highly efficient approach for quantifying the host antibody response [3, 4].

As with other pathogens, high-throughput DNA sequencing and proteomics of schistosomes [5, 6] have provided rich sets of genomic, transcriptomic, and protein data. These data, coupled with DNA microarray technologies and analysis methods, have enabled the development of schistosome and other parasite protein microarrays. However, compared with nucleic acid microarrays, there are available a wide and diverse range of protein microarray systems each with its own inherent strengths and weaknesses [7]. The variables to be considered for construction of a protein microarray include the source of protein used (e.g., native extract, recombinant cellular or cell-free synthesis); whether the microarray is printed or is of in situ construction; the type of detection system; the microarray surface chemistry; whether an analytical, functional, or reverse phase microarray is produced; and whether the microarray is manufactured commercially or in the laboratory.

Currently, we suggest researchers consider a purpose built microarray fabricated within the laboratory or with the assistance of a collaborator. The protein microarray can be made using standard 96-well format laboratory equipment, a commercial cell-free protein expression kit, nitrocellulose-coated glass slides, and a microarray contact printer. We encourage researchers to seek specialist help when using a contact microarrayer although there are comprehensive reviews available [8]. Here we present, as an example, details of our development and application of the first schistosome protein microarray [2]. While the procedures described may require some modification dependent on, for example, the parasite species under consideration or the size of the protein microarray to be produced, the general methods we present are likely applicable for the construction of most parasite protein microarrays. The chapter describes the development of a parasite protein microarray comprising four steps: (1) Selection of target genes of interest for protein expression; (2) Production of the cDNA template; (3) Manufacture of the microarray; and (4) Scanning and probing of the microarray (Fig. 1).

2 Materials

2.1 Protein Selection, RNA Purification, and cDNA Amplification

1. Data-mining software: e.g., TMHMM (http://www.cbs.dtu.dk/services/TMHMM/), SignalP (http://www.cbs.dtu.dk/services/SignalP/), BLAST (http://blast.ncbi.nlm.nih.gov/Blast.cgi), interProScan (http://www.ebi.ac.uk/Tools/pfa/iprscan/).
2. Protein, genomic, and transcriptomic datasets; e.g., NCBI (http://www.ncbi.nlm.nih.gov/genbank/), *Schistosoma japonicum* Genome Project (http://www.chgc.sh.cn/japonicum/), and SchistoDB (www.schistodb.net).

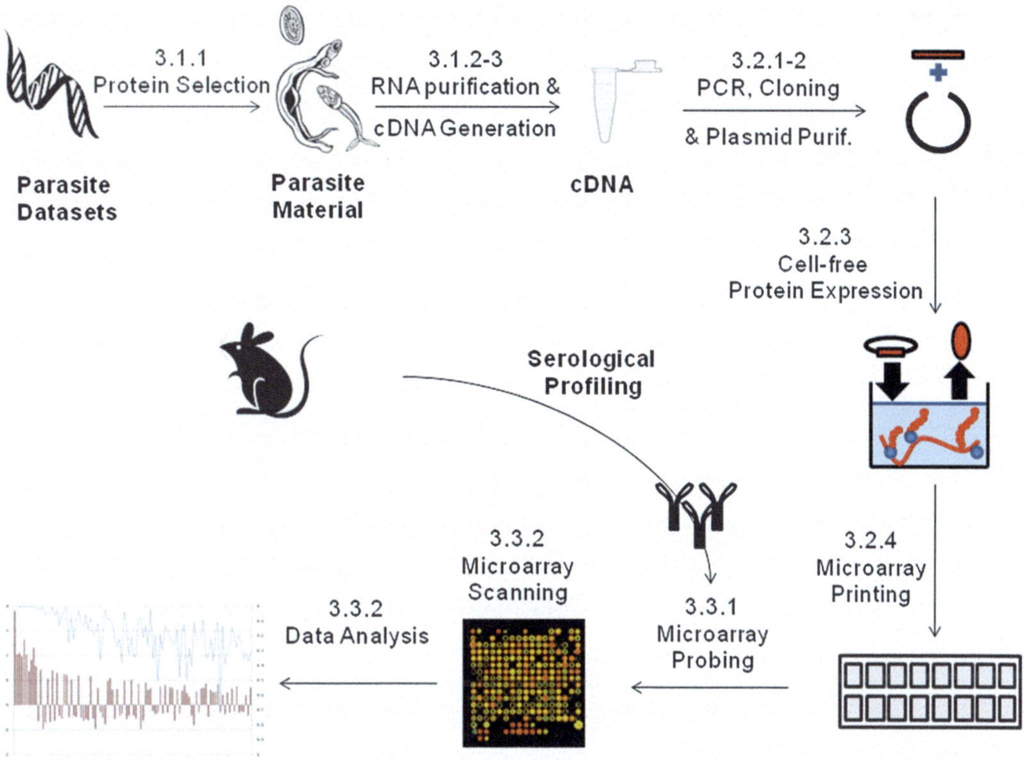

Fig. 1 Parasite protein microarray workflow

3. *Schistosoma japonicum* and *S. mansoni* lifecycle stages.
4. Sterile homogenizer tips and cordless motor (Kimble 749521-1590).
5. TRIZOL (Invitrogen 15596-026).
6. RNeasy Minikit (Qiagen 74104).
7. Chloroform.
8. Ethanol.
9. 3 M sodium acetate, pH 5.2.
10. 1 M Tris–HCl, pH 7.4.
11. RNase-free water.
12. RQ1 RNAse-free DNAse (Promega M6101) or equivalent.
13. Bioanalyzer RNA Pico Kit (Agilent 5067-1513).
14. Bioanalyzer 2100 (Agilent).
15. QuantiTect Whole Genome Amplification kit (Qiagen 207043) or equivalent.
16. Sensiscript Reverse Transcription kit (Qiagen 205211).

2.2 Polymerase Chain Reaction (PCR), Recombination Cloning, Plasmid Purification, Cell-Free Protein Expression, and Microarray Printing

1. PCR primers specific for target sequences with adaptor sequences for homologous recombination in 96-well plate format (from a commercial supplier).
2. 96-Well PCR plates.
3. Reagents for PCR.
4. 96-Well PCR thermocycler.
5. Reagents and equipment for agarose gel electrophoresis.
6. Multichannel pipettes.
7. Commercial or lab prepared chemically competent cells (e.g., DH5α or Top10).
8. Linearized vector for cell-free expression (e.g., pXT7, pXi, pIVEX, or similar) (*see* **Note 1**).
9. Adhesive plastic sheet for 96-well plates.
10. Super Optimal Catabolic (SOC) media (2 % tryptone, 0.55 % yeast extract, 10 mM NaCl, 10 mM KCl, 10 mM $MgCl_2$, 10 mM $MgSO_4$, and 20 mM glucose).
11. Luria-Bertani (LB) Media (1 % tryptone, 1 % NaCl, and 0.05 % yeast extract).
12. 96-Well flat bottom blocks for bacterial culture –2 mL (19579 Qiagen).
13. Airpore Tape Sheets (19571 Qiagen).
14. QIAprep 96 Turbo Miniprep kit with vacuum manifold or equivalent.
15. 96-Well round bottom plates.
16. 384-Well round bottom plates.
17. Cell-free protein expression kit (e.g., rapid translation system (RTS) 100 *E. coli* HY kits, 5 PRIME; or similar system such as Expressway, Invitrogen; EasyXpress, Qiagen).
18. Complete, mini, EDTA-free protease inhibitor cocktail tablets (Roche) or equivalent.
19. Tween 20 (P1379 Sigma-Aldrich).
20. Microarray spotting robot including printing pins (e.g., OmniGrid family of microarrayers, Digilab).
21. 3, 8, or 16 pad nitrocellulose coated slides (Whatman or GraceBio Labs).

2.3 Protein Microarray Probing, Scanning, and Data Analysis

1. Array Blocking Buffer (Whatman 10485356).
2. *E. coli* lysate (from a commercial supplier).
3. FAST frame (Whatman 10486001).
4. FAST slide incubation chambers (Whatman 10486046).

5. TBST (20 mM Tris–HCl pH 7.5, 0.5 M NaCl, 0.05 % Tween 20) and TBS (20 mM Tris–HCl pH 7.5, 0.5 M NaCl) washing buffers.
6. EBNA1 recombinant protein or antibodies reactive against test sera for use as primary antibody positive control (from a commercial supplier).
7. Mixed species IgG for use as secondary antibody positive control (from a commercial supplier).
8. Biotin-conjugated antibody reactive against antibodies or subtypes to be measured in test sera (from a commercial supplier).
9. Streptavidin-conjugated Cy5 fluorophore (Surelight P3, Columbia Biosciences).
10. Laser microarray scanner (e.g., Genepix 4300A, Molecular Devices, or ScanArray, Perkin Elmer).
11. Scanner imaging software (e.g., Genepix Pro, Molecular Devices, or ScanArray Express, Perkin Elmer).
12. Data analysis software (e.g., Excel, Microsoft; Bioconductor packages/R project for statistical computing, www.bioconductor.org/ www.r-project.org).

3 Methods

3.1 Protein Selection, RNA Purification, and cDNA Generation

It is a prerequisite that the researcher has access to genomic, transcriptomic, or proteomic data for the parasite species under consideration from which to bioinformatically select a subset of protein-coding genes for microarray printing. A cDNA PCR template is prepared by isolating total RNA from one or more parasite lifecycle stages, using standard techniques. The isolated total RNA is affinity purified, treated with DNAse and the quality is assessed using a Bioanalyzer. Finally, cDNA is produced using the whole transcriptome amplification and reverse transcriptase kits. While there are many advantages to using cDNA as the PCR template, it is also possible to use genomic DNA (*see* **Note 2**) or phage libraries (*see* **Note 3**) with some modification of the methods we describe here.

3.1.1 Protein Selection

The ideal protein microarray would have full proteome coverage; however, given the thousands of proteins expressed by most parasites, this is currently technologically and financially unfeasible. Therefore, genes must be selected using criteria suited to the particular aim of the project and the parasite species used. For vaccine discovery, as was the case with the schistosome protein microarray, genes were selected from available *Schistosoma japonicum* and *S. mansoni* gene and protein datasets on the basis of protein localization, lifecycle stage expression, and sequence homology within schistosome and host species (*see* **Note 4**). Well-established schistosome vaccine candidates and other characterized antigens were also included. Our final selection process was completed using standard bioinformatic tools and spreadsheet software; the details of this selection process, and other potential strategies, are beyond the scope of this chapter. The reader should refer to recent reviews on vaccine bioinformatics [9] for further information.

3.1.2 RNA Purification

For specific details regarding parasite RNA isolation and purification, readers should refer to the excellent chapter in a previous volume of this series by Hoffmann and Fitzpatrick [10]; note that a protocol that is best suited to individual need and the characteristics of the specific parasite should be selected. In brief, for schistosomes, we use a method combining guanidinium thiocyanate-phenol-chloroform extraction (TRIZOL reagent) and affinity column purification (Qiagen RNeasy Minikit). The purified total RNA is DNAse-treated (Promega) and then checked for quality by Bioanalyzer (Agilent). As many selected genes may not be universally expressed in male or female parasites or across all development stages, we recommend extracting and combining total RNA from several parasite lifecycle stages of mixed sex to provide complete cDNA coverage for PCR amplification (*see* **Note 5**). The basic procedure is as follows:

1. Mechanically homogenize pooled freshly collected parasites using a hand-held motor (Kimble) and disposable tips in TRIZOL as recommended by the manufacturer. It is important to ensure that there are sufficient parasites to provide a suitable yield of total RNA for cDNA amplification (at least 10 ng of total RNA).
2. Complete remaining steps for TRIZOL total RNA isolation as per Hoffmann and Fitzpatrick [10] but with some modifications if necessary (*see* **Note 6**).
3. Remove the aqueous phase containing total RNA and gently mix in a separate tube with an equal volume of 70 % ethanol.
4. Transfer to a Qiagen affinity column. Centrifuge at room temperature for 15 s. Repeat for several aliquots if volume is larger than column capacity.
5. Complete remaining steps of total RNA purification as described in Hoffmann and Fitzpatrick [10] and in the Qiagen kit instructions.
6. Resuspend total RNA pellet in RNAse-free water following ethanol precipitation.
7. Check total RNA concentration spectroscopically and store aliquots at −80 °C if not needed immediately.
8. Treat total RNA samples with DNAse to remove all contaminating DNA as directed in the RQ1 RNAse-free DNAse protocol (Promega).

9. Mix 1–8 μL of total RNA sample with 1 μL RQ1 10× reaction buffer, RQ1 RNAse-free DNAse at 1 U/μg RNA, and RNAse-free water to a final volume of 10 μL. If sample contains less than 1 μg total RNA include only 1 unit of RQ1 DNAse.
10. Incubate at 37 °C for 30 min.
11. Add 1 μL of RQ1 DNAse stop solution and incubate at 65 °C for 10 min.
12. Assess the quality of the isolated and purified total RNA using a Bioanalyzer RNA Pico kit and 2100 Bioanalyzer (*see* **Note 7**). This method only requires 1 μL of RNA sample.
13. Follow the detailed instructions for the RNA Pico kit carefully for consistent results, taking particular care to avoid bubbles when loading the Pico chip.
14. Ensure that all total RNA samples have high RNA integrity number (RIN) values as determined by the Bioanalyzer software (*see* **Note 8**).

3.1.3 cDNA Generation

If parasite total RNA is limited or precious (e.g., in our case, total RNA samples from schistosome miracidia, eggs, or schistosomula lifecycle stages) a cDNA amplification step can be useful. Whole transcriptome amplification kits (QuantiTect WTA) use a combination of random and oligo-DT primers to amplify up to 40 μg of cDNA from as little as 10 ng of RNA (*see* **Note 9**). For higher yield or less precious total RNA samples we used conventional reverse transcription kits (Sensiscript RT, Qiagen) with oligo-DT primers to generate cDNA (protocol not described here). Finally, equal concentrations of each cDNA source were mixed for the final PCR template. The summarized protocol for whole transcriptome amplification is as follows (*see* the QuantiTect manual for further details):

1. Prepare fresh RT mix as directed and add 5 μL to ≥10 ng total RNA in 5 μL nuclease-free water/TE buffer. Vortex and centrifuge.
2. Incubate at 37 °C for 30 min, stop reaction at 95 °C for 5 min, and cool to 22 °C.
3. Prepare fresh ligation mix as directed and add 10 μL to RT reaction. Vortex and centrifuge.
4. Incubate at 22 °C for 2 h.
5. Prepare fresh amplification mix as directed and add 30 μL to the ligation reaction. Vortex and centrifuge.
6. Incubate at 30 °C for 8 h (high yield reaction) and stop reaction at 95° for 5 min.
7. Quantify cDNA, diluting if necessary, and store at −20 °C.

3.2 PCR, Recombination Cloning, Plasmid Purification, Cell-Free Protein Expression, and Microarray Printing

In contrast to the PCR amplification and plasmid purification methods that are standard, variations are possible for the cloning (*see* **Note 10**) and protein expression steps. While the DNA template for cell-free protein expression can be generated using other methods, cloning based on homologous recombination is efficient and suitable for high-throughput workflow [2, 3]. In addition, as the highly efficient *E. coli* based cell-free protein expression system can potentially cause the loss of post-translational modifications and disulfide bonds important for epitope formation, disulfide kits and cell-free systems using wheat germ cells, rabbit reticulocytes, and human cells are available for use. Readers are encouraged to examine the suitability of these other methods for construction of their particular parasite protein microarray.

3.2.1 PCR Amplification

1. Design PCR primers for each of the selected parasite genes including 20 base pairs complementary to the expression vector of choice to allow for homologous recombination cloning.
2. Order primers from a commercial supplier in 96-well format.
3. Prepare the PCR template by diluting amplified cDNA to 50 ng/μL and mixing equal volumes from each source.
4. Amplify using a standard PCR protocol in a 25 μL volume (*see* **Note 11**).
5. Check 3 μL of the PCR reaction by electrophoresis in a 1 % (w/v) agarose gel.

3.2.2 Recombination Cloning and Plasmid Purification

1. Combine 1 μL PCR product with 4 μL of linear vector in a new 96-well plate chilled on ice (*see* **Note 12**).
2. Add 10 μL of thawed DH5α cells to each well with care to avoid contamination between wells.
3. Cover with adhesive plastic sheet and store on ice for 30 min.
4. Heat shock in a 42 °C water bath for 1 min.
5. Chill on ice for 2 min.
6. Dispense 200 μL SOC media to each well, cover plate with adhesive plastic sheet, and incubate for 1 h at 37 °C.
7. Add 1.1 mL of LB media (with 50 μg/mL kanomyocin) into flat-bottom well blocks and transfer transformation mixture.
8. Cover with Airpore Tape Strips and incubate overnight at 37 °C with 600 rpm shaking.
9. Check for turbid media in cells—high turbidity indicates a successful transformation while slight turbidity suggests background from an empty vector. If necessary, repeat cloning steps for missing inserts with the option of picking single colonies.

10. Make glycerol stocks by mixing 80 μL of cell culture with an equal volume of 50 % (v/v) glycerol. Store at −80 °C in 96-well round bottom plates (*see* **Note 13**).
11. Pellet remaining cells by centrifuging at 3,000 rpm for 8 min.
12. Discard supernatant and proceed with QIAprep 96 turbo Minikit protocol (including optional purification step) using a vacuum manifold.
13. Elute plasmid in 100 μL of EB buffer (supplied in kit).
14. Run agarose gel electrophoresis to check size of plasmids and inserts compared with empty vector. Quality control (QC)-PCR can be used to check for the presence of the insert (*see* **Note 14**).
15. It is recommended that all or a subset of the purified plasmids or previously stored glycerol stocks are sequenced prior to protein expression.

3.2.3 Cell-Free Protein Expression

1. On the basis of the PCR and cloning gels, the QC-PCR and sequencing results select plasmids for cell-free protein expression and printing.
2. Prepare the RTS reaction mix as directed.
3. Transfer 10 μL (>0.5 μg DNA) of miniprep DNA into 96-well round bottom plates, add 40 μL of the reaction mix, cover with an adhesive plastic sheet, and briefly spin the plate (*see* **Note 15**).
4. Incubate for 5 h at 30 °C and shake at 300 rpm.
5. Stop the reaction with the addition of 16 μL 4× stop solution (0.2 % v/v Tween 20 and 5 complete, mini, EDTA-free protease inhibitor cocktail tablets (Roche) per 10 mL) (*see* **Note 16**).
6. Cover plate and centrifuge at 3,000 rpm for 3 min and store on ice ready for printing.

3.2.4 Microarray Printing

Load the cell-free protein solution onto printing plates including controls and recombinant proteins. We recommend including dilutions of recombinant cellularly expressed parasite proteins with known antigenicity that are also printed as cell-free extracts (comparison control), secondary antibody positive controls (mixed species IgG), host-specific positive control (e.g., Epstein-Barr virus nuclear antigen 1 (EBNA1) or anti-host antibodies), parasite extract antigen (e.g., schistosome soluble worm antigen preparation (SWAP)), no plasmid DNA negative control (protein expression mix only), and buffer only negative control (*see* **Note 17**). Setup the microarrayer as directed by the manufacturer and load the gal file defining feature location. Print protein microarrays and allow drying. Store at room temperature in a desiccator cabinet.

3.3 Protein Microarray Probing, Scanning, and Data Analysis

3.3.1 Protein Microarray Probing

Probe the protein microarrays with sera or with antibodies directed against protein expression tags. However, when probing with sera pre-absorption with *E. coli* lysate is required to reduce background signal from antibodies reactive with bacterial antigens present in the RTS protein extract. A quality control probe to check the print quality is also recommended. Select slides from the start, the middle, and the end of the print run for probing with antibodies directed against the N- and C-terminal expression tags (in pXi/pXT7 the N- and C-terminal tags are 10× His and HA, respectively). All incubation steps use a platform rocker.

1. Fit the incubation chambers over the slide and mount on the slide frame.
2. Hydrate each microarray chamber with blocking buffer (BB) and leave at room temperature for 30–60 min with gentle rocking.
3. Dilute sera in 1:100 with BB and 10 % (w/v) *E. coli* lysate and incubate at room temperature for 30–60 min with gentle rocking.
4. For each microarray chamber, aspirate BB. Add pre-absorbed sera or QC antibodies diluted (1:500 typically) in BB alone. Take care not to let microarray pads dry out.
5. Incubate overnight in a humidified box at 4 °C with gentle rocking.
6. Aspirate and wash three times with TBST.
7. Add diluted (typically 1:1,000 in BB) biotin-conjugated secondary antibody. Incubate at room temperature for 1 h with rocking.
8. Aspirate and wash five times with TBST.
9. Add diluted streptavidin-conjugated Cy5 fluorophore (typically 1:200 in BB), and incubate for 1 h at room temperature. Then aspirate and wash three times with TBST and three times with TBS.
10. Remove slides from incubation chambers and frame. Wash with purified water.
11. Centrifuge slide for 5 min at 500 × *g*. Store slide in the dark until scanned.

3.3.2 Scanning and Data Analysis

The image acquisition and basic data analysis techniques presented here, although still under development, are applicable to most projects; however, the reader is advised to consult the scanner and imaging software manuals as well as specialist references for further details of image acquisition and data analysis. Scan the probed slides using a confocal laser microarray scanner (e.g., Genepix 4300A). Adjust the laser and photomultiplier tube (PMT) values to maximize signal while not over-saturating features. Quantify the signals with image analysis software (e.g., Genepix Pro 7) and calculate the final feature intensity by subtracting the local background from the signal. Use a standard spreadsheet package (Excel) or more specialized software (R project, Bioconductor) for further data analysis. Transform the signal intensities and normalize between the microarrays using the No DNA negative controls [11]. Positive features are defined as having a signal greater than the average of the No DNA negative controls plus 2–3 standard deviations. Confirm the probing protocol was successful and sample integrity was maintained by checking the printed control features within each protein microarray. Similarly, comparison controls should have comparable reactivity to the cell-free protein equivalent features. Ensure that over 90 % of features have full-length protein expression in the quality control probed microarrays across the entire print run. Full-length proteins will have positive signal intensities for both the N- and C-terminal expression tags for each feature. When completing serological profile studies, positive antigens can be statistically compared between infection-resistant and -susceptible host groups for selection of vaccine targets. Diagnostic antigens are selected by comparing different cohorts of infected and uninfected sera samples.

4 Notes

1. Theoretically, any T7 vector capable of in vitro expression may be used. For the schistosome protein microarray the proprietary pXi vector, similar to the pXT7 vector [3], was used. Linear vector is generated by digesting the multiple cloning site in the circular vector with specific restriction enzymes and PCR amplifying further linear vector.
2. When using genomic DNA instead of cDNA, PCR primers must be designed to include coding regions only and it may be necessary to express and print several polypeptide fragments for a protein with introns. Similarly, due to the efficiency limitations of PCR coding sequences longer than 3,000 base pairs must also be split into fragments. Open reading frames can be found using online ORF/gene finder or prediction programs.
3. In our hands, λ phage libraries were not as successful as cDNA when PCR amplifying large (>1 kb) sequences
4. To allow for some losses during the PCR amplification, cloning, and protein expression stages, select >20 % more genes than required for protein microarray printing [2].
5. This may depend on the genes selected and the transcriptional differences within the parasite. In our case, numerous genes

were only expressed in male or female schistosome worms or a particular lifecycle stage.

6. Our group routinely extracts total RNA from schistosome parasites; however we have made several minor modifications in the Hoffmann and Fitzpatrick [10] protocol regarding homogenization, incubation times, and purification. Therefore, for optimal results, each laboratory is encouraged to empirically determine the optimal method.
7. RNA quality of samples can also be assessed using electrophoresis on a denaturing agarose gel.
8. The Bioanalyzer literature notes that a particular RIN is no guarantee of experimental success but it is a good indication of the quality of the RNA samples. For the schistosome protein microarray, we used samples with RIN values above 4–5 before attempting cDNA synthesis.
9. Due to random priming, the QuantiTect WTA kit cannot guarantee full-length sequences. However, in practise our template, consisting of cDNA prepared from multiple lifecycle stages and male and female worms using the WTA and conventional RT kits, was successful in amplifying most sequences.
10. It is possible to use other methods adaptable for high-throughput cloning including restriction site cloning, ligation independent cloning, or sequence and ligation independent cloning [12]. The 5 Prime RTS manual also suggests using overlap extension PCR to generate linear DNA template.
11. Using 1 μL of cDNA template, amplify in a typical cycling protocol: denature for 2 min at 94 °C; followed by 30 cycles of 94 °C for 30 s; 55 °C for 15 s and 68 °C for 1 min/kb; and a final extension for 10 min at 68 °C.
12. Cloning efficiency may be improved by varying PCR product volumes between 0.5 and 1 μL.
13. It is useful to take glycerol stocks at this stage for repeat printing or sequencing checks.
14. A quality control PCR is also recommended. Perform a standard PCR reaction using previously designed primers and purified plasmids and check for presence of the correct insert on an agarose gel.
15. The volume required depends on the size of the microarray to be printed and the printing protocol used.
16. The final concentration of Tween 20 will determine the viscosity of the printing solution. Viscosity has a large impact on the quality of the final protein microarray and can cause spot bleeding or incomplete spot printing [8]. Users are encouraged to empirically determine the best concentration,

dependant on the printing pins used, the humidity of the microarrayer, and the print surface.

17. Prior to the final print, it is recommended that a range of dilutions of recombinant proteins and parasite antigen extracts is printed and probed to determine their optimal concentration.

Acknowledgements

We want to thank all the staff at Antigen Discovery Incorporated and the Protein Microarray Laboratory, University of California, Irvine. We also wish to thank Hamish McWilliam for his illustrations of schistosome lifecycle stages. PLF's research was supported by National Institute of Allergy and Infectious Disease Grants U54AI065359 and U01AI078213. DLD, AL, and DPM receive support from the National Health and Medical Research Council of Australia.

References

1. Vigil A, Davies DH, Felgner PL (2010) Defining the humoral immune response to infectious agents using high-density protein microarrays. Future Microbiol 5:241–251
2. Driguez P, Doolan DL, Loukas A, Felgner PL, McManus DP (2010) Schistosomiasis vaccine discovery using immunomics. Parasit Vectors 3:4
3. Davies DH, Liang X, Hernandez JE, Randall A, Hirst S, Mu Y, Romero KM, Nguyen TT, Kalantari-Dehaghi M, Crotty S, Baldi P, Villarreal LP, Felgner PL (2005) Profiling the humoral immune response to infection by using proteome microarrays: high-throughput vaccine and diagnostic antigen discovery. Proc Natl Acad Sci U S A 102: 547–552
4. Davies DH, Wyatt LS, Newman FK, Earl PL, Chun S, Hernandez JE, Molina DM, Hirst S, Moss B, Frey SE, Felgner PL (2008) Antibody profiling by proteome microarray reveals the immunogenicity of the attenuated smallpox vaccine modified vaccinia virus ankara is comparable to that of Dryvax. J Virol 82:652–663
5. Liu F, Lu J, Hu W, Wang SY, Cui SJ, Chi M, Yan Q, Wang XR, Song HD, Xu XN, Wang JJ, Zhang XL, Zhang X, Wang ZQ, Xue CL, Brindley PJ, McManus DP, Yang PY, Feng Z, Chen Z, Han ZG (2006) New perspectives on host-parasite interplay by comparative transcriptomic and proteomic analyses of Schistosoma japonicum. PLoS Pathog 2:e29
6. Liu F, Zhou Y, Wang ZQ, Lu G, Zheng H, Brindley PJ, McManus DP, Blair D, Zhang QH, Zhong Y, Wang S, Han ZG, Chen Z (2009) The Schistosoma japonicum genome reveals features of host-parasite interplay. Nature 460:345–351
7. Chandra H, Reddy PJ, Srivastava S (2011) Protein microarrays and novel detection platforms. Expert Rev Proteomics 8:61–79
8. Austin J, Holway AH (2011) Contact printing of protein microarrays. Methods Mol Biol 785:379–394
9. He Y, Rappuoli R, De Groot AS, Chen RT (2010) Emerging vaccine informatics. J Biomed Biotechnol 2010:218590
10. Hoffmann KF, Fitzpatrick JM (2004) Gene expression studies using self-fabricated parasite cDNA microarrays. Methods Mol Biol 270: 219–236
11. Trieu A, Kayala MA, Burk C, Molina DM, Freilich DA, Richie TL, Baldi P, Felgner PL, Doolan DL (2011) Sterile protective immunity to malaria is associated with a panel of novel P. falciparum antigens. Mol Cell Proteomics 10(M111):007948
12. Li MZ, Elledge SJ (2007) Harnessing homologous recombination in vitro to generate recombinant DNA via SLIC. Nat Methods 4:251–256

Chapter 14

A Transposon-Based Tool for Transformation and Mutagenesis in Trypanosomatid Protozoa

Jeziel D. Damasceno, Stephen M. Beverley, and Luiz R.O. Tosi

Abstract

The ability of transposable elements to mobilize across genomes and affect the expression of genes makes them exceptional tools for genetic manipulation methodologies. Several transposon-based systems have been modified and incorporated into shuttle mutagenesis approaches in a variety of organisms. We have found that the Mos1 element, a DNA transposon from *Drosophila mauritiana*, is suitable and readily adaptable to a variety of strategies to the study of trypanosomatid parasitic protozoa. Trypanosomatids are the causative agents of a wide range of neglected diseases in underdeveloped regions of the globe. In this chapter we describe the basic elements and the available protocols for the in vitro use of Mos1 derivatives in the protozoan parasite *Leishmania*.

Key words Mariner, *Leishmania*, Transposon, Mutagenesis, In vitro transposition

1 Introduction

Trypanosomatid parasitic protozoa of the *Leishmania* and *Trypanosoma* genera are the causative agents of leishmaniasis, Chagas disease, and African trypanosomiasis. These major neglected diseases of humans and domestic animals cause high rates of morbidity in underdeveloped areas of the globe [1, 2]. These parasites efficiently circumvented the defense strategies mounted by their hosts and chemotherapy and vaccination strategies are either inadequate or unavailable [3]. The interaction between trypanosomatids and their diverse environments is deeply influenced by the parasite peculiar control of gene expression, which is not fully understood. Canonical promoter sequences have not been described and functionally unrelated genes are transcribed as large polycistrons and independently regulated mainly at the post-translational level [4–7]. A better comprehension of these parasites unique gene expression apparatus will not only shed light on the evolutionary history of trypanosomatids but also contribute to the rational design of more effective therapeutic strategies. The

Christopher Peacock (ed.), *Parasite Genomics Protocols*, Methods in Molecular Biology, vol. 1201,
DOI 10.1007/978-1-4939-1438-8_14, © Springer Science+Business Media New York 2015

collection of tools available for the manipulation of trypanosomatids has a bearing on the kind of biological question that can be addressed. In fact, many advances in our understanding of the basic biology of these protozoa have come from studies of the sequence, organization and expression of these organisms' genomes using a relatively limited repertoire of genetic tools [8–11].

In this chapter, we focus on the use of transposon-based mutagenesis in the trypanosomatid *Leishmania*. Transposable elements (TEs) are mobile DNA sequences with the ability to relocate and entail changes in genome structure and the expression [12]. Such remarkable feature makes TEs exceptional tools for genetic manipulation methodologies. The structure and transposition mechanisms of these mobile DNA sequences are highly diverse and determine their classification [13]. TEs are normally divided into Class I, or retrotransposons and Class II, which are the DNA transposons. Current transposon technology is mostly based on Class II DNA TEs due to the relative simplicity of transposon structure and mode of transposition. Nonetheless, the design of TE mutagenesis protocols heavily depends on the transposon system used, as well as on the intended outcome.

The most common application of TEs as a genetic tool is found in the determination of gene function where transposon mutagenesis mediating gain or loss of function can be easily explored either in gene-to-gene strategies or in genome-wide approaches. TE-mediated insertional mutagenesis and/or protein tagging can be adapted and explored in the study of trypanosomatids. Among the main classes of TEs found in eukaryotes, only retroelements have been described in trypanosomatids [14], making them amenable to transposon-based mutagenesis approaches. Heterologous TE systems can be imported into parasites and mobilization can be carried out in vivo [15, 16]. In this approach, the expression of the transposase activity within the parasite and the introduction of modified transposons tend to be less manageable due to difficulties in attaining high transfection efficiencies and in controlling the level of transposase expression. An efficient alternative for the constraints of in vivo transposition in trypanosomatids is the use of shuttle mutagenesis strategies, in which the transposon is mobilized prior to the introduction into the parasite [17]. The majority of shuttle mutagenesis protocols make use of non-autonomous TEs in which the transposase gene has been replaced with reporter genes or selectable markers and the transposase activity is supplemented in *trans* [18, 19]. Shuttle mutagenesis can be carried out in *Escherichia coli* or in in vitro reactions. The in vitro mobilization reaction not only avoids some of the shortcomings that are inherent of in vivo transposition systems, but also constitutes a manageable and practical strategy to introduce genetic alterations into protozoa parasites.

Several TEs have been manipulated and tailored to the point that they are easily incorporated into shuttle mutagenesis strategies

[18, 20, 21]. One of such example is the *mariner/Tc1* superfamily of transposons, which is widely distributed in nature and includes various Class II TEs. The in vivo mobilization of mariner-based TEs has been described in a wide range of organisms [15, 22–25]. The heterologous mobilization of the mariner element Mos1 from *Drosophila mauritiana* within the *Leishmania* genome set the bases for the use of transposon shuttle mutagenesis strategies in trypanosomatids [15]. We have found that this element is suitable and readily adaptable to be employed in a wide range of in vitro approaches to investigate gene function in *Leishmania*.

A major characteristic of the element Mos1 from *D. mauritiana*, which is a defining member of the *mariner/Tc1* transposon family, is its minimal *cis* requirements for transposition [26]. This feature makes Mos1 an especially useful element in transposon-based approaches. The terminal inverted repeats (TIRs) that define the boundaries of mariner transposons contain the binding sites for the transposase and are essential for mobilization [27, 28]. However, the 28 bps TIRs of Mos1 alone does not suffice for optimal transposition of modified versions of the element in vitro. The retention of a few base pairs internal to the TIRs is necessary for proper trans-mobilization by the active transposase [19]. Tolerance for cargo DNA length varies among different TEs and greatly affects their functionality. Different from other TEs that can carry longer sequences [29], *mariner* elements can be rendered unmovable by the increase in cargo length [18, 30].

The Mos1 mobilization in *Leishmania* emphasizes its usefulness as a tool for probing gene function in this parasitic protozoan. However, the estimate frequency of in vivo transposition of Mos1 within the *Leishmania* genome is as low as 10^{-6} for a single allele inactivation. Besides the difficulties in modulating the levels of expression of the heterologous transposase within the parasite, the diploid nature of the *Leishmania* genome plays an important part in the intrinsic limitations of an in vivo transposition approach. Considering that the efficiency of Mos1 in vitro transposition can be as high as 10^{-3}/target DNA molecule [19], the in vitro mobilization of Mos1-derived elements constitutes a fine alternative for the in vivo strategy. Higher mobilization efficiency and the possibility to control the transposition reaction not only facilitate the construction of insertion libraries into a variety of targets but also expand the applicability of these exceptional tools.

The *mariner* in vitro transposition reaction developed for use in trypanosomatids includes the recombinant Mos1 transposase and a variety of modified elements cloned into a donor plasmid. The modified *mariner* elements available for use in *Leishmania* promote the inactivation of the target gene upon insertion and can also be expressed in bacteria. Some of them mediate the expression of translational or transcriptional fusions and, therefore, are adequate for subcellular localization studies or gene trapping strategies. As illustrated in Fig. 1 and revised elsewhere [21],

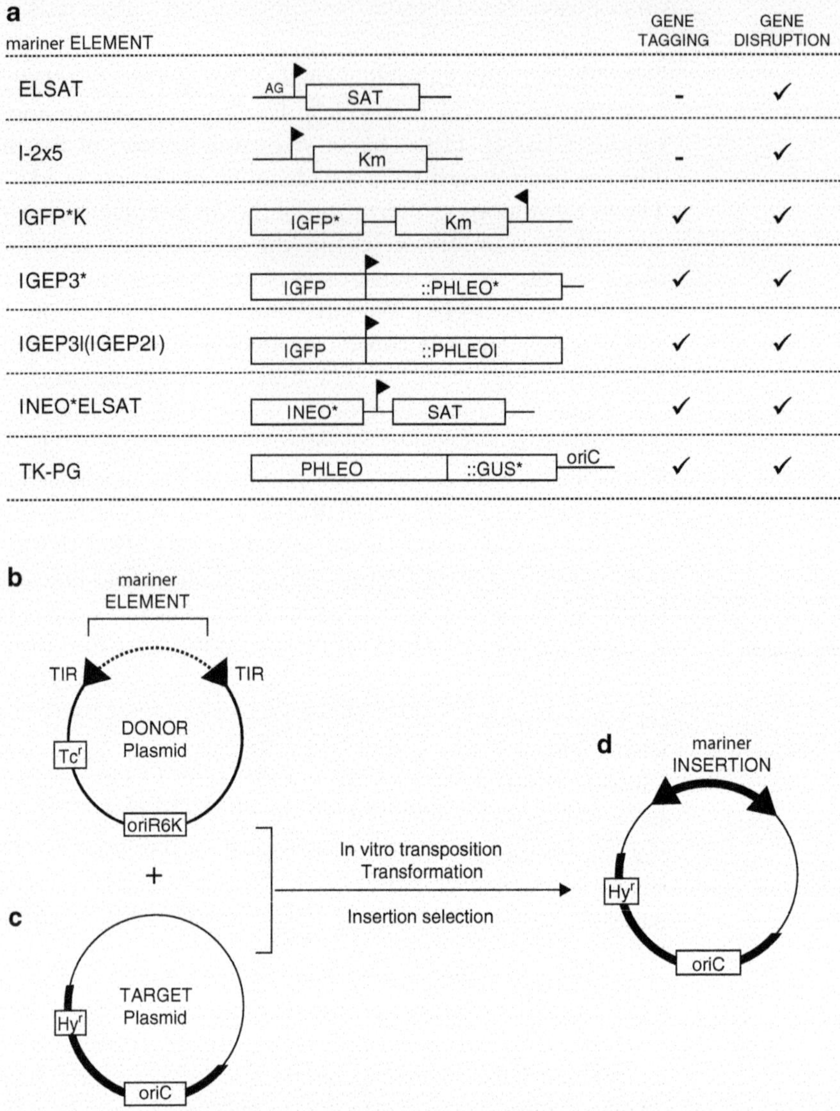

Fig. 1 The in vitro *mariner* transposition assay. The major components of the transposition reaction include a modified version of the mariner element (**a**), which is carried in the donor plasmid (**b**), the target plasmid (**c**) and the recombinant Mos1 transposase. Following the reaction, the transposition reaction products are transformed into bacteria and selection is carried out with the appropriate drugs according to the selection markers in target plasmid and in the modified Mos1 element (**d**). (**a**) The Mos1 modified transposons available for use in Gene Tagging and/or Gene disruption protocols in *Leishmania* [21]. These elements contain variable elements according to the application intended; all modified elements bear a drug resistance marker for selection of integration events. The selection markers include SAT (Nourseothricin resistance marker), Km (Kanamycin resistance marker), PHLEO (Phleomycin/zeocin resistance marker) and NEO (G418 resistance marker). The generation of transcriptional or translational fusions is mediated by tag protein genes such as GFP (Green Fluorescent Protein), GUS (β-glucuronidase) or NEO (Neomycin Phosphotransferase II); *black arrowheads*, *E. coli* promoters; "/," indicates that the gene lacks a start or stop codon; "*" the gene contains a stop codon; AG, trans splice acceptor site. (**b**) The transposon donor plasmid contains an R6K replication origin (*ori*R6K) and will not propagate in *pir*– bacteria strains used to select the transposition production. The *arrowheads* represent the Terminal Inverted Repeats (TIR) and the internal sequence at 5′ and 3′-ends that contain the required *cis-elements* for transposition and define the minimal Mos1 element; the *dashed line* represents the different markers and/or reporter genes that may compose the modified elements

some TEs bear eukaryotic selectable markers such as Neomycin Phosphotransferase (NPT), Streptothricin Acetyl Transferase (SAT), and the Bleomycin binding protein (PHLEO) and their use is limited to gene disruption protocols. Other elements mediate the selection of protein fusions in the parasite. Most of these elements use the Green Fluorescent Protein (GFP) as reporter of translational fusions. Some of these elements lack the reporter stop codon and maintain an open reading frame across the element sequence and allow the recovery of products that preserve both amino and carboxy termini of the target protein [18]. Other trapping reporters, such as β-glucuronidase or NPT are also available [20]. In this chapter we describe the available protocol for in vitro use of modified Mos1 elements in *Leishmania* studies.

2 Materials

2.1 The Mos1 Transposase Expression and Purification

1. Ca^{2+}-competent *E. coli* strain BL21 (DE3)/pLysS.
2. The pET3a vector bearing the *D. mauritiana* Mos1 transposase (pET3a-TPase construct; [19]).
3. LB medium, 1 % tryptone, 0.5 % yeast extract, 1 % NaCl.
4. 1 M isopropyl-β-D-thio-galactopyranoside (IPTG).
5. Cell Resuspension Buffer, 20 mM Tris–HCl (pH 7.6), 2 mM $MgCl_2$, 25 % sucrose, 0.6 mM phenylmethylsulfonyl fluoride (PMSF), 1 mM benzamidine (BZA), 1 mM dithiothreitol (DTT).
6. Liquid nitrogen.
7. Cell Lysis Buffer, 20 mM Tris–HCl (pH 7.6), 4 mM EDTA, 200 mM NaCl, 1 % deoxycholate, 1 % nonylphenoxy polyethoxy ethanol (NP-40), 0.6 mM PMSF, 1 mM BZA, 1 mM DTT.
8. DNAse I.
9. 1 M $MgCl_2$.
10. Lysozyme.
11. Inclusion Bodies Wash Buffer, 100 mM Tris–HCl (7.6), 4 M deionized urea.
12. Column Buffer, 20 mM Tris–HCl (pH 7.6), 50 mM NaCl, 4 M guanidine-HCl, 1 mM PMSF, 1 mM BZA, 5 mM DTT.
13. DEAE-Sephadex equilibrated in Column Buffer.
14. SDS-PAGE apparatus.
15. 10 kDa cut-off Slide-A-Lyzer dialysis cassette (Pierce).
16. Dialysis Buffer A, 10 % glycerol, 25 mM Tris–HCl (pH 7.6), 50 mM NaCl, 5 mM $MgCl_2$, 2 mM DTT.
17. Dialysis Buffer B, 10 % glycerol, 25 mM Tris–HCl (pH 7.6), 50 mM NaCl, 5 mM $MgCl_2$, 0.5 mM DTT.

2.2 The In Vitro Transposition Assay and Selection

1. 10× Transposition Reaction Buffer, 250 mM HEPES (pH 7.9), 1 M NaCl, 20 mM DTT, 50 mM $MgCl_2$.
2. 100 % glycerol.
3. 10 mg/ml purified acetylated BSA.
4. Reaction Stop Buffer, 50 mM Tris–HCl (pH 7.6), 0.5 mg/ml proteinase K, 10 mM EDTA, 250 mg/ml yeast tRNA.
5. 25:24:1 Phenol–chloroform–isoamyl alcohol.
6. 3 M sodium acetate.
7. 100 % ethanol (EtOH).
8. 70 % EtOH.
9. 10 mM Tris–HCl pH 7.5.
10. *Pir– E. coli* electrocompetent cells (DH10B).
11. Electroporator.
12. Liquid and semisolid LB medium.
13. Selection drugs at appropriate concentration, Ampicillin (100 mg/ml), hygromycin (30 mg/ml), nourseothricin (50 mg/ml), and Zeocin (100 mg/ml). *See* **Note 1**.

3 Methods

3.1 The Expression of the Mos1 Transposase

1. Transform the pET3a-TPase construct into Ca^{2+}-competent *E. coli* strain BL21 (DE3)/pLysS, plate transformed cells onto semisolid LB medium containing ampicillin (100 μg/ml) and incubate overnight at 37 °C. *See* **Note 2**.
2. Pick one colony and inoculate into LB medium containing 100 μg/ml ampicillin and incubate overnight at 37 °C under vigorous shaking.
3. Make a 1:100 dilution of the saturated culture into 100 ml of fresh LB medium containing 100 μg/ml ampicillin. Further incubate at 37 °C under vigorous shaking to an OD_{600} of 0.7–0.8.
4. Add IPTG to a final concentration of 0.5 mM in order to induce expression of the transposase. Incubate at 37 °C with vigorous shaking for 1 h. *See* **Note 3**.
5. Harvest cells by centrifugation at 1,300 × *g* for 10 min at 4 °C.
6. Resuspend cell pellet in 1/100 of induced cell culture volume in Cell Resuspension Buffer.
7. Quick freeze in liquid nitrogen. *See* **Note 4.**

3.2 The Purification of the Mos1 Transposase

1. Thaw the sample at room temperature.
2. Add lysozyme to a final concentration of 1 mg/ml and incubate for 5 min at room temperature with gentle agitation.
3. Add 1 ml of Cell Lysis Buffer and incubate for 15 min at room temperature with gentle agitation.
4. Add 60 μg of DNAseI and $MgCl_2$ to a final concentration of 10 mM. Pipette up and down until the sample is no longer viscous and Incubate for 20 min at room temperature.
5. Pellet inclusion bodies by centrifuging for 15 min at 14,000 × *g* at 4 °C. Discard the supernatant.
6. Resuspend pellet in 1 ml of ice cold Wash Buffer by vortexing (or pipetting up and down). Centrifuge for 15 min at 14,000 × *g* at 4 °C and discard the supernatant.
7. Repeat **step 5** two more times.
8. Resuspend the final inclusion bodies-containing pellet in 0.5 ml of ice cold Column Buffer. Vortex vigorously to completely dissolve the pellet.
9. Centrifuge 3 min at 14,000 × *g* at 4 °C. Take the supernatant, save a 50 μl aliquot and proceed to the next step.
10. Apply the sample from the previous step onto a 10 ml DEAE-Sephadex column previously equilibrated with ice-cold Column Buffer. Carry on this step at cold-room temperature.
11. Elute the DEAE-Sephadex column using the Column Buffer and collect up to ten 0.5 ml fractions. Carry on this step at cold-room temperature.
12. Analyze 20 μl of each eluted fraction by SDS-PAGE. Also include an aliquot with equivalent volume of the pre induced control (**step 4**; Subheading 3.1) and input material before loaded into DEAE-Sephadex column (**step 8**; this section). *See* **Note 5**.
13. Pool together the transposase-containing fractions up to 2 ml and dilute the sample to 12 ml using ice cold Column Buffer. *See* **Note 6**.
14. Transfer the sample to the dialysis slide. Perform dialysis against 1 l of Dialysis Buffer A for 6 h at 4 °C.
15. Discard Dialysis Buffer A and perform a second round of dialysis at 4 °C overnight, using 1 l of Dialysis Buffer B. *See* **Note 7**.
16. Discard the dialysis buffer and remove insoluble material from dialyzed sample by centrifugation at 10,000 × *g* for 10 min at 4 °C.
17. Take the supernatant, add glycerol to a final concentration of 50 % and estimate protein concentration by BCA method.
18. Store 100 μl aliquots at −80 °C. *See* **Note 8**.

3.3 The In Vitro Transposition Reaction

1. Prepare a typical transposition reaction, which is carried out in 20 μl in 0.6 ml microfuge tubes and contains 2 μl of 10× Transposition Reaction Buffer, 2 μl of 100 % glycerol, 0.5 μl of purified acetylated BSA, 30 fmol of donor plasmid, and 10 fmol of target plasmid. *See* **Note 9**.
2. Add 100 ng of recombinant Mos1 transposase. *See* **Note 10**.
3. Incubate reaction for 1 h at 25 °C.
4. Add 80 μl of Reaction Stop Buffer and incubate for 30 min at 30 °C.
5. Add 100 μl of 25:24:1 phenol–chloroform–isoamyl alcohol and vortex vigorously.
6. Centrifuge at 14,000 × *g* for 15 min and transfer 90 μl of the aqueous phase (upper layer) into a 1 ml microfuge tube.
7. Add 10 μl of 3 M sodium acetate, 250 μl of 100 % EtOH and incubate at −80 °C for 1 h.
8. Precipitate DNA by centrifugation at 14,000 × *g* for 30 min at 4 °C. Discard the supernatant.
9. Wash precipitated DNA with 1 ml of 70 % EtOH and centrifuge at 14,000 × *g* for 15 min at 4 °C. Discard the supernatant and remove residual liquid with a pipette.
10. Resuspend the pellet in 10 μl of 10 mM Tris–HCl pH 7.5.
11. Transform 2 μl of the purified transposition reaction into *pir– E. coli* DH10B electrocompetent cells. *See* **Note 11**.
12. Plate transformed cells onto semisolid LB medium containing the appropriated selective drugs and incubate overnight at 37 °C. *See* **Note 12**.

4 Notes

1. The target DNA and modified elements used in the transposition reaction will determine the appropriate selection drug and/or the adequate combination of drugs, as exemplified in Fig. 1.
2. In our hands, the use of electrocompetent *E. coli* BL21 strain did not allow the selection of the pET3a-TPase transformant.
3. Save an aliquot of cell suspension before addition of IPTG. Process this aliquot according to the described protocol scaling down buffers and reagents. The final protein lysate will be used as the pre-Induction control.
4. This step can be omitted if the samples are to be processed immediately. In this case proceed to **step 2** in the next section. When performing large scale induction, divide induced resuspended cells into 1 ml aliquots, freeze in liquid nitrogen and

stored at −80 °C. Take one aliquot at a time to perform the protein purification.

5. Unbound transposase will normally be eluted between fractions 3 and 7 from a 10 ml column. However, it is advisable to analyze all eluted fraction using SDS-PAGE. The exclusion of the ion exchange chromatography step yielded inactive transposase after refolding suggesting that the column purification eliminates an inhibitory factor.
6. Omission of the dilution step resulted in precipitation of inactive protein during the following dialysis step.
7. This is a limiting step in refolding active transposase. The recovering of active enzyme is very sensitive to the refolding conditions used. Eventually, It may be necessary empirically determine the optimal conditions for this step.
8. Transposase activity is sensitive to freeze and thaw. Therefore, it is important to aliquot the sample before freezing. Use one aliquot at a time for transposition reaction.
9. Transposition efficiency may vary depending on the amount and purity of target and donor DNA. The maximum activity can be reached using 150 ng of donor plasmid. Efficiency of transposition can also be improved by using DNA preparations containing a high proportion of supercoiled DNA.
10. Due to variations in the refolding process, the transposition efficiency may vary among different transposase preparations. It is advisable to test each preparation prior to conducting transposition reactions. Excessive transposase (above 100 nM) does not increase transposition efficiencies.
11. Incubate transformed cells for 1 h at 37 °C with vigorous agitation. Plate 10 μl of a 1:100 dilution onto medium containing the appropriate antibiotic for selection of the target DNA to determine the transformation efficiency. Plate the undiluted suspension onto medium containing the antibiotics required to double-select the expression of resistance markers found on both target and donor plasmids. Determine transposition efficiency by dividing the number of colonies grown in double-selection medium by the transformation efficiency. Control transposition efficiencies should range from 10^{-4} to 10^{-3}.
12. Depending on the planned application, insertions events can be further characterized and used for functional studies within the parasite. These include the transfection of either tagged genes for subcellular localization of its product, or interrupted versions of genes for knockout generation. This can be done using individual insertion events or a library of transposition product in mass transfection into the parasite.

References

1. Alvar J, Velez ID, Bern C, Herrero M, Desjeux P, Cano J, Jannin J, den Boer M (2012) Leishmaniasis worldwide and global estimates of its incidence. PLoS One 7:e35671
2. Blum JA, Neumayr AL, Hatz CF (2012) Human African trypanosomiasis in endemic populations and travellers. Eur J Clin Microbiol Infect Dis 31:905–913
3. Croft SL, Sundar S, Fairlamb AH (2006) Drug resistance in leishmaniasis. Clin Microbiol Rev 19:111–126
4. Martinez-Calvillo S, Yan S, Nguyen D, Fox M, Stuart K, Myler PJ (2003) Transcription of Leishmania major Friedlin chromosome 1 initiates in both directions within a single region. Mol Cell 11:1291–1299
5. Johnson PJ, Kooter JM, Borst P (1987) Inactivation of transcription by UV irradiation of T. brucei provides evidence for a multicistronic transcription unit including a VSG gene. Cell 51:273–281
6. Mottram JC, Murphy WJ, Agabian N (1989) A transcriptional analysis of the Trypanosoma brucei hsp83 gene cluster. Mol Biochem Parasitol 37:115–127
7. Holzer TR, McMaster WR, Forney JD (2006) Expression profiling by whole-genome interspecies microarray hybridization reveals differential gene expression in procyclic promastigotes, lesion-derived amastigotes, and axenic amastigotes in Leishmania mexicana. Mol Biochem Parasitol 146:198–218
8. Cruz A, Beverley SM (1990) Gene replacement in parasitic protozoa. Nature 348: 171–173
9. Kapler GM, Coburn CM, Beverley SM (1990) Stable transfection of the human parasite Leishmania major delineates a 30-kilobase region sufficient for extrachromosomal replication and expression. Mol Cell Biol 10:1084–1094
10. ten Asbroek AL, Ouellette M, Borst P (1990) Targeted insertion of the neomycin phosphotransferase gene into the tubulin gene cluster of Trypanosoma brucei. Nature 348: 174–175
11. Kolev NG, Tschudi C, Ullu E (2011) RNA interference in protozoan parasites: achievements and challenges. Eukaryot Cell 10: 1156–1163
12. Burns KH, Boeke JD (2012) Human transposon tectonics. Cell 149:740–752
13. Ivics Z, Li MA, Mates L, Boeke JD, Nagy A, Bradley A, Izsvak Z (2009) Transposon-mediated genome manipulation in vertebrates. Nat Methods 6:415–422
14. Bringaud F, Ghedin E, El-Sayed NM, Papadopoulou B (2008) Role of transposable elements in trypanosomatids. Microbes Infect 10:575–581
15. Gueiros-Filho FJ, Beverley SM (1997) Trans-kingdom transposition of the Drosophila element mariner within the protozoan Leishmania. Science 276:1716–1719
16. Leal S, Acosta-Serrano A, Morita YS, Englund PT, Bohme U, Cross GA (2001) Virulence of Trypanosoma brucei strain 427 is not affected by the absence of glycosylphosphatidylinositol phospholipase C. Mol Biochem Parasitol 114:245–247
17. Garraway LA, Tosi LR, Wang Y, Moore JB, Dobson DE, Beverley SM (1997) Insertional mutagenesis by a modified in vitro Ty1 transposition system. Gene 198:27–35
18. Goyard S, Tosi LR, Gouzova J, Majors J, Beverley SM (2001) New Mos1 mariner transposons suitable for the recovery of gene fusions in vivo and in vitro. Gene 280:97–105
19. Tosi LR, Beverley SM (2000) Cis and trans factors affecting Mos1 mariner evolution and transposition in vitro, and its potential for functional genomics. Nucleic Acids Res 28:784–790
20. Augusto MJ, Squina FM, Marchini JF, Dias FC, Tosi LR (2004) Specificity of modified Drosophila mariner transposons in the identification of Leishmania genes. Exp Parasitol 108:109–113
21. Damasceno JD, Beverley SM, Tosi LR (2010) A transposon toolkit for gene transfer and mutagenesis in protozoan parasites. Genetica 138:301–311
22. Coates CJ, Jasinskiene N, Morgan D, Tosi LR, Beverley SM, James AA (2000) Purified mariner (Mos1) transposase catalyzes the integration of marked elements into the germ-line of the yellow fever mosquito, Aedes aegypti. Insect Biochem Mol Biol 30:1003–1008
23. Lampe DJ, Akerley BJ, Rubin EJ, Mekalanos JJ, Robertson HM (1999) Hyperactive transposase mutants of the Himar1 mariner transposon. Proc Natl Acad Sci U S A 96: 11428–11433
24. Sherman A, Dawson A, Mather C, Gilhooley H, Li Y, Mitchell R, Finnegan D, Sang H (1998) Transposition of the Drosophila element mariner into the chicken germ line. Nat Biotechnol 16:1050–1053

25. Zhang JK, Pritchett MA, Lampe DJ, Robertson HM, Metcalf WW (2000) In vivo transposon mutagenesis of the methanogenic archaeon Methanosarcina acetivorans C2A using a modified version of the insect mariner-family transposable element Himar1. Proc Natl Acad Sci U S A 97:9665–9670
26. Miskey C, Papp B, Mates L, Sinzelle L, Keller H, Izsvak Z, Ivics Z (2007) The ancient mariner sails again: transposition of the human Hsmar1 element by a reconstructed transposase and activities of the SETMAR protein on transposon ends. Mol Cell Biol 27: 4589–4600
27. Auge-Gouillou C, Hamelin MH, Demattei MV, Periquet M, Bigot Y (2001) The wild-type conformation of the Mos-1 inverted terminal repeats is suboptimal for transposition in bacteria. Mol Genet Genomics 265:51–57
28. Auge-Gouillou C, Hamelin MH, Demattei MV, Periquet G, Bigot Y (2001) The ITR binding domain of the Mariner Mos-1 transposase. Mol Genet Genomics 265:58–65
29. Thibault ST, Singer MA, Miyazaki WY, Milash B, Dompe NA, Singh CM, Buchholz R, Demsky M, Fawcett R, Francis-Lang HL, Ryner L, Cheung LM, Chong A, Erickson C, Fisher WW, Greer K, Hartouni SR, Howie E, Jakkula L, Joo D, Killpack K, Laufer A, Mazzotta J, Smith RD, Stevens LM, Stuber C, Tan LR, Ventura R, Woo A, Zakrajsek I, Zhao L, Chen F, Swimmer C, Kopczynski C, Duyk G, Winberg ML, Margolis J (2004) A complementary transposon tool kit for Drosophila melanogaster using P and piggyBac. Nat Genet 36:283–287
30. Balciunas D, Wangensteen KJ, Wilber A, Bell J, Geurts A, Sivasubbu S, Wang X, Hackett PB, Largaespada DA, McIvor RS, Ekker SC (2006) Harnessing a high cargo-capacity transposon for genetic applications in vertebrates. PLoS Genet 2:e169

Chapter 15

Separation of Basic Proteins from *Leishmania* Using a Combination of Free Flow Electrophoresis (FFE) and 2D Electrophoresis (2-DE) Under Basic Conditions

Marie-Christine Brotherton, Gina Racine, and Marc Ouellette

Abstract

Basic proteins, an important class of proteins in intracellular organisms such as *Leishmania*, are usually underrepresented on 2D gels. This chapter describes a method combining basic proteins fractionation using Free flow electrophoresis in isoelectric focusing mode (IEF-FFE) followed by protein separation using two-dimensional gel electrophoresis (2-DE) in basic conditions. The combination of these two techniques represents a great improvement for the visualization of *Leishmania* proteins with basic pI using 2D gels.

Key words Basic proteins, *Leishmania*, Two-dimensional gel electrophoresis (2-DE), Free flow electrophoresis (FFE), DeStreak

1 Introduction

Intracellular organisms, such as *Leishmania*, are predicted to have a more basic proteome than free-living cells [1]. For instance, 57.8 % of *L. infantum* proteins are predicted to harbor a pI higher than 7.0 and 23.0 % are predicted to have a pI greater than 9.0, which is considered as highly basic proteins (www.tritrypdb.org v. 4.2). However, these proteins tend to be poorly represented on classical 2D gels compared to the acidic ones [2].

The first step of our protocol, IEF-FFE, consists in a liquid-based isoelectric focusing technique where the sample is injected continuously into a thin film of carrier ampholytes establishing a pH gradient in the separation chamber (reviewed in ref. 3) (*see* Fig. 1). By applying an electric field perpendicular to the buffer flow direction, the proteins are separated according to their respective pI and collected at the end of the separation chamber into a 96-well plate (*see* Fig. 1). This fractionation step allows the enrichment of basic proteins and ensures a better representation of the less abundant ones [4].

Christopher Peacock (ed.), *Parasite Genomics Protocols*, Methods in Molecular Biology, vol. 1201,
DOI 10.1007/978-1-4939-1438-8_15,

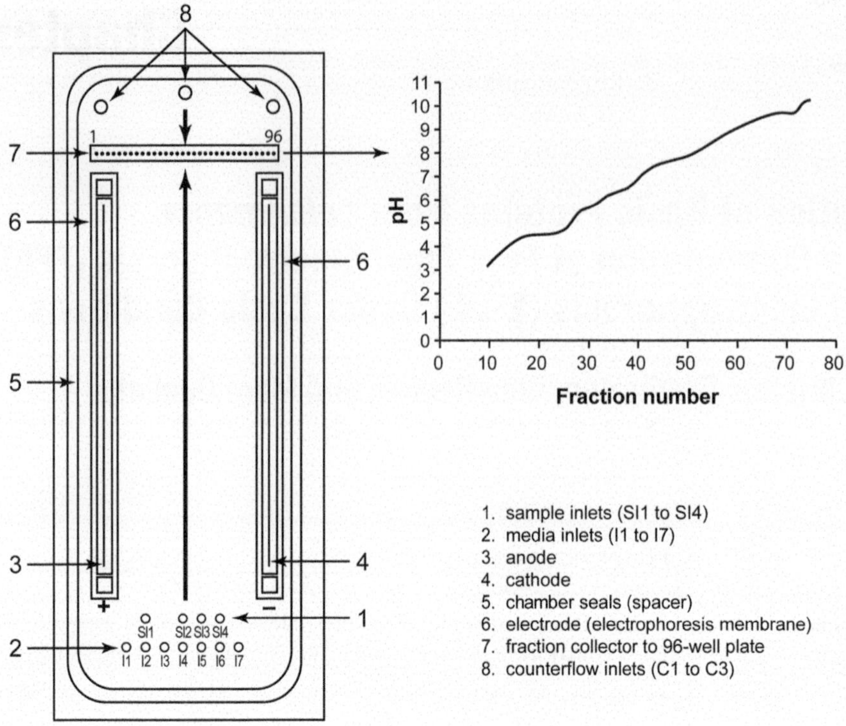

Fig. 1 Schematic representation of IEF-FFE separation chamber (modified from ref. 10). For details *see* Subheading 1

The second step of the protocol consists in a modified 2-DE to allow a better separation of basic proteins. The 2-DE consists of a protein separation according to their pI in an immobilized pH gradient in the first dimension followed by a separation according to their molecular weight in the second dimension [5]. In basic conditions, the water and the reducing agent dithiothreitol (DTT) tend to be transported towards the anode during the isoelectric focusing step [6, 7]. The loss of DTT leads to the oxidation of the protein thiol groups, intra- and inter-chain disulfide bonds formation and, consequently, protein aggregation [6, 8]. Finally, the presence of extra spots and spot trains are observed on basic 2D gels and are caused by the variation of the number of oxidized thiol group in the proteins [9]. To circumvent these problems, hydroxyethyl disulfide (DeStreak) instead of DTT is used in the rehydration step of the isoelectric focusing [4, 9].

2 Materials

2.1 *Leishmania* Culture

1. MAA/20: to prepare 1 L, dissolve 5.95 g of Hepes, 0.35 g of $NaHCO_3$, 100 mg of L-glutamine, 2.5 g of dextrose and 5.0 g of tryptic soy broth in deionized H_2O, adjust the volume to 800 mL and sterilized by autoclaving. Store at −20 °C in the

dark. Before use, complete the medium with 200 mL of inactivated FBS 20 %, 10 mL of L-glutamine 200 mM, 10 mL of penicillin–streptomycin 10 mg/mL, and 3 mL of Hemin solution 5 mg/mL. Adjust the pH using HCl at 7.0 for promastigotes and at 5.8 for amastigotes. Store the supplemented medium at 4 °C (*see* **Note 1**).

2.2 Proteins Extraction

1. Hepes–NaCl buffer: to prepare 500 mL, dissolve 2.5 g of Hepes, 4 g of NaCl, 0.1875 g of KCl, 0.05 g of Na_2HPO_4, and 0.54 g of dextrose in deionized H_2O, adjust the pH at 7.05 using NaOH 10 N and then adjust the volume to 500 mL. Sterilize by autoclaving and store at room temperature.
2. 2D lysis buffer: to prepare 25 mL, dissolve 10.51 g of urea, 3.81 g of thiourea, 0.75 g of CHAPS, 0.075 g of DTT, 0.036 g of TCEP, 125 μL of IPG pH 4–7, and 62.5 μL of IPG pH 3–10 in deionized H_2O and adjust the volume to 25 mL. Aliquot in 1 mL and store at −20 °C (*see* **Note 2**).
3. Tris-Base 50 mM.
4. Protease inhibitors cocktail (Sigma): dissolve all the bottle in 10 mL of deionized H_2O, aliquot in 1 mL and store at −20 °C.
5. 2-D Quant kit (GE Healthcare).

2.3 FFE Tests Prior to Protein Separation

2.3.1 Free Flow Apparatus

The FFE device commercialized by FFE Weber GmbH (Kirchheim, Germany) was used to perform the described experimental procedures.

2.3.2 FFE High HGP Buffers (See **Note 3**)

1. Sulfanilic acid azochromotrop (SPADNS): to prepare 100 mL, dilute 1 mL of SPADNS in 99 mL of distilled H_2O. Store at room temperature.
2. pI mix (BD IEF-FFE).
3. Stock solution HPMC/glycerol: mix 210.0 g of glycerol, 210.0 g of ProMetHEUS hydroxypropylmethylcellulose (HPMC) 0.8 %, and 210.0 g of distilled H_2O. Store at 4 °C (*see* **Note 4**).
4. Anodic stabilization medium high HGP: mix 7.0 g of H_2SO_4 1 M, 10.5 g of distilled H_2O, and 52.5 g of stock solution HPMC/glycerol. Store at 4 °C.
5. Separation medium 1 high HGP: mix 7.5 g of distilled H_2O, 52.5 g of stock solution HPMC/glycerol, and 10.0 g of Prolyte 3–10 FFE reagent 1. Store at 4 °C (*see* **Note 5**).
6. Separation medium 2 high HGP: mix 10.5 g of distilled H_2O, 157.5 g of stock solution HPMC/glycerol, and 42.0 g of Prolyte 3–10 FFE reagent 2. Store at 4 °C (*see* **Note 5**).

7. Separation medium 3 high HGP: mix 7.5 g of distilled H_2O, 52.5 g of stock solution HPMC/glycerol, and 10.0 g of Prolyte 3–10 FFE reagent 3. Store at 4 °C (*see* **Note 5**).
8. Cathodic stabilization medium high HGP: mix 7.0 g of NaOH 1 M, 10.5 g of distilled H_2O, and 52.5 g of stock solution HPMC/glycerol. Store at 4 °C.
9. Counterflow medium high HGP: mix 82.5 g of distilled H_2O and 247.5 g of stock solution HPMC/glycerol. Store at 4 °C.
10. Anode: mix 360.0 g of distilled H_2O and 40.0 g of H_2SO_4 1 M. Store at room temperature.
11. Cathode: mix 360.0 g of distilled H_2O and 40.0 g of NaOH 1 M. Store at room temperature.

2.4 Protein Fractionation Using IEF-FFE (See Note 6)

1. Anodic stabilization medium: mix 7.5 g of H_2SO_4 1 M, 42.5 g of distilled H_2O, 31.5 g of urea, 11.5 g of thiourea, and 3.4 g of mannitol. Store at 4 °C.
2. Separation medium 1: mix 34.85 g of distilled H_2O, 31.5 g of urea, 11.5 g of thiourea, 3.4 g of mannitol, and 15.15 g of Prolyte 3–10 FFE reagent 1. Store at 4 °C (*see* **Note 5**).
3. Separation medium 2: mix 50.0 g of distilled H_2O, 63.0 g of urea, 23.0 g of thiourea, 6.8 g of mannitol, 1.543 g of CHAPS, and 50.0 g of Prolyte 3–10 FFE reagent 2. Store at 4 °C (*see* **Note 5**).
4. Separation medium 3: mix 21.0 g of distilled H_2O, 21.0 g of urea, 7.7 g of thiourea, 2.3 g of mannitol, and 12.3 g of Prolyte 3–10 FFE reagent 3. Store at 4 °C (*see* **Note 5**).
5. Cathodic stabilization medium: mix 15.0 g of NaOH 1 M, 85.0 g of distilled H_2O, 63.0 g of urea, 23.0 g of thiourea, and 6.8 g of mannitol. Store at 4 °C.
6. Counterflow medium: mix 250.0 g of distilled H_2O, 157.5 g of urea, 57.5 g of thiourea, and 16.88 g of mannitol. Store at 4 °C.
7. Anode: mix 360.0 g of distilled H_2O and 40.0 g of H_2SO_4 1 M. Store at room temperature.
8. Cathode: mix 360.0 g of distilled H_2O and 40.0 g of NaOH 1 M. Store at room temperature.
9. Amicon Ultra-15 columns (Millipore).

2.5 Protein Separation Using 2D Gels in Alkaline Conditions

2.5.1 2D Gels Apparatus

1. Ettan IPGphor II isoelectric focusing system (GE Healthcare).
2. Ettan DALTtwelve system separation unit (GE Healthcare).
3. ProXPRESS 2D Proteomic Imaging System (PerkinElmer Life Sciences).
4. Progenesis SameSpots software (Nonlinear Dynamics).
5. ProXcision robot (PerkinElmer Life Sciences).

2.5.2 2D Gels Buffers and Materials

1. 2-D Quant kit (GE Healthcare).
2. 18-cm Immobiline DryStrips pH 6–9 or 6–11.
3. DeStreak Rehydration Solution (GE Healthcare).
4. Mineral oil (DryStrip cover fluid).
5. Equilibration solution: to prepare 200 mL, mix 6.67 mL of Tris–HCl 1.5 M pH 8.8, 72.07 g of urea, 69.0 mL of glycerol 87 % v/v, 4.0 g of SDS, trace of bromophenol blue and adjust the volume to 200 mL using deionized H_2O. Aliquot in 40 mL and store at −20 °C.
6. Reduction solution: dissolve 1.0 g of DTT in 100 mL of equilibration solution prior to use.
7. Alkylation solution: dissolve 2.5 g of iodoacetamide in 100 mL of equilibration solution prior to use.
8. Low molecular weight calibration kit for SDS electrophoresis (GE Healthcare): 1 vial of 576 μg diluted in 1 mL de Laemmli buffer and heat for 5 min at 100 °C. Store at −20 °C.
9. SDS electrophoresis buffer 2×: to prepare 3 L, dissolve 18.2 g of Tris-Base, 86.5 g of glycine and 6.0 g of SDS in deionized H_2O and adjust the volume to 3 L. Store at room temperature.
10. SDS electrophoresis buffer 1×: to prepare 20 L, dissolve 60.5 g of Tris-Base, 288 g of glycine, and 20.0 g of SDS in deionized H_2O and adjust the volume to 20 L. Store at room temperature.
11. Agarose solution: dissolve 125.0 mg of agarose and trace of bromophenol blue in 25 mL of SDS electrophoresis buffer 1×. Heat in a microwave oven until agarose is dissolved. Store at room temperature.
12. Acrylamide–Bisacrylamide solution (30:0.8 %): to prepare 1 L, dissolve 300 g of acrylamide and 8 g of bis-acrylamide in deionized H_2O and adjust the volume to 1 L. Store in the dark at 4 °C.
13. Tris–HCl 1.5 M pH 8.8: to prepare 1 L, dissolve 181.7 g of Tris-Base in deionized H_2O and adjust the pH to 8.8 using 6 M HCl. Adjust the volume to 1 L. Store at room temperature.
14. Bind-Silane solution: to prepare 20 mL, mix 16 mL of ethanol, 400 μL of acetic acid, 20 μL of Bind-Silane, and 3.6 mL of deionized H_2O. Must be prepared freshly.
15. Acrylamide solution 12 %: to prepare 12 gels of 20×24 cm, mix 385 mL of acrylamide–bisacrylamide solution (30:0.8 %), 240.6 mL of Tris–HCl 1.5 M pH 8.8, 315 mL of deionized H_2O, and 9.63 mL of SDS 10 %. Must be prepared freshly.

16. Displacing solution: mix 100 mL of Tris–HCl 1.5 M pH 8.8, 200 mL of glycerol, and 100 mL of deionized H_2O. Store at room temperature.
17. Fixation solution: to prepare 1 L, mix 400 mL of methanol, 70 mL of acetic acid, and 530 mL of deionized H_2O. Store at room temperature.
18. SYPRO Ruby Protein Gel Stain (Invitrogen).
19. Destain solution: to prepare 1 L, mix 100 mL of methanol, 70 mL of acetic acid, and 830 mL of deionized H_2O. Store at room temperature.

3 Methods

3.1 Growth of Leishmania Cultures

1. Inoculate 10 mL of MAA/20 medium with *Leishmania*. For promastigotes, incubate at 25 °C and for amastigotes, incubate at 37 °C in ventilated flask in the presence of 5 % CO_2.
2. Isolate proteins from the culture when it reaches exponential phase of growth according to the OD_{600}.

3.2 Proteins Extraction

1. Harvest the cells by centrifugation at 1,300×g for 5 min at room temperature.
2. Wash the cells twice in Hepes–NaCl.
3. Resuspend the pellet in 300 μL of 2D lysis buffer supplemented with 10 μL of protease inhibitors cocktail and 10 μL of Tris-Base 50 mM.
4. Incubate at room temperature for 2 h with frequent vortexing.
5. Centrifuge at 9,400×*g* for 2 min at room temperature.
6. Collect the supernatant containing the proteins and keep on ice (*see* **Note 7**).
7. Quantify the proteins using the 2-D Quant kit as specified by the manufacturer (GE Healthcare).

3.3 FFE Test Prior to Separation

3.3.1 FFE Setup

1. Turn on the cooler and set the temperature at 10 °C.
2. Put the filter papers (0.6 mm) in distilled H_2O.
3. Put the separation chamber in vertical position and install the spacer (0.4 mm).
4. Superimpose the electrode membrane drained of excess glycerol/isopropanol and the filter paper and place carefully the smooth side of the appropriate membrane on each electrode. Check for the appropriate alignment of the membrane on the electrode and wipe any glycerol at the bottom of the electrodes.
5. Close the separation chamber.

6. Close the middle two clamps simultaneously and close the other pairs in order to finish with both extremities of the chamber. Tighten each pairs of clamps beginning in the middle.
7. Put the tubes in distilled H_2O and open the media tubes (I1–I7). Make sure that the valves at the end of the separation chamber and the sample pump clamp are opened.
8. Turn on the media pump and fill the chamber with distilled H_2O. Make sure to remove all the air bubbles in the chamber by putting the media pump reverse and then forward again.
9. When the entire chamber is filled with distilled H_2O with no more air bubbles, open the counterflow tubes (C1–C3). When these tubes are filled with distilled H_2O, close the valves at the end of the separation chamber.
10. Check for blocked tubes among the 96 collection tubes. To unblock a tube, apply a negative pressure using a syringe until the water droplets are released freely in a constant manner.
11. Put the separation chamber in horizontal position.

3.3.2 Stripes Test

This test is used to verify the laminar flow, the tightness of the separation chamber, and the delivery by the media tubes and the media pump. This test should be done before each FFE fractionation day.

1. Close the sample pump clamp.
2. Stop the media pump and put the media tubes I2, I4, and I6 in the SPADNS solution and let the other tubes in distilled H_2O (*see* **Note 8**).
3. Turn on the media pump at a flow rate of 200 mL/h. The three pink stripes in the chamber should be of identical width and parallel. The fractions could be collected in a 96-well plate.
4. Stop the media pump and put back the media tubes I2, I4, and I6 in distilled H_2O.
5. Restart the media pump and wait until all the dye disappeared from the chamber.
6. Correct the setup if the stripe test was incorrect.

3.3.3 Pump Calibration

The media and sample pumps can be calibrated simultaneously. It must be done daily.

1. Fill a bottle and an eppendorf tube with distilled H_2O and determine the actual weight of each.
2. Stop the media pump, put the media tubes (I1–I7) in the bottle and the sample tube in the eppendorf tube.
3. Start simultaneously the media pump at 200 mL/h and the sample pump at 1,000 μL/h.

4. After 10 min, stop both pumps and determine again the weight of the bottle and the eppendorf tube.
5. Determine the new Calfac using the following formula: New Calfac = (Old Calfac/(Weight difference/10 min)) × Flow rate.
6. Set the new Calfac on the FFE.

3.3.4 High HGP Test

This test is used to verify the pI marker separation. Furthermore, the buffers used in this test contribute to prepare the chamber for the subsequent protein separation. This must be done daily.

1. Stop the media pump and open the clamp of the sample pump.
2. Put the media tube I1 in anodic stabilization medium high HGP, I2 in separation medium 1 high HGP, I3 to I5 in separation medium 2 high HGP, I6 in separation medium 3 high HGP, I7 in cathodic stabilization medium high HGP, and C1–C3 in counterflow medium high HGP (*see* **Note 9**).
3. Turn on the media pump at 57 mL/h and fill the separation chamber with separation media.
4. Put the anode and cathode tubes in the proper solutions.
5. When the chamber is filled with media, set the voltage at 1,500 V, the current at 50 mA, the power limit at 60 W and switch on the high voltage (*see* **Note 10**).
6. Wait approximately 10 min for the stabilization of the current.
7. Close the clamp of the sample pump and inject at 1,000 μL/h the pI markers (60 μL of pI markers diluted in 240 μL of separation medium 3 high HGP) by the sample inlet 4.
8. Monitor the separation of the pI marker by collecting the fractions in a 96-well plate. One sharp pink line at the left followed by 6 sharp yellow lines should be observed.

3.3.5 FFE Fractionation Medium Test

This test is used to verify the pI marker separation in the protein separation media. This test can also be used to calculate how much time after the beginning of the injection we must collect the sample in a 96-well plate.

1. Stop the media pump and the high voltage and open the clamp of the sample pump.
2. Put the media tube I1 in anodic stabilization, I2 in separation medium 1, I3 and I4 in separation medium 2, I5 in separation medium 3, I6 and I7 in cathodic stabilization, and C1–C3 in counterflow medium (*see* **Note 9**).
3. Turn on the media pump at 57 mL/h and fill the separation chamber with separation media.

4. Set the voltage at 750 V, the current at 50 mA, the power limit at 60 W and switch on the high voltage.
5. Wait for approximately 10 min for the stabilization of the current.
6. Close the clamp of the sample pump and inject at 1,000 μL/h the pI markers (60 μL of pI markers diluted in 240 μL of separation medium 2) by the sample inlet 2.
7. Monitor the separation of the pI marker by collecting the fractions in a 96-well plate. One sharp pink line at the left followed by six sharp yellow lines should be observed.
8. If all the tests are correct, the protein fractionation can begin.

3.4 Protein Fractionation Using IEF-FFE

1. Wash the remaining pI markers in the chamber for approximately 10 min.
2. Prepare the sample by diluting 3 mg of protein in separation medium 2 to obtain a concentration of 1 mg/mL. Add 2 μL/mL of non-diluted SPADNS to each sample (*see* **Note 11**).
3. Inject the sample at 1,000 μL/h and start collecting in FFE 96-well plates when the marker starts to leak from the collection tubes (*see* **Note 12**).
4. Measure the pH of each fraction between the pink and the yellow markers in order to pool the appropriate fractions for the following 2D gel separation. At this step, pooled fractions can be stored at −20 °C (*see* **Note 13**).
5. Wash the chamber with separation media for 10 min between each sample.
6. Concentrate the pooled samples using Amicon Ultra-15 columns as specified by the manufacturer until it reach approximately 250 μL and wash three times with 1 mL of 2D lysis buffer.
7. Quantify the proteins using the 2-D Quant kit as specified by the manufacturer.
8. At the end of the working day, carry out the active and passive washes of the FFE as specified in the user guide.

3.5 Protein Separation Using 2D Gels in Alkaline Conditions

3.5.1 Isoelectric Focusing (First Dimension)

1. Prepare the sample by diluting 150 μg of proteins in DeStreak Rehydration Solution to obtain a total of 345 μL. Add 5 μL of IPG buffer of the appropriate pH (6–9 or 6–11).
2. Load the sample onto the strip holder in the Ettan IPGphor II isoelectric focusing system. Avoid the introduction of air bubbles.
3. Remove the plastic protector and apply the appropriate Immobiline DryStrips (pH 6–9 or 6–11) on the sample and cover it with mineral oil (*see* **Note 14**).

4. Start the isoelectric focusing in the Ettan IPGphor II isoelectric focusing system according to the following program (*see* **Note 15**):
   (a) Rehydration: 30 V for 12 h.
   (b) Step-*n*-hold: 500 V for 1 h.
   (c) Gradient: 1,000 V for 1 h.
   (d) Gradient: 8,000 V for 3 h (*see* **Note 16**).
   (e) Step-*n*-hold: 8,000 V for 48 000 Vh.
5. Put each strip in a tube. Store at −20 °C until further use.

3.5.2 SDS-PAGE (Second Dimension)

1. Spread 400 μL of Bind-Silane solution on each small glass and polish using 20 % ethanol (*see* **Note 17**).
2. Prepare the gel caster as described in the Ettan DALTtwelve system manual.
3. Immediately before casting the gels, complete the acrylamide solution 12 % (for 12 gels) with 9.63 mL of APS 10 % and 1.36 mL of TEMED 10 %.
4. Cast the gels as described in the Ettan DALTtwelve system manual.
5. Put 10 mL of reduction solution in each tube containing the strip and incubate on a rocking platform for 15 min at room temperature.
6. Drain the tube and put 10 mL of alkylation solution in each tube and incubate on a rocking platform for 15 min at room temperature.
7. Prepare molecular marker by adding 10 μL of Low molecular weight markers for SDS electrophoresis followed by 40 μL of agarose on an IEF paper.
8. Rinse the gel cassettes in hot running tap water and drain upside down.
9. Remove all the water at the gel surface using Whatman paper.
10. Place and push the strip on the gel with the positive side on the left and the plastic side on the glass and push an IEF paper containing the molecular weight beside (*see* **Note 18**).
11. Pour agarose on the gel cassettes to cover the strip and the IEF paper and let polymerise.
12. Fill the Ettan DALTtwelve tank with 1× SDS electrophoresis buffer until the first line (approximately 7.5 L) and set the temperature at 25 °C.
13. Put the gel cassettes (and blanks if necessary) in the tank.
14. Fill the top of the tank with 2× SDS electrophoresis buffer until it reaches the minimum fluid line.

15. Run the electrophoresis at 5 W/gel for 30 min followed by 17 W/gel for 4 h (or until the blue font line reaches the bottom of the gels).
16. Stop the electrophoresis and remove the cassette from the tank.
17. Open each cassette gently.
18. Put the glass plate with the gel in the fixation solution overnight.

3.5.3 Gel Staining, Imaging and Analysis

1. Stain the gels with Sypro Ruby for 5 h with agitation in the dark (*see* **Note 19**).
2. Wash 3× 1 h with destain solution.
3. Rinse briefly with distilled H_2O and store the gel individually in a well-closed plastic bag in the dark at 4 °C.
4. Image the gels using the ProXPRESS 2D Proteomic Imaging System.
5. Perform the gel analysis using Progenesis SameSpots software.
6. Cut the spots of interest using the ProXcision robot and send them for MS/MS identification.

4 Notes

1. Our protocol was optimized using *L. infantum* and modifications might be necessary for other species.
2. Do not heat any buffer containing urea, because heating urea leads to the formation of isocyanate, which can lead to carbamylation of proteins and, consequently, affect subsequent MS/MS identification.
3. The high HGP media can be stored for a few weeks at 4 °C.
4. To prepare the stock solution HPMC/glycerol, it is important to add only small amount of HPMC at the time in a stepwise manner with vigorous stirring. Then, the solution must be agitated overnight to ensure complete dissolution.
5. As a quality control, the conductivity and the pH of each separation medium must be measured prior to the experiment. For the High HGP media, the values must be around 295 μS and pH 4.00 for separation 1, 495 μS and pH 6.98 for separation 2 and 302 μS and pH 9.85 for separation 3. For the protein fractionation media, the values must be around 356 μS and pH 4.45 for separation 1, 633 μS and pH 7.38 for separation 2, and 451 μS and pH 9.74 for separation 3.
6. The FFE separation buffer must be done freshly. Furthermore, for a better precision accuracy, even the liquid must be measured by weighting.

7. For longer period, protein samples must be stored at −80 °C.
8. Each time the media is changed, the media pump must be stopped to avoid the introduction of air bubbles in the chamber.
9. Keep all the media on ice during the entire procedure.
10. The high voltage of the FFE can cause severe injury. In case of problem, always switch off the high voltage before doing anything.
11. To avoid protein degradation, keep the sample on ice during all the injection.
12. Monitor closely the level of media and sample to avoid the introduction of air bubbles in the separation chamber.
13. As the pH of the fractions must slightly differ from the pH of the Immobiline DryStrip, it is recommended to do a test run to determine which pH fractions correspond to which pH on the Immobiline DryStrip.
14. Use a forceps to manipulate the gel strip. Be cautious to put the strip in the right position (plus and minus ends) in the strip holder.
15. All the strip holders must be parallel in the apparatus.
16. When the voltage reaches 8,000 V, put a filter paper on each electrode in the bottom of the strip holder. If the voltage has difficulties to reach 8,000 V, put the filter paper on each electrode and change them regularly until the voltage reaches 8,000 V.
17. This step is used to stick the gel onto the glass plate which is really useful for the staining and spot picking steps.
18. The gel strip and the IEF paper must not be in contact. Furthermore, avoid the presence of air bubble between the gel strip and the separation gel.
19. The Sypro Ruby staining is light sensitive. From this step, always work in the dark.

Acknowledgments

We would like to thank Drs. Aude Foucher and Jolyne Drummelsmith for previous optimization with our FFE. MO is member of the CIHR Group on Host–Pathogen Interactions and of the Centre for Host–Parasite Interactions "Programme Regroupements Statégiques" of the Fonds du Québec pour la Recherche sur la Nature et les Technologies. This work was funded by a CIHR grant to MO. MO holds the Canada Research Chair in Antimicrobial Resistance.

References

1. Kiraga J et al (2007) The relationships between the isoelectric point and: length of proteins, taxonomy and ecology of organisms. BMC Genomics 8:163
2. Gorg A et al (2009) 2-DE with IPGs. Electrophoresis 30(Suppl 1):S122–S132
3. Nissum M, Foucher AL (2008) Analysis of human plasma proteins: a focus on sample collection and separation using free-flow electrophoresis. Expert Rev Proteomics 5: 571–587
4. Brotherton MC et al (2010) Analysis of stage-specific expression of basic proteins in Leishmania infantum. J Proteome Res 9: 3842–3853
5. Bjellqvist B et al (1982) Isoelectric focusing in immobilized pH gradients: principle, methodology and some applications. J Biochem Biophys Methods 6:317–339
6. Altland K et al (1988) Isoelectric focusing of basic proteins: the problem of oxidation of cysteines. Electrophoresis 9:474–485
7. Gorg A et al (1997) Very alkaline immobilized pH gradients for two-dimensional electrophoresis of ribosomal and nuclear proteins. Electrophoresis 18:328–337
8. Herbert B et al (2001) Reduction and alkylation of proteins in preparation of two-dimensional map analysis: why, when, and how? Electrophoresis 22:2046–2057
9. Olsson I et al (2002) Organic disulfides as a means to generate streak-free two-dimensional maps with narrow range basic immobilized pH gradient strips as first dimension. Proteomics 2:1630–1632
10. Zischka H et al (2008) Purification of Saccharomyces cerevisiae mitochondria by zone electrophoresis in a free flow device. Methods Mol Biol 432:51–64

Chapter 16

Proteomic Analysis of Posttranslational Modifications Using iTRAQ in *Leishmania*

Dan Zilberstein

Abstract

iTRAQ is a high coverage quantitative proteomics technique identifies and quantitates abundance changes of multiple (up to eight) distinct protein samples. To date, one iTRAQ-MS/MS assay can identify up to quarter of cells proteome. Each of the eight tags covalently binds to the N-terminus as well as arginine and lysine side chains of peptides, enabling labeling of the entire peptide population in each sample. Following tagging, the various protein samples are mixed and subjected to LC-MS/MS analysis. In the first round identical peptides from the different protein populations focus in a single pick. Subsequently, sequence of each peptide is determined. The tags whose *m/z* is similar to that of natural amino acids are used to determine relative abundance. To date, iTRAQ enabled identification of almost 2,000 *Leishmania* proteins. Here, we provide protocols for protein abundance changes and for phosphoproteomics analysis in *Leishmania* parasites.

Key words *Leishmania*, iTRAQ, Affinity tag, Proteomics, Phosphoproteomics, Protein expression, Quantitative proteomics

1 Introduction

Genome wide analysis of gene expression enables panoramic view on transcriptome abundance changes. However, studies of last decade revealed that at any given time, only half of mRNA molecules are translated into proteins [1–3]. The level of correlation varies according to gene function, i.e., proteins involved in signaling and metabolic pathways demonstrated stronger correlation, while those in large complexes showed weaker correlation [4]. In *Leishmania*, an organism that lack transcriptional regulation of protein coding genes [5], the correlation is even worse, around 0.2 [6, 7]. Therefore, at least in *Leishmania* (as well as all other trypanosomatid parasites) high coverage quantitative proteome expression must be determined in order to study phenotypic expression.

Christopher Peacock (ed.), *Parasite Genomics Protocols*, Methods in Molecular Biology, vol. 1201,
DOI 10.1007/978-1-4939-1438-8_16, © Springer Science+Business Media New York 2015

To date, four high coverage quantitative methods are available; Isobaric Tags for Relative and Absolute Quantification (iTRAQ) [8], Isotope-Coded Affinity Tag (ICAT) [9], Stable Isotope Labeling by Amino acids in Cell culture (SILAC) and dimethylation [10]. iTRAQ and ICAT are employed for multiple samples whereas the latter compare abundance of only 2–3 distinct samples. iTRAQ is in my opinion the better technique for quantitative relative abundance measurement of multiple samples because the tags covalently bind N-termini of all peptides, not to selective SH groups (as do ICAT).

We started to use iTRAQ in 2006, soon after it became available, and published the first papers on protein abundance changes during *L. donovani* differentiation 2 years later [11]. iTRAQ detected 1,700 proteins of which 920 identified at all differentiation time points. This high coverage facilitated systems analysis of several pathways. For example, we found that during transition from promastigotes to amastigotes, parasites undergone metabolic retooling in a highly regulated and coordinated manner. We further observed that changes in posttranslational modifications, such as phosphorylation, methylation, acetylation, and glycosylation also occur during differentiation [12]. Further analysis of the *L. donovani* phosphoproteomic revealed stage-specific phosphorylation motifs [13]. Recently, we employed iTRAQ to quantitate differentiation-derived changes in the phosphorylation state of *L. donovani* proteome [13]. This analysis revealed signal-specific phosphorylation of protein kinases.

To date, only two other laboratories have published the use of iTRAQ for proteome analyses of *Leishmania* parasites. Sardar et al. [14] used it to assess effect of oxygen stress inducing agents such as menadione (ROS) and S-nitroso-*N*-acetylpenicillamine (RNS) and their combined effect on protein abundance in *L. donovani* promastigotes. iTRAQ detected ~20 % of promastigotes proteome, i.e., 1,653 proteins, which is comparable to Rosenzweig et al. [11]. Of these, the abundance of about a quarter of these proteins changed after exposure to oxygen stresses. Lynn et al. [15] used iTRAQ to compare protein abundance changes between amastigotes and promastigotes of *L. mexicana*. The analysis identified only a few hundred proteins in each life stage.

The aim of this chapter is to share with the *Leishmania* research community protocols and experience for iTRAQ analysis. We hope that more research groups will use this method. It is relatively expensive, but the information that comes out of this analysis is worthwhile.

This chapter is dedicated to Professor Emeritus Robert (Bob) W. Olafson of Victoria University who founded The UVic Proteomics Center. Bob introduced me to iTRAQ and thanks to his great vision we have been able to show how well *Leishmania* differentiation is regulated.

2 Materials

2.1. *Parasites growth media*: *L. donovani* 1SR promastigotes are grown in EARL's-based medium 199 supplemented with 10 % heat inactivated fetal calf serum (FCS) at pH 7 and amastigotes in medium 199 with untreated 25 % FCS, titrated to pH 5.5 with 10 mM Tris/succinate [16]. Our experience with other *Leishmania* strains is that changing media have minimal effect on protein repertoire or abundance.

2.2. *Cell wash buffers and storage*: In order to minimize unnecessary stresses on parasites we recommend that cell harvest from medium and subsequent washes be carried out using the medium salt solution. For example, in accordance with our growth medium, M199, we harvest and wash parasites with EARL's salt solution at pH similar to growth conditions. If the samples are to be used for phosphoproteomics, phosphatase inhibitors should be added to the washing buffers. Before solubilizing cells and for storage, cells were pelleted and kept on ice until they were subjected to protein extraction. Storing cell pellet but not extracts is critical for long-term maintenance of phosphorylation sites. We found that when we store cell extracts we lose phosphorylation sites within a few weeks, whereas cell pellets maintain them for months. For long-term storage (>month) it is recommended to keep the cell pellets in liquid nitrogen.

2.3. For phosphopeptides enrichment we used 10 μm diameter TiO_2 beads from GL Science (Tokyo, Japan). iTRAQ tags were from Applied Biosystems (Applied Biosystems, Inc., Foster City, CA, USA). Trypsin was from Promega (USA).

3 Methods

3.1 iTRAQ Labeling, Assay, and Analysis with Focus on Protein Phosphorylation

3.1.1 Principles of iTRAQ Labeling

Isobaric Tags for Relative and Absolute Quantification (iTRAQ) utilizes amine-reactive isobaric tags to label all peptides in a particular protein digest. An advantage of this method is that four to eight protein digest samples can each be tagged differently, allowing direct comparison of up to eight samples. Samples are combined at an equal ratio and subjected to LC-MS/MS, where, on fragmentation, every fragmented peptide tag produces distinct signature ions differing by an m/z value of 114–117 in the 4 plex and in addition 113, 118, 119, and 121 in the 8 plex. The relative intensities of these signals represent the relative abundance of the analyzed peptide in each sample. Relative abundance values of all peptides attributed to each specific protein are averaged to represent the relative abundance of the entire protein [8, 17].

3.1.2 Perspectives

iTRAQ is not the only quantitative proteomic technique available to date. Other methods are Isotope-Coded Affinity Tag (ICAT) [9], Stable Isotope Labeling by Amino acids in Cell culture (SILAC) and dimethylation [10]. iTRAQ and ICAT are employed for multiple samples whereas the latter compare abundance of only two samples. iTRAQ is the better technique for quantitative relative abundance measurement of multiple samples because the tags bind N-termini, not selective SH groups (ICAT). Our experience with iTRAQ focused on time course changes in protein abundance (this section) and posttranslational changes during *L. donovani* differentiation (Subheading 3.1.5). When we started using iTRAQ in 2006 only four-flex iTRAQ tage were available. The eight tags appeared in late 2008, but we found that addition of four tags to the same assay affected sensitivity as well as repertoire of proteins identified in the assays.

3.1.3 Optimization of Assay

Optimization is a pre requisite for successful use of iTRAQ; it is important to define the optimal extraction procedure, trypsinization conditions (this is even more important for phosphoproteomics), tag to protein ratio, protein concentration loaded to mass spectrometer. In most cases (at least in ours) proteomic centers start with frozen cells. Nonetheless, you should insist on optimizing (with them) treatments and assay to your experimental system.

3.1.4 Cell Harvest and Protein Extraction

Cell growth conditions should be specific to the strain used and/or specific experiment. Our assays were done using axenic promastigotes and amastigotes of a cloned line of *L. donovani* 1SR [16]. Differentiation of promastigotes to amastigotes in axenic culture is carried out as follows; transfer late-log promastigotes from promastigote medium at 26 °C to amastigote medium (*see* content **item 2.1**) at 37 °C in 5 % CO_2 incubator. Split cells 1:10 24 h after initiation of differentiation, using pre-warmed amastigote medium. Maturation of axenic amastigotes completed within 120 h [18]. In our analyses, aliquots of cells (total of ~3×10^9/ml, equals 8 mg cell protein/ml) were collected at seven differentiation time points, washed twice in ice-cold EARL's salt solution and finally pelleted and kept frozen until used.

Reduce 1 mg of protein from each sample with dithiothreitol (30 min at 37 °C), then alkylate cysteine sulfhydryls were with iodoacetamide (30 min at 37 °C in darkness). Add 20 μg of trypsin to each sample (enzyme–sample ratio of 50:1) and digest at 37 °C for 16 h. Subsequently, acidify each sample with formic acid (final concentration of 0.5 % v/v) prior to reversed phase solid phase extraction (SPE) for desalting and sample cleanup (sodium deoxycholate should be removed by centrifugation as it precipitates under acidic conditions). Waters HLB Oasis reversed phase SPE cartridges (10 mg) is recommended to be used to desalt peptides.

For phosphopeptides enrichment, following binding and washing, elute peptides from Oasis HLB with TiO_2 binding buffer (70 % v/v acetonitrile, 5 % v/v trifluoroacetic acid (TFA), and 300 mg/ml lactic acid).

3.1.5 TiO_2 Phosphopeptide Enrichment

To each sample, added 10 mg of TiO_2 beads (GL Science, 10 μm diameter), then incubated the mixture with end-over-end rotation for 30 min at 4 °C. Wash TiO_2 beads five times with 200 μl TiO_2 binding buffer (30 s mix, centrifuge 60 s at 2,000×*g*). Remove liquid gently, be careful not to disturb TiO_2 beads. Subsequently, wash five times with 200 μl buffer B (80 % ACN, 0.1 % TFA). Remove liquid gently, be careful not to disturb TiO_2 beads.

Transfer TiO_2 beads to StageTips with C_8 frits for phosphopeptides for elution. Elution should be performed in steps: 40 μl 0.5 % NH_4OH, then 40 μl 0.5 % NH_4OH/30 % ACN, then 40 μl 0.5 % NH_4OH/50 % ACN. Formic acid (12 μl) should then be added to the eluent to lower the pH for HLB Oasis SPE cleanup prior to iTRAQ labeling. Freeze eluent at −80 °C and lyophilized to dryness.

3.1.6 Peptide Labeling (4 Plex) and LC-MS/MS Analysis

Labeling of peptides with the iTRAQ reagent should be carried out essentially as instructed by the manufacturer (Applied Biosystems, Inc., Foster City, CA, USA). Briefly, Rehydrate peptides (100 μg protein) with 30 μl of 0.5 M triethylammonium carbonate buffer (TEAB; pH 8.5). iTRAQ label (10 μl in 100 % acetonitrile) is then diluted with 70 μl of 100 % ethanol prior to addition of the 80 μl volume to the sample. Samples are then vortexed to mix and centrifuged. Labeling reaction conducted for 60 min at room temperature (~23 °C). Up to four separately labeled samples are pooled and vacuum concentrated to remove the organic solvent component prior to LC-MS/MS analysis. In our analyses, when 4 plex iTRAQ tags was used, two labeled peptide mixes were created; one mix included equal amounts of protein from promastigote, 2.5, 5, and 10 h of differentiation (early differentiation), and the other included promastigotes, 15 and 24 h of differentiation and mature amastigotes (120 h) [11].

Samples are acidified with formic acid prior to LC-MS/MS analysis with 25–50 % of each iTRAQ sample is injected per LC-MS/MS analysis using an LC Packings Famos Autosampler with an LC Packings Ultimate Nanoflow HPLC coupled to a QSTAR Pulsar I. Samples are analyzed by LC-MS/MS two times with the second analysis employing an exclusion list generated from the first analysis as outlined in next section (Subheading 3.1.7).

A desalting column (0.3×5 mm) HPLC plumbing configuration is used to protect the analytical column (Magic C_{18} resin, 150 mm×75 μm column diameter). Samples are then loaded onto the desalting column at 30 μl/min with 100 % solvent A (2 % v/v acetonitrile, 0.1 % formic acid) for 5 min. 2 h HPLC analytical separation at 300 nl/min (60 min linear gradient from 0 to 20 % solvent B (98 % v/v acetonitrile, 0.1 % formic acid), 30 min linear gradient from 20 to 40 % solvent B, 12 min linear gradient from 40 to 80 % solvent B). MS/MS spectra are then acquired in a data dependent manner selecting the top two most intense eluting ions in the 400–1,600 *m/z* range with a 2+ to 5+ charge state. Following selection for MS/MS analysis, precursor ions should be excluded from selection for MS/MS analysis for 180 s.

3.1.7 Data Acquisition, Processing, and Analysis

Raw MS/MS data from the first LC-MS/MS analysis should be converted to peak lists using Mascot Script (we used version 1.6b21; Matrix Science). Peak lists are queried against the *L. infantum* ver. 3 (i.e., latest version available) using latest version of Mascot with the following parameters:

- Allowing one missed trypsin cleavage site.
- Fixed/constant modifications:
  - Methylthio (C), iTRAQ4-plex (N-term), iTRAQ4-plex (K).
- Variable modifications:
  - Deamidated (NQ), Phospho (S/T), Phospho (Y).
- Peptide Mass Tolerance of ±0.3 Da.
- Fragment Mass Tolerance of ±0.15 Da.

Mascot assignments of MS/MS to peptide sequences with an *ion score >30* should considered to be of high quality and should be used to generate an exclusion list of peptide masses to exclude from selection for MS/MS in the second LC-MS/MS analysis.

The MS analysis results in separate data files (*.wiff) for each LC-MS/MS analysis. All data files are searched for protein identification and relative abundance using ProteinPilot [19]. Data should be searched against the available *L. infantum* database.

To view data, we recommend to use ProteinPilot viewer that can be downloaded for free from Applied Biosystems (http://download.appliedbiosystems.com/proteinpilot). Once installed, the ".group" result file can be opened using ProteinPilot. To view data with peaks represented as observed, the ".wiff files" must be placed in the same directory structure that they were analyzed in. The directory structure can be viewed on the "Summary Statistics" tab in the Protein Pilot result window.

3.2 Other Posttranslational Modifications

Mass spectrometers that read iTRAQ labeled peptides also detect changes in *m/z* made by amino acid methylation, acetylation and glycosylation. ProteinPilot than determines the type of modification identified. Unlike protein phosphorylation, no extra care of protein samples is required for the other PTMs. Hence, one can use any type of protein analysis to detect PTMs. We did so with our first iTRAQ time course proteomics analysis of differentiation.

In addition to protein abundance changes we were able to detect significant changes in protein the above indicated PTMs [12]. Furthermore, using these assays detected for the first time protein fucosylation in *Leishmania* promastigotes and amastigotes.

3.3 Validation with Nano-LC-Monitoring (MRM) MS Analysis

To validate PTM expression profile it is recommended to use as nano-LC-monitoring MS analysis (MRM) [20] that employs synthetic phosphopeptides that are 10 Da heavier than the endogenous peptides to quantitate expression profile.

A heavy version (+10 kDa) of phosphopeptides of choice (for example, EGIIPYTEV(pT)R, the phosphorylation site of the alpha subunit of eIF2 in *L. donovani* [20]) was spiked into samples and the resulting peptide mixes were mixed with TiO_2 beads and phosphopeptides eluted in two steps, using 30 and 50 % ACN in 0.5 % NH_4OH. In our case, the enriched phosphopeptides were subjected to MRM-MS analysis at the Genome BC Proteomics Centre at the University of Victoria. All data was analyzed using MultiQuant 1.1 (Applied Biosystems). The ratio of endogenous EGIIPYTEV(pT) R phosphopeptide levels in the samples to those of the heavy phosphopeptide (averaged from five MRM transitions) is then normalized.

Acknowledgments

I thank Dr. Polina Tsigankov for critical reading. I thanks Drs. Christoph ** and *** for providing me their protocols. This work was supported by U.S.-Israel Binational Foundation grant 2009226.

References

1. Lu P, Vogel C, Wang R, Yao X, Marcotte EM (2007) Absolute protein expression profiling estimates the relative contributions of transcriptional and translational regulation. Nat Biotechnol 25:117–124
2. Greenbaum D, Colangelo C, Williams K, Gerstein M (2003) Comparing protein abundance and mRNA expression levels on a genomic scale. Genome Biol 4:117
3. Garcia-Martinez J, Gonzalez-Candelas F, Perez-Ortin JE (2007) Common gene expression strategies revealed by genome-wide analysis in yeast. Genome Biol 8:R222
4. Schmidt MW, Houseman A, Ivanov AR, Wolf DA (2007) Comparative proteomic and transcriptomic profiling of the fission yeast Schizosaccharomyces pombe. Mol Syst Biol 3:79
5. Clayton C, Shapira M (2007) Post-transcriptional regulation of gene expression in trypanosomes and leishmanias. Mol Biochem Parasitol 156:93–101
6. Lahav T, Sivam D, Volpin H, Ronen M, Tsigankov P, Green A, Holland N, Kuzyk M, Borchers C, Zilberstein D, Myler PJ (2011) Multiple levels of gene regulation mediate differentiation of the intracellular pathogen Leishmania. FASEB J 25:515–525
7. Haile S, Papadopoulou B (2007) Developmental regulation of gene expression in trypanosomatid parasitic protozoa. Curr Opin Microbiol 10:569–577
8. Melanson JE, Avery SL, Pinto DM (2006) High-coverage quantitative proteomics using amine-specific isotopic labeling. Proteomics 6:4466–4474

9. Gygi SP, Rist B, Gerber SA, Turecek F, Gelb MH, Aebersold R (1999) Quantitative analysis of complex protein mixtures using isotope-coded affinity tags. Nat Biotechnol 17:994–999
10. Ong SE, Blagoev B, Kratchmarova I, Kristensen DB, Steen H, Pandey A, Mann M (2002) Stable isotope labeling by amino acids in cell culture, SILAC, as a simple and accurate approach to expression proteomics. Mol Cell Proteomics 1:376–386
11. Rosenzweig D, Smith D, Opperdoes FR, Stern S, Olafson RW, Zilberstein D (2008) Retooling *Leishmania* metabolism: from sandfly gut to human macrophage. FASEB J 22:590–602
12. Rosenzweig D, Smith D, Myler PJ, Olafson RW, Zilberstein D (2008) Post-translational modification of cellular proteins during Leishmania donovani differentiation. Proteomics 8:1843–1850
13. Tsigankov P, Gherardini PE, Helmer-Citerich M, Spath GF, Myler PJ, Zilberstein D (2014) Regulation dynamics of Leishmania differentiation: deconvolution signals and identifying phosphorylation trend. Mol Cell Proteomics 13:1787–1799
14. Sardar AH, Kumar S, Kumar A, Purkait B, Das S, Sen A, Kumar M, Sinha KK, Singh D, Equbal A, Ali V, Das P (2013) Proteome changes associated with Leishmania donovani promastigote adaptation to oxidative and nitrosative stresses. J Proteomics 81:185–199
15. Lynn MA, Marr AK, McMaster WR (2013) Differential quantitative proteomic profiling of Leishmania infantum and Leishmania mexicana density gradient separated membranous fractions. J Proteomics 82:179–192
16. Saar Y, Ransford A, Waldman E, Mazareb S, Amin-Spector S, Plumblee J, Turco SJ, Zilberstein D (1998) Characterization of developmentally-regulated activities in axenic amastigotes of *Leishmania donovani*. Mol Biochem Parasitol 95:9–20
17. Mahoney DW, Therneau TM, Heppelmann CJ, Higgins L, Benson LM, Zenka RM, Jagtap P, Nelsestuen GL, Bergen HR, Oberg AL (2011) Relative quantification: characterization of bias, variability and fold changes in mass spectrometry data from iTRAQ-labeled peptides. J Proteome Res 10:4325–4333
18. Barak E, Amin-Spector S, Gerliak E, Goyard S, Holland N, Zilberstein D (2005) Differentiation of *Leishmania donovani* in host-free system: analysis of signal perception and response. Mol Biochem Parasitol 141:99–108
19. Shilov IV, Seymour SL, Patel AA, Loboda A, Tang WH, Keating SP, Hunter CL, Nuwaysir LM, Schaeffer DA (2007) The Paragon Algorithm, a next generation search engine that uses sequence temperature values and feature probabilities to identify peptides from tandem mass spectra. Mol Cell Proteomics 6:1638–1655
20. Gosline SJ, Nascimento M, McCall LI, Zilberstein D, Thomas DY, Matlashewski G, Hallett M (2011) Intracellular eukaryotic parasites have a distinct unfolded protein response. PLoS One 6:e19118

Chapter 17

Large-Scale Differential Proteome Analysis in *Plasmodium falciparum* Under Drug Treatment

Judith Helena Prieto, Elisabeth Fischer, Sasa Koncarevic, John Yates, and Katja Becker

Abstract

Here, we establish a methodology for large-scale quantitative proteomics using SIL (stable isotope labeling) to examine protein expression changes in trophozoite stages of the malaria parasite *Plasmodium falciparum* following drug treatment. For this purpose, exposure to $^{13}C_6{}^{15}N_1$-isoleucine was optimized in order to obtain 99 % atomic enrichment. Proteome fractionation with anion exchange chromatography was used to reduce sample complexity and increase quantitative coverage of protein expression. Tryptic peptides of subfractions were subjected to SCX/RP separation, measured by LC-MS/MS, and quantified using the software tool Census. In drug-treated parasites, we identified a total number of 1,253 proteins, thus increasing the overall number of proteins so far identified in the trophozoite stage by 30 % in the previous literature. A relative quantification was obtained for more than 800 proteins. About 5 % of proteins showed a clear up- or downregulation upon drug treatment.

Key words Quantitative proteomics, SIL, Census, Malaria, *Plasmodium*, Drug effects

1 Introduction

Since publication of the first malaria parasite genome sequence [1] and concurrent publication of the proteome of *Plasmodium falciparum* [2], the field of systems biology has acquired a wealth of tools to integrate data about the proteome, transcriptome, and metabolome in an attempt to gain insight into cellular processes and functions [3–5]. The aim of systems biology is to evaluate entire protein networks of a given cell at a given spatial or temporal point. Current proteome technologies are still unable to accomplish such a task, but improvements in mass spectrometry, bioinformatics, and protein separation, as well as labeling techniques are allowing scientists to gain a step forward in this understanding [6–9]. In this chapter we present one of the quantitative analytical tools currently available for proteomics of the *Plasmodium* parasite.

Christopher Peacock (ed.), *Parasite Genomics Protocols*, Methods in Molecular Biology, vol. 1201,
DOI 10.1007/978-1-4939-1438-8_17, © Springer Science+Business Media New York 2015

Proteomics is defined as the analysis of the entire proteome complement expressed in any biological sample at a given time under specific conditions [10]. Two avenues are the driving forces in the field: functional and expression proteomics. Functional proteomics seeks to characterize the components of signaling pathways, subcellular structures, and multiprotein complexes, among others. In expression proteomics, the differences in abundance of proteins in two different samples are measured. This is referred to as a differential proteome and is used to identify proteins that are significantly altered in two or more different samples. It is used extensively to catalog proteins present in one stage of development such as the protein repertoires of *Cryptosporidium parvum* sporozoites [11] or the different stages of development of the *Plasmodium falciparum* proteome, where its very complex life cycle was dissected [2, 12], allowing identification of clusters of expression. In these cases a qualitative approach was used to make conclusions of abundance of proteins in the sample. Absolute quantification would be ideal for comparing results across laboratories and experiments, and progress is being made in this regard [13]. Nonetheless if a quantitative study is to be carried out for a large number of proteins, an internal standard has to be introduced. A good "internal standard" in a bottom-up proteomics experiment would be the same peptide being quantified but slightly heavier so as to distinguish it from the target of interest. This approach allows a higher amount of peptides to be analized and an increase in precision in the quantification [14].

In the methodology described, we used MudPIT (multidimensional protein identification technology), where complex mixtures of proteins are digested and loaded onto a biphasic microcapillary column, which is then interfaced to the mass spectrometer. In order to have all proteins labeled, we made use of the organism cell machinery to incorporate the label by replacing the isotope of interest in the growth media. Depending on the organism, up to three generations are needed for up to 99 % incorporation of the labeled isotope. The samples are mixed 1:1 with an identical sample grown in unlabeled media. The idea is that all peptides detected in the mass spectrometer have internal control peptides that went through the same digestion treatment and further downstream steps. The ratio of the heavy (labeled) over light (unlabeled) peptides should be equal to one if no expression change has occurred in the growth conditions. The peptides that exhibit larger or smaller ratios than the average can be studied in detail. Incorporation of ^{15}N for protein quantification has been used in many model organisms, including yeast, *Arabidopsis*, *Drosophila*, and rat [15–17]. It is useful in microbiology since culture conditions to grow the organism of choice can be adjusted to incorporate just labeled media. However, there are examples where an organism requires a medium that cannot be readily adjusted to fully incorporate ^{15}N such as *P. falciparum*, which requires red blood cells for its growth. Nonetheless it can incorporate heavy isoleucine ($^{13}C^{15}N$-Ile). *P. falciparum* can be metabolically labeled using SILAC (single amino acids in cell culture) [18]. SILAC is a method that allows metabolic incorporation of "heavy" and "light" forms of an amino acid into proteins. In a SILAC experiment, two different cell populations are grown in media that is identical except that one contains a ^{13}C-substitute amino acid in place of the naturally occurring ^{12}C version. The labeled amino acid must be one that is not synthesized by the parasite but supplied solely from the media. After a number of cell divisions, one population of cells has completely incorporated the "heavy" amino acid, while the other has incorporated the "light." Using a SILAC labeling approach in *P. falciparum*, labeled isoleucine was found in all but one protein of the 5,000 predicted possibilities, and the observed mass difference of 7 Da allowed for discernible isotopic envelopes in the mass spectrometer. This was beautifully suggested by Hyde and co-workers [19] and used in our laboratory to quantify the changes of up to 800 proteins upon treatment with two different antimalarial drugs [14]. The overall and detailed methodology is presented here.

2 Materials

2.1 Preparation of Cell Culture Medium with and Without Labels (Heavy and Light)

1. Medium RPMI-1640 without isoleucine was special ordered through Servichem (in Germany) directly from Cell Culture Technologies GmbH (Switzerland). The formula is from GIBCO #52400 but lacks isoleucine, L-glutamine, and $NaHCO_3$, comes as dry powder, and needs to be prepared before usage. The instructions come with the packaging, the dry powder, and five vials. The kit is kept at −20 °C. Once thawed, mix thoroughly. Add to 3 l MilliQ water until everything is solubilized. Solutions 1 and 2 and 8 g $NaHCO_3$ are added, and subsequently vials 3–5 are mixed in as well as 1.2 g of L-glutamine, 400 ml Albumax II from GIBCO (5 % solution), 14.55 ml glucose (45 % solution), 218 μl gentamicin (50 mg/ml), and 80 ml hypoxanthin (10 mM; solubilize 136 mg in 1.5 ml NaOH and dissolve to 100 ml).
2. Add unlabeled isoleucine to final concentration of 50 mg/L to "light" media and labeled isoleucine to "heavy" media to an end value of 52.7 mg/L. Make a stock solution for each medium. There is no need to set pH, as HEPES is in the solution as a buffering agent. Labeled $^{13}C_6$,$^{15}N_1$-isoleucine is from Cambridge Isotopes (Andover, MA).
3. Bring volume to 4 l and filter-sterilize medium. Store media at 4 °C in the dark up to a maximum of 4 months in 0.5 l bottles.

2.2 Cell Culture

1. Blood (from blood bank type A+).
2. Incubator (CO_2, N_2).
3. Sterile bank, laminal flow hood.
4. *Plasmodium falciparum* strain 3D7.
5. Sorbitol (5 %) for synchronization.
6. For magnetic sorting (MACS, magnetic activated cell sorting, commercialized Miltenyi Biotec GmbH, Bergisch Gladbach, Germany).

2.3 Protein Extraction, Subfractionation, and Digestion

1. Urea, Triton X-100, Tris, EDTA, $MgSO_4$ (Roth), benzonase (Merck).
2. Use for fractionation Vivapure MiniH spin columns. (Weak anion exchange, diethylamine (D).) Catalog IX01DH24, www.sartorius.de. Capacity of 4 mg, 400 μl.
3. OrgoSol Detergent-OUT Kit (Calbiochem, catalog number 496950-1MEDI) was used to precipitate protein after fractionation.
4. Trypsin (Gold-Promega) and LysC (Wako Chemicals USA), IAA (iodoacetamide), TCEP (tris(2-carboxyethyl)phosphine) (Sigma).

2.4 Proteomics

1. The work can be carried out on different instruments. Because of the need of high mass accuracy (mass difference at the higher charge states) and a need for speed, an Orbitrap instrument (Thermo Scientific) is suggested.
2. For search and quantification there are different packages available. Integrated Proteomics Applications (http://integrated-proteomics.com/products/ip2/) was used in our study. However, Maxquant [20] and Proteome Discover Software (Thermo Scientific) are alternatives.

3 Methods

3.1 Cultivation of Plasmodium falciparum

Blood stages of the *P. falciparum* strain 3D7 (chloroquine-(CQ) sensitive) were maintained in culture using a modified protocol of Trager and Jensen [21, 22]. Labeled and unlabeled RPMI medium 1640 supplemented with $NaHCO_3$ and HEPES, pH 7.4, 22 μg/ml gentamicin sulfate, 2.1 mM L-glutamine, 0.2 mM hypoxanthine, 0.2 % Albumax II, and 0.16 % glucose. Washed human erythrocytes of blood group A positive were added to a hematocrit of 5 %. Parasites were maintained at a parasitemia of 1–10 % in an atmosphere of 94 % N_2/3 % O_2/3 % CO_2 at 37 °C and synchronized to the ring stage via the sorbitol method [23].

3.1.1 Synchronization of Parasite Culture

A tight window (>90 % of parasites at the same stage) can be accomplished with a combination of magnet purification, which selects for trophozoites and schizonts, and repeated treatment with sorbitol 3–5 times. *See* **Note 1**. This window will grow as parasites continue to multiply. Parasite growth and parasitemia were monitored by assessing Giemsa-stained blood smears under the microscope to establish the appropriate time point for drug treatment and cell harvesting. For determining IC_{50} values on the parasites, the semiautomated microdilution technique based on ^{3}H-hypoxanthine incorporation was applied.

1. Flow cells through a magnet (MACS) under sterile conditions. Only late-stage parasites are selected. LD columns and LS adapter were used in our laboratory. *See* **Note 1**.
2. Spin down cells at 500 × *g* at room temperature for 3 min after the next schizogony has taken place (4–5 h after magnetizing) and the first rings have invaded the new erythrocytes.
3. Discard medium (supernatant).
4. Add 5 ml 5 % sorbitol for 5–8 min at room temperature.
5. Spin down and discard sorbitol supernatant, and wash with RPMI medium twice.
6. Continue growing in medium. Only rings will survive treatment.
7. Repeat after approximately 6 h and once every life cycle.

3.1.2 Determination of IC_{50} Values on *P. falciparum*

Isotopic drug sensitivity assays were employed to investigate the susceptibility of the malaria parasites to various compounds. Incorporation of radioactive [^{3}H] hypoxanthine was carried out with the modifications of Fivelman et al. [24].

1. Parasites were incubated at a parasitemia of 0.25 % (>70 % ring forms) and 1.25 % hematocrit in hypoxanthine-free medium.
2. After 48 h, 0.5 μCi [^{3}H] hypoxanthine was added into each well, and the plates were incubated for another 24 h.
3. Harvest the cells of each well on a glass fiber filter (Perkin-Elmer, Rodgau-Jügesheim, Germany), wash, and dry cells on said filter.
4. Consider the radioactivity measured, in counts per minute, to be proportional to the respective growth of *P. falciparum* in the well. Calculate IC_{50} values.

3.2 Heavy Isotope Labeling of *P. falciparum* Proteins

In order to gain the highest protein yield, parasites at the trophozoite stage were harvested and grown in specialized cell culture conditions (*see* Subheading 2). Procedures published by Nirmalan et al. [19] and Koncarevic et al. [25] were combined. Human serum was replaced by Albumax. Custom RPMI-1640 medium devoid of isoleucine was supplemented with $^{13}C_6$,$^{15}N_1$-isoleucine to yield 7 Da mass shifts per isoleucine in a peptide.

1. Preincubate synchronized cultures for 24 h in isoleucine-free medium.
2. Add heavy ($^{13}C_6$, $^{15}N_1$) isoleucine (52.7 mg/ml) or light isoleucine (50 mg/ml) to parallel cultures. Isoleucine incorporation was applied for three complete life cycles (3 × 48 h) before adding the drugs.
3. Incubate blood-stage cultures with a parasitemia of 8–10 % at the *late ring/early trophozoite stage* (20–24 h post-infection) with the drug of choice, chloroquine (CQ) in our case (2 × IC_{50} values = 17 nM for CQ) or the solvent used for control cultures. *See* **Note 2**.
4. Harvest cells after 12 h of drug exposure, at which point the parasites reach the *late trophozoite/schizont* stage in parallel with the controls. *Note*: The window of the sorbitol-induced synchrony was approximately 4 h. No reinvasion takes place, as medium change would be needed, during the drug exposure.

3.3 Preparation of Parasite Extracts

1. To avoid as much as possible contaminating red blood cell material, isolate only infected erythrocytes.
2. Use a larger magnet (SuperMACS) than the one used for synchronization for collecting the trophozoite (column D in our case). The column has ferromagnetic fibers coated with a cell-friendly coating kept previously in ethanol. *See* **Note 3** for preparation of column and parasite isolation.
3. Harvest parasites, after column enrichment, by lysing the red blood cells in saponin-containing buffer (7 mM K_2HPO_4, 1 mM NaH_2PO_4, 11 mM $NaHCO_3$, 58 mM KCl, 56 mM NaCl, 1 mM $MgCl_2$, 14 mM glucose, 0.02 mM saponin, pH 7.5). Resuspend and wash twice with PBS buffer. *See* **Note 4** for details.
4. Disrupt parasites with three cycles of freezing and thawing and ultrasonication in digestion buffer (4 M urea, 0.4 % Triton X-100, 50 mM Tris–HCl, 5 mM EDTA, 10 mM $MgSO_4$, pH 8.0) in the presence of protease inhibitors.
5. Perform RNA/DNA digestion with benzonase® (Merck) for 30 min at 4 °C. A first centrifugation step at 15,000 × *g* removes contaminating hemozoin.
6. Centrifuge cell extract at 100,000 × *g* for 30 min. The supernatant is used as "parasite" extract for analysis preceded by a fractionation step. The insoluble pellet was used as an "insoluble fraction" directly for the analyses. *See* **Note 5**.

3.4 Proteome Subfractionation

In order to increase the number of identifications, a pre-fractionation was carried out. Weak anion exchanger chromatography was used to subfractionate the soluble proteins of *P. falciparum*.

1. Mix samples to be compared (light and heavy) in a protein-to-protein ratio of 1:1 wt/wt.
2. Equilibrate vivapure Mini H columns with a buffer containing 2 M urea, 0.2 % Triton X-100, 25 mM Tris–HCl, 2.5 mM EDTA, and 5 mM $MgSO_4$, pH 8.0 and then load with 1:1 mixed protein extract.
3. Elute bound proteins by stepwise increasing salt concentrations (200–500 mM NaCl in the same equilibration buffer 2 M urea, 0.2 % Triton X-100, 25 mM Tris, 2.5 mM EDTA, 5 mM $MgSO_4$, pH 8.0). In our case three samples were collected: flow through, 200 mM NaCl, and 500 mM NaCl.
4. Protein concentration in eluted samples was measured; fractions were precipitated by OrgoSol™-DETERGENT-OUT™ detergent removal kit (Calbiochem, Darmstadt, Germany). Any other precipitation protocol should work as well.

3.5 Protein Digestion

In order to analyze the protein mixture, a bottom-up proteomics approach is used. The peptides need to be first digested with two enzymes that will cut at all arginine and lysine sites. This allows for an easier data analysis than if using unpredictable digestion protocols.

1. Dissolve protein extract in 60 μl 8 M urea and 100 mM Tris pH 8.5 (freshly made).
2. Add 0.3 μl 1 M TCEP (5 mM final concentration) and incubate at RT for 20 min, to reduce all disulfide bridges.
3. Add 1.2 μl 500 mM IAA (10 mM final concentration, freshly made) and incubate at RT for 15 min in the dark to alkylate all free cysteines.
4. Digest protein mixture with Lys-C by adding 1.5 μl Lys-C (0.5 μg/μl) enzyme, (1/20 to 1/100) and incubate in the dark for 3–4 h at 37 °C.
5. Prepare and optimize solution for digestion with trypsin by adding 180 μl 100 mM Tris pH 8.5 to dilute urea by factor of 4 (2 M final concentration). Add 2.4 μl 100 mM $CaCl_2$ (1 mM final concentration).
6. Add 1.5 trypsin (0.5 μg/μl) (1/20 to 1/100) and incubate in the dark overnight at 37 °C, for optimal digestion.
7. Stop reaction by adding 13.3 μl 90 % formic acid (5 % final concentration).
8. Spin at top speed for 15 min to collect all insoluble matter, transfer the supernatant to a new tube, and freeze at −80 °C (or −20 °C).

3.6 Multidimensional Protein Identification Technology (MudPIT)

The digested protein mixture and the resulting peptides were analyzed via mass spectrometry. In order to introduce the large amount of peptides, these were separated in a freshly prepared two-phase capillary column that had two distinguishable resins and separated peptides by hydrophobicity and charge. The peptide mixture was loaded onto the column and connected to a nano-HPLC where increasing salt concentration steps and a linear organic buffer gradient eluted the peptides from the multidimensional column into the mass spectrometer. Approximately 100 μg of protein was used for MudPIT on an LTQ-Orbitrap (ThermoElectron). All samples were analyzed in triplicate. The MudPIT column and experiment were carried out as follows:

1. Package sample loading column before use by having a total of 6 cm, 250 μm i.d. (inside diameter) fused silica, and loading with 3 cm Aqua 5 μm C18 material and 3 cm Luna SCX 5 μm material. The first 3 cm will hold the peptide mixture, and the second phase (SCX) will separate peptides by charge.
2. Pack analytical column, for separation of peptides by reverse phase, by having 10 cm 100 μm i.d. fused silica column, loaded with Aqua 5 μm C18 material and an orifice of 5 Å.
3. Run the liquid chromatography gradient, between increasing salt steps, with Buffer B (90–95 % organic) 0.2 ml/min (300 nl/min at tip). As a guideline the method can look similar to this: 0–10 % B in approx. 10 min, 10–60 % B in 95 min, 60–100 % B in approx. 10 min, and finally 100 % B for approx. 10 min and back to low Buffer B with 100–0 % B in 5 min and equilibrate before next salt step with 0 % for 10 min. Depending on the complexity of the sample, the amount of salt steps has to be established. In our case a 6-step MudPIT was carried out for each of the soluble samples after fractionation and a full 12 step for the insoluble sample.
4. Set up the mass spectrometer to run MS/MS data-dependent method; as an example the settings we used were as follows: 1 full scan 400–1,600 m/z (centroid mode) with 5 ms/ms scan "Most intense if no parent masses found," checked repeat count: 2 repeat duration: 30 s, exclusion list size: 100, and exclusion duration: 60 s.

3.7 Data Processing

Once the data is collected it has to be searched against a database, and further filtered to confident identifications, and the peptides containing isoleucine have to be quantified.

1. Analyze the obtained MS/MS spectra with IPA (Integrated Proteomics Applications Inc.) using a non-redundant *Plasmodium* database (PlasmoDB.org, as of February 2012, version 8.2 available). Keep false-positive rate under 5 %. The software package has two optional search engines (SEQUEST

or ProLUCID); the results are filtered with DTASelect2 and quantified with Census. *See* **Note 6**.

2. Acquire relative quantification data of proteins by calculating a ratio of ratios. The output from Census is ratio 1 of the unlabeled and labeled sample for the drug-treated sample (unlabeled drug treated$_1$/labeled control$_1$) as well as ratio 2 for the control sample (unlabeled control$_2$/labeled control$_2$) of a different MudPIT run (normalization run). Through calculation of a ratio of ratios (1/2) for each quantified peptide, we filtered out eventual labeling deficiencies and received a ratio that corresponds to (unlabeled drug treated/unlabeled control). The changes observed by this calculation identify proteins with differential expression status upon drug treatment. *See* **Note 2**.

4 Notes

1. The synchronization can be carried out more easily with a preliminary magnet purification step before sorbitol treatment. We accomplished this with a small column that was washed and equilibrated before infected red blood cells flow through. The optimal flow rate is one drop every 2–3 s. This enrichment for late trophozoites and early schizonts needs to be carried out under the sterile bank with sterile solutions, as the parasite will need to keep on growing under these conditions.
2. Four samples were run in parallel: labeled control, unlabeled control, labeled sample, and unlabeled sample. The ratio of ratios guarantees that any changes because of cell growth under labeling conditions are accounted for.
3. Wash three times with 3 column volumes of sterile water. Washing the column as thoroughly as possible is important, otherwise cells might be damaged because of the presence of ethanol. The magnet is previously equilibrated with 3–4 column volumes of RPMI medium. A SuperMACS magnet is used with a D column that allows for 10^9 magnetically labeled cells. The sample is in RPMI medium and loaded on the column at a flow rate not to exceed 7 ml/min, which we accomplish with a needle with orifice G-21. We wash with 3–4 column volumes of RPMI medium. The magnet is removed, and cells are eluted with the amount of column volumes depending on the amount of plates loaded (e.g., 4 big plates, 4 column volumes for elution). The limit in our case for the D column is 4 big plates with a parasitemia of 8 % (trophozoites). For elution, wash with 40 ml and collect at the bottom of the column. Wash with 2 column volumes and collect at the top of the column with pipette and by using a three-way spigot and a 50 ml syringe to push RPMI medium through. The elution is

collected at the top of the column as some red blood cells will adhere to the top and will be lost if elution is taken by the classical drip method. Each column volume is collected separately as separate fractions and ought to be checked visually by Giemsa staining of thin films in order to assess where the enrichment is the highest. Cells are centrifuged without breaks for 5 min (it takes approx. 20 min for centrifuge to stop without breaks). All pellets are collected.

4. For 4 big plates the yield is about 500 μl parasitized erythrocytes with an 80–90 % enrichment. Pellet is resuspended in 10 ml buffer with saponin and incubated for 10 min at 37 °C (20 × volume of pellet). The color will change to a strong red wine color. Pellet is centrifuged at 2,300 rpm for 5 min. A second wash in saponin is carried out with no incubation. The pellet is washed twice with PBS and centrifuged at 1,500 × *g* in a small centrifuge.
5. For the freezing-thawing cycle, make sure that the cell extract is completely frozen. Once centrifugation steps are carried out, the insoluble pellet can be collected by freezing with N_2 the outside of tube and picking the insoluble fraction with a spatula.
6. Peptides were evaluated after first taking the union of search results (DTASelect2.0 output files). One of the strengths of SILAC is the quantification of proteins when only a single peptide is identified. The identification of the labeled peptide and its counterpart speaks for its presence.

Acknowledgement

The authors wish to thank the Alexander von Humboldt Foundation for funding Dr. Judith Helena Prieto. The work was supported by the Deutsche Forschungsgemeinschaft SPP 1710, BE1540/23-1 (to K.B.).

References

1. Gardner MJ, Hall N, Fung E et al (2002) Genome sequence of the human malaria parasite Plasmodium falciparum. Nature 419:498–511
2. Florens L, Washburn MP, Raine JD et al (2002) A proteomic view of the Plasmodium falciparum life cycle. Nature 419:520–526
3. Winzeler EA (2008) Malaria research in the post-genomic era. Nature 455:751–756
4. Lakshmanan V, Rhee KY, Daily JP (2011) Metabolomics and malaria biology. Mol Biochem Parasitol 175:104–111
5. malERA Consultative Group on Basic Science and Enabling Technologies (2011) A research agenda for malaria eradication: basic science and enabling technologies. PLoS Med 8:e1000399

6. Olsen JV, Schwartz JC, Griep-Raming J et al (2009) A dual pressure linear ion trap Orbitrap instrument with very high sequencing speed. Mol Cell Proteomics 8:2759–2769
7. Sharma K, Weber C, Bairlein M et al (2009) Proteomics strategy for quantitative protein interaction profiling in cell extracts. Nat Methods 6:741–744
8. Park SK, Venable JD, Xu T et al (2008) A quantitative analysis software tool for mass spectrometry-based proteomics. Nat Methods 5:319–322
9. Kuss C, Gan CS, Gunalan K et al (2012) Quantitative proteomics reveals new insights into erythrocyte invasion by Plasmodium falciparum. Mol Cell Proteomics 11(M111):010645
10. Dierick JF, Dieu M, Remacle J et al (2002) Proteomics in experimental gerontology. Exp Gerontol 37:721–734
11. Sanderson SJ, Xia D, Prieto H et al (2008) Determining the protein repertoire of Cryptosporidium parvum sporozoites. Proteomics 8:1398–1414
12. Lasonder E, Ishihama Y, Andersen JS et al (2002) Analysis of the Plasmodium falciparum proteome by high-accuracy mass spectrometry. Nature 419:537–542
13. Southworth PM, Hyde JE, Sims PF (2011) A mass spectrometric strategy for absolute quantification of Plasmodium falciparum proteins of low abundance. Malar J 10:315
14. Prieto JH, Koncarevic S, Park SK et al (2008) Large-scale differential proteome analysis in Plasmodium falciparum under drug treatment. PLoS One 3:e4098
15. Krijgsveld J, Ketting RF, Mahmoudi T et al (2003) Metabolic labeling of C. elegans and D. melanogaster for quantitative proteomics. Nat Biotechnol 21:927–931
16. Wu WW, Wang G, Baek SJ et al (2006) Comparative study of three proteomic quantitative methods, DIGE, cICAT, and iTRAQ, using 2D gel- or LC-MALDI TOF/TOF. J Proteome Res 5:651–658
17. McClatchy DB, Dong MQ, Wu C et al (2007) 15N metabolic labeling of mammalian tissue with slow protein turnover. J Proteome Res 6:2005–2010
18. Ong SE, Blagoev B, Kratchmarova I et al (2002) Stable isotope labeling by amino acids in cell culture, SILAC, as a simple and accurate approach to expression proteomics. Mol Cell Proteomics 1:376–386
19. Nirmalan N, Sims PF, Hyde JE (2004) Quantitative proteomics of the human malaria parasite Plasmodium falciparum and its application to studies of development and inhibition. Mol Microbiol 52:1187–1199
20. Cox J, Matic I, Hilger M et al (2009) A practical guide to the MaxQuant computational platform for SILAC-based quantitative proteomics. Nat Protoc 4:698–705
21. Trager W, Jensen JB (1976) Human malaria parasites in continuous culture. Science 193: 673–675
22. Cranmer SL, Magowan C, Liang J et al (1997) An alternative to serum for cultivation of Plasmodium falciparum in vitro. Trans R Soc Trop Med Hyg 91:363–365
23. Lambros C, Vanderberg JP (1979) Synchronization of Plasmodium falciparum erythrocytic stages in culture. J Parasitol 65: 418–420
24. Fivelman QL, Adagu IS, Warhurst DC (2004) Modified fixed-ratio isobologram method for studying in vitro interactions between atovaquone and proguanil or dihydroartemisinin against drug-resistant strains of Plasmodium falciparum. Antimicrob Agents Chemother 48:4097–4102
25. Koncarevic S, Bogumil R, Becker K (2007) SELDI-TOF-MS analysis of chloroquine resistant and sensitive Plasmodium falciparum strains. Proteomics 7:711–721

Chapter 18

Use of ^{13}C Stable Isotope Labelling for Pathway and Metabolic Flux Analysis in *Leishmania* Parasites

Eleanor C. Saunders, David P. de Souza, Jennifer M. Chambers, Milica Ng, James Pyke, and Malcolm J. McConville

Abstract

This protocol describes the combined use of metabolite profiling and stable isotope labelling to define pathways of central carbon metabolism in the protozoa parasite, *Leishmania mexicana*. Parasite stages are cultivated in standard or completely defined media and then rapidly transferred to chemically equivalent media containing a single ^{13}C-labelled nutrient. The incorporation of label can be followed over time or after establishment of isotopic equilibrium by harvesting parasites with rapid metabolic quenching. ^{13}C enrichment of multiple intracellular polar and apolar (lipidic) metabolites can be quantified using gas chromatography-mass spectrometry (GC-MS), while the uptake and secretion of ^{13}C-labelled metabolites can be measured by ^{13}C-NMR. Analysis of the mass isotopomer distribution of key metabolites provides information on pathway structure, while analysis of labelling kinetics can be used to infer metabolic fluxes. This protocol is exemplified using *L. mexicana* labelled with ^{13}C-U-glucose. The method can be used to measure perturbations in parasite metabolism induced by drug inhibition or genetic manipulation of enzyme levels and is broadly applicable to any cultured parasite stages.

Key words *Leishmania* spp., Metabolomics, Stable isotope, Central carbon metabolism, Gas chromatography, Mass spectrometry

1 Introduction

The parasitic protozoa comprise an important group of human pathogens that are the cause of devastating diseases, including malaria, toxoplasmosis, human African trypanosomiasis, Chagas disease, and the leishmaniases [1]. Current drug therapies for all of these diseases are limited and, in many cases, their utility has been severely undermined by the emergence of drug resistance in clinical isolates. As metabolic enzymes represent a major class of targets for antiparasitic drugs, new methods for quantitatively analyzing parasite metabolism are needed. While genome-based annotations have added considerably to our understanding of parasite metabolism and provide a global view of the metabolic potential of these

Christopher Peacock (ed.), *Parasite Genomics Protocols*, Methods in Molecular Biology, vol. 1201,
DOI 10.1007/978-1-4939-1438-8_18,

pathogens, they are limited in that many protein-encoding genes (often more than 50 %) in these pathogens cannot be assigned a function based on homology. Moreover, existing genomic, transcriptomic, and proteomic approaches provide limited information on the precise structure of metabolic networks, metabolic fluxes, and the relative contribution of nutrient salvage versus de novo metabolic pathways to parasite growth. Metabolomic approaches are increasingly being used to fill some of these gaps and to complement other -omic approaches [2–4]. In particular, metabolite profiling, combined with stable isotope labelling experiments, has the potential to (a) define the structure and operation of canonical as well as novel or unanticipated metabolic pathways in different developmental stages and (b) directly measure the contribution of different carbon sources to parasite growth in vitro and in vivo. These approaches are also widely applicable for monitoring metabolic processes and the physiological state of parasites during infection and/or in response to pharmacological treatments.

Leishmania spp. are the causative agents of a spectrum of diseases in humans that range from self-resolving, cutaneous lesions to the disseminating mucocutaneous and life-threatening visceral forms of disease [5]. Symptomatic diseases occur in more than 12 million people annually, resulting in 50,000 deaths, principally from visceral leishmaniasis [6]. There are currently no defined vaccines for human leishmaniasis and existing front-line drug treatments are inadequate due to toxicity, expense, and/or the emergence of drug-resistant strains [7]. Information on *Leishmania* metabolism is primarily derived from genetic and biochemical studies that are often limited in the number of metabolic pathways they investigate. More recently, genomic approaches have been used to reconstruct global and organelle-specific metabolic networks in *Leishmania* parasites [8], such as the curated genome-wide metabolic databases, LeishCyc (BioCyc) and KEGG [9], and the predicted proteome of glycosomes [10]. Genome-scale stoichiometric models of *Leishmania* metabolism have also been generated using flux balance analysis [11]. However, as mentioned above, these latter approaches are limited by the presence of many protein-encoding genes with no annotated function and recent studies that suggest that species-specific differences in parasite biology may be primarily regulated by differences in gene copy number rather than the expression of species-specific genes [12]. *Leishmania* spp. also constitutively transcribe most protein-encoding genes as polycistronic mRNA messages and lack conventional transcriptional regulation (or transcription factors), suggesting that the major regulatory processes occur after transcription and possibly even after protein translation [13, 14]. Finally, any attempt to model parasite metabolism in vivo requires detailed information on the availability of carbon sources and other essential nutrients in the relevant host niches that, in the case of intracellular stages of *Leishmania* spp., are still poorly defined [15]. For these reasons, it is clear that the direct measurement of metabolite levels and metabolic fluxes provides unique insights into parasite metabolism as well as being complementary to other -omics approaches.

Here, we describe a method for mapping key pathways in *Leishmania* central carbon metabolism using metabolite profiling and stable-isotope labelling approaches. This method can be generically applied to any microbial pathogen [16, 17], and even used to measure metabolic pathways in intracellular stages [18]. We have recently used this approach to dissect the role of glycosomal succinate metabolism and the mitochondrial TCA cycle in *L. mexicana* promastigotes [19]. In this method, relevant parasite stages are metabolically labelled with ^{13}C-labelled carbon sources (i.e., ^{13}C-U-glucose, $^{13}C_1$-glucose, ^{13}C-U-glutamate, ^{13}C-U-aspartate) under standard (non-perturbed) growth conditions. Rapid sampling of labelled parasites at early time points is then used to measure the kinetics of labelling. Alternatively, parasites can be sampled at a single final time point after isotopic equilibrium has been reached. In both protocols, it is essential that parasites are harvested under conditions that result in effective metabolic arrest [19, 20]. Following metabolite extraction, the incorporation of ^{13}C label into intracellular metabolite pools is detected using either gas chromatography (GC)- or liquid chromatography (LC)-mass spectrometry (MS). The level of ^{13}C enrichment can be determined after correcting for natural abundance of naturally occurring stable isotopes and, when combined with analysis of the culture supernatant using ^{13}C-NMR, can be used to determine absolute fluxes.

2 Materials

2.1 *Leishmania mexicana* Culture and Stable Isotope Labelling

1. Tissue culture hood.
2. Culture medium, such as RPMI supplemented with 10 % iFBS or completely defined medium (CDM) supplemented with 0.5 % bovine serum albumin (non-delipidated) [29] (*see* **Note 1**).
3. Labelling medium, RPMI or CDM medium containing 13 mM ^{13}C-U-glucose (Cambridge Isotope Laboratories) instead of unlabelled glucose. Labelling medium is supplemented with 0.5 % BSA (non-delipidated) instead of iFBS (*see* **Note 2**). Media can be stored at 4 °C and warmed to 27 °C before use.
4. 80 % ethanol (v/v) spray.
5. Tissue culture flasks (25 cm^2).
6. Incubator (27 °C) for culturing parasites.
7. Auto pipette, sterile plastic pipettes (10 ml), and sterile pipette tips (20–200 μl).

8. Centrifuge (RT, capable of spinning 15 ml tubes).
9. Screw cap tubes (15 ml, sterile).

2.2 Metabolic Quenching and Extraction

1. Dry ice-ethanol bath (half fill a plastic container with ethanol and slowly add dry ice, until the rapid bubbling slows and dry ice is still visible).
2. Methanol:water (3:1 v/v) containing 5 μM *scyllo*-inositol (3 ml of methanol (99.9 % purity, GC-MS grade) + 980 μl water (ultrapure water) and 20 μl of a 1 mM *scyllo*-inositol stock).
3. 1× phosphate-buffered saline (PBS) (approximately 30 ml), *chilled*.
4. Chloroform (99.8 % purity, liquid chromatography grade).
5. Ultrapure water.
6. Thermometer (−10 to 50 °C range, *see* **Note 17**). An electronic temperature probe can also be used.
7. Medium-sized Styrofoam box, filled with ice.
8. Refrigerated centrifuges, *chilled*: This method utilizes two refrigerated centrifuges (a large benchtop centrifuge for 15 ml conical screw cap tubes and a microcentrifuge for 1.5 ml microtubes) although one is sufficient.
9. Water bath, 60 °C.
10. 15 ml conical screw cap tube (*see* **Note 17**).
11. 1.5 ml safe-lock microtubes (*see* **Note 17**).

2.3 GC-MS Derivatization and Instrumentation

1. Methoxyamine hydrochloride (Aldrich, 226904) in pyridine (20 mg/ml) is prepared just prior to use. Weigh 10 mg of methoxyamine hydrochloride into a clean GC-MS vial. Add 500 μl of pyridine using a positive displacement pipette with a clean glass capillary. Cap vial and vortex, ensuring that no white crystals remain. This solution is hydroscopic and should be used immediately to prevent uptake of moisture.
2. Methanol (99.9 % purity grade details).
3. N,O-bis(trimethylsilyl)trifluoroacetamide (BSTFA) + 1 % trimethylchlorosilane (TMCS) (1 ml vials under argon atmosphere, e.g., Thermo Fisher Scientific, TS-38831).
4. Microtubes (safe-lock).
5. Tweezers (*see* **Note 17**).
6. GC-MS vials (screw cap vials, clear, Agilent), vial inserts (250 μl pulled point glass, Agilent), and screw caps (Agilent) (*see* **Note 17**).
7. Centrifugal evaporator, capable of spinning 1.5 ml microtubes, e.g., Christ RVC vacuum concentrator.

Table 1
GC-MS settings for the analysis of central carbon metabolites in *L. mexicana*

	Electron impact (EI)	Chemical ionization (CI)
Injection	• One wash of the syringe with hexane (discarded) • Four washes with sample (not discarded) • Sample injection, 1 μl (no pre- or post-injection delay) • Three washes of the syringe with methanol (discarded) • Five washes of the syringe with hexane (discarded)	
Inlet	General purpose split/splitless liner with glass wool, tapered and deactivated (for example Agilent part number 5183-4711), held at 250 °C	
Carrier gas	Helium (ultrahigh purity). Constant column flow rate of 1 ml/min. Inlet purged at 20 ml/min for 60 s gas saver at 15 ml/min after 60 s	
Capillary column	A multipurpose, low bleed column suitable for GC-MS, for example VF5ms capillary column (Agilent CP9013; 0.25 mm i.d., 0.25 μm film thickness) containing a 10 m EZ guard section	
Oven program (run time 24 min)	• 70 °C, 1 min hold • Ramp from 70 to 295 °C at 12.5 °C/min • Ramp from 295 to 320 °C at 25 °C/min • 320 °C, 2 min hold	
Transfer line	250 °C	
Ionization source	Inert EI ion source (Agilent part number G3170-65760), 250 °C	CI ion source complete (Agilent part number G3170-65403), 300 °C. CI reagent gas, methane 14 %
Detector	Full scan, range 50–500 *m/z*. SIM can also be performed	Full scan, range 50–700 *m/z*. SIM can also be performed

8. GC-MS system—We use an Agilent 6890 Series GC coupled with a 5973 mass selective detector, and a 7683 series automatic liquid sampler. GC is performed with a 30 m. MS is performed using electron impact (EI) or chemical ionization (CI) (Table 1). MSD ChemStation (ChemStation D.01.02.16, Agilent Technologies) is used for instrument control, editing/running the sample sequence, and data analysis.

2.4 NMR Analysis

1. Internal standards (1 ml volumes).

   5 mM stock solution of 3-(trimethylsilyl)-1-propanesulfonic acid-d_6 (DSS-d_6 with 0.2 % w/v NaN_3 in D_2O).

   21.4 mM ^{13}C-U-glycerol with 0.2 % w/v NaN_3 in D_2O.

   21.4 mM imidazole with 0.2 % in NaN_3 in D_2O.

2. ^{13}C-spectra are collected at 200 MHz using an 800 MHz Bruker-Biospin Avance NMR fitted with a cryoprobe. NMR data is analyzed using the TopSpin™ software package, version 2.

3 Methods

A workflow for the following steps is given in Fig. 1.

3.1 L .mexicana Culture and Stable Isotope Labelling

1. *L. mexicana* promastigotes are cultivated in RPMI containing 10 % iFBS at 27 °C in standard 75 cm^2 tissue culture flasks (*see* **Note 2** for alternative medium). Promastigotes (2.4×10^8 total) are harvested while in mid-log phase (1×10^7 parasite/ml) by centrifugation ($805 \times g$, 10 min, room temperature) and the cell pellet resuspended in completely defined medium (CDM) containing 0.5 % BSA. Following incubation for 1 h at 27 °C, parasites (1.6×10^8 total for 3–4 replicate analyses) are harvested by centrifugation and then resuspended in labelling CDM in which individual unlabelled carbon sources are replaced by ^{13}C-U-labelled carbon sources at a density of 2×10^7 parasites/ml. These steps must be performed quickly to ensure that pelleted parasites do not become nutrient limited

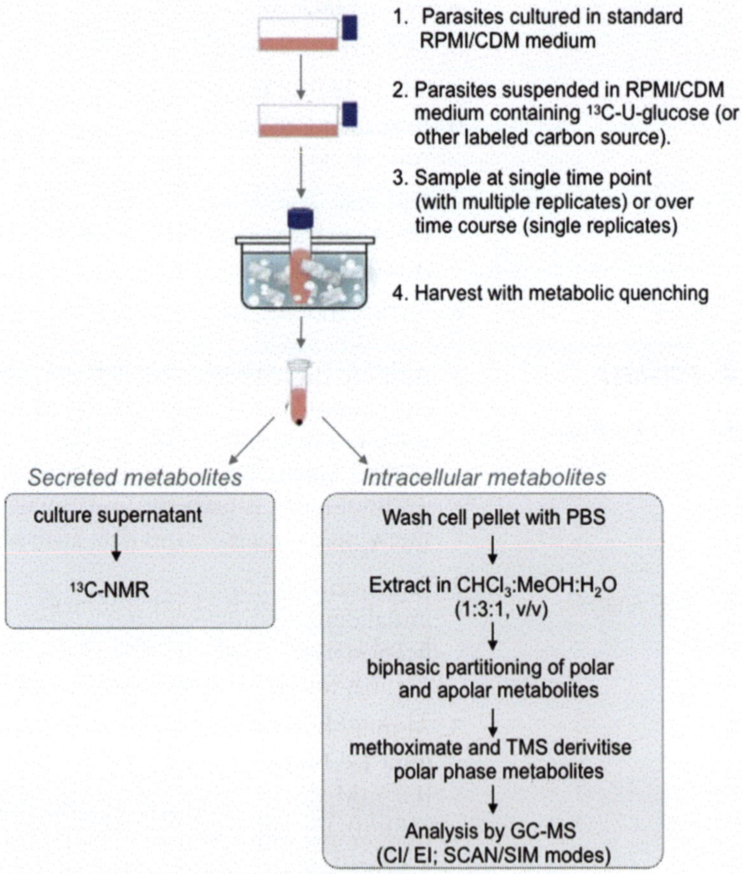

Fig. 1 Workflow for ^{13}C-labeling experiments

(*see* **Note 3**). As a control, an equivalent aliquot of parasites (2×10^7 parasites/ml) is suspended in unlabelled CDM and then harvested to provide a zero time point. Parasite suspensions are transferred to 75 cm^2 culture flasks and incubated at standard growth temperature of 27 °C prior to harvest.

3.2 Metabolic Quenching and Extraction

1. Single aliquots can be harvested at multiple time points or multiple replicates can be generated at a single time point (*see* **Note 4**). Before quenching ensure that centrifuges, 1× PBS, and consumables are chilled to 0–4 °C (*see* Subheading 3). Prepare the dry ice-ethanol bath (see Subheading 3). Prepare the methanol:water extraction solution (*see* Subheading 3).
2. When parasites are harvested at multiple time points, individual aliquots of parasite suspension (containing 4×10^7 parasites total) are rapidly transferred to a 15 ml tube that is immediately immersed in the dry ice-ethanol bath with gentle agitation. As the culture suspension reaches 4 °C (after approximately 10 s, as monitored by insertion of a thermometer) the tube is transferred to an ice bath (*see* **Note 5**). This procedure ensures that non-perturbed cells are chilled rapidly *without* freezing. The chilled cell suspension (*see* **Note 6**) is subsequently centrifuged (805 × *g*, 10 min, 0 °C) and the culture supernatant transferred to a separate tube and stored on ice for subsequent ^{13}C-NMR analysis (*see* **Note 7**, *see* Subheading 3.4). For experiments that involve the sampling of multiple replicates at a single time point, the entire culture flask (1.6×10^8 cells total, for 3–4 replicates) is immersed in the dry ice-ethanol bath to bring the entire culture suspension to 4 °C. The flask contents can then be transferred to a pre-chilled 15 ml conical screw cap tube and centrifuged (805 × *g*, 10 min, 0 °C) with the resulting culture supernatant transferred to a separate tube as above.
3. The metabolically quenched cell pellets (single sample, 4×10^7 parasites total) are resuspended in pre-chilled PBS (1 ml) and transferred to 1.5 ml safe-lock microtubes. For multiple replicates, the larger cell pellet (1.6×10^8 parasites total) is resuspended in pre-chilled PBS (approximately 3 ml) and aliquots of the cell suspension (4×10^7 parasites total) are transferred to separate 1.5 ml safe-lock microtubes to generate experimental replicates. Cell pellets are immediately centrifuged (10,000 × *g*, 1 min, 0 °C) in a microcentrifuge and the resulting pellets are washed a further two times with pre-chilled PBS (*see* **Note 8**).
4. Chloroform ($CHCl_3$, 50 μl) is added to each sample and the cell pellet disrupted by vortexing and/or sonication in a bath sonicator. The prepared methanol:water (3:1 v/v) solution containing the internal standard (200 μl) is added to make a monophasic solution of chloroform:methanol:water (final 1:3:1 v/v) and the suspension is vortex mixed. Sonication in a

water bath (4 × 20 s) can be used to aid sample dispersal and metabolite extraction. Samples are then incubated in a water bath (60 °C, 15 min) to facilitate metabolite extraction. The solution should remain in a single phase throughout.

5. After centrifugation in a microcentrifuge (16,100 × *g*, 0 °C, 5 min), the supernatants are transferred to a new 1.5 ml safe-lock microtube and the pellet discarded (*see* **Note 9**). Water (100 μl) is added to the extracts to form a two-phase system (final 1:3:3 v/v), vortex mixed, and centrifuged (16,100 × *g*, 0 °C, 1 min) to facilitate phase separation.
6. The upper aqueous phase, containing polar metabolites, is transferred to a new 1.5 ml safe-lock microtube using a 200 μl pipette with plastic tips, ensuring that the interface is not disturbed (*see* **Note 10**). Transfer the entire upper phase extract to GC-MS glass vial insert as aliquots (75–100 μl at a time) with intermediate drying in vacuo in a Christ RVC vacuum concentrator (55 °C). In parallel, transfer a 10 μl aliquot of the stock solution of metabolite mix (*see* **Note 11**) to glass GC-MS inserts and dry under the same conditions. Wash all samples twice with anhydrous methanol (40 μl followed by 20 μl) with intermediate drying to ensure that samples are completely dry.

3.3 GC-MS Analysis of Intracellular Metabolites

1. Place the glass inserts into 2.5 ml GC-MS vials (use clean tweezers to avoid any contamination) and add prepared methoxyamine chloride solution (20 μl) using a positive displacement glass pipette. Cap the vials, vortex mix (*see* **Note 12**), and incubate with gentle agitation overnight on a plate rocker or orbital shaker at room temperature.
2. Add BSTFA containing 1 % TMCS (20 μl) (*see* **Note 13**). Vortex mix and incubate at room temperature. The derivatization time should be kept constant by adding BSTFA reagent 1 h prior to each of the GC/MS runs. Sample runs should be randomized and a standard metabolite mixture and a reagent blank (comprising pure hexane) run after every four sample runs (*see* **Note 14**).
3. GC is performed using a DB5ms capillary column (J&W Scientific, 30 m, 250 Table 1). The GC inlet and GC-MS transfer line temperatures are maintained at 270 °C and 250 °C, respectively. The oven temperature gradient is programmed as follows: 70 °C (1 min); 70–295 °C at 12.5 °C/min; 295–320 °C at 25 °C/min; 320 °C for 2 min. MS can be performed in either electron ionization (EI, Fig. 2a) or chemical ionization (CI) mode. In EI mode, metabolites are ionized with high-energy electrons at an applied energy of 70 eV, resulting in substantial fragmentation and the generation of complex mass spectral fingerprints (Fig. 2b). Metabolite identification is achieved by comparison with commercial or in-house-

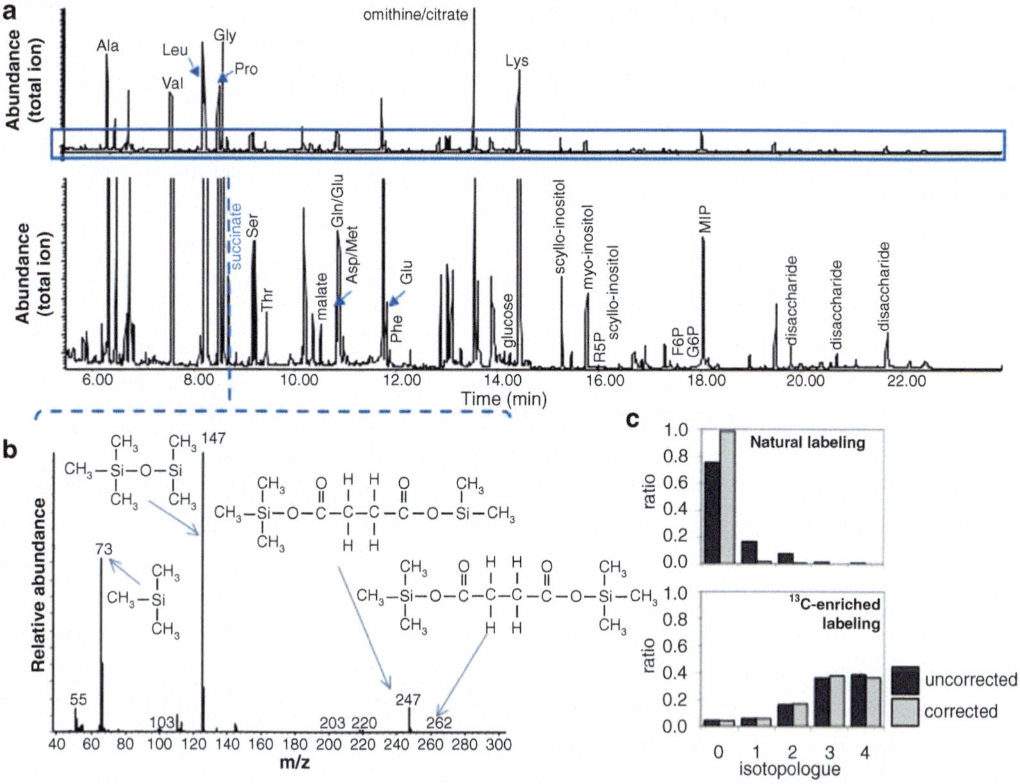

Fig. 2 GC-MS analysis of *L. mexicana* polar metabolites. Total ion chromatogram (TIC) of *L. mexicana* promastigote polar metabolites after methoximation and trimethylsilyl (TMS) derivatization. The *lower* panel shows expanded TIC scale and selected detected metabolites. (**b**) EI mass spectrum of the TMS derivative of succinate and chemical structure of the diagnostic fragments, *m/z* 247 and *m/z* 262. The structure of generic TMS fragments at *m/z* 147 and *m/z* 73 is also shown. (**c**) The molar ratios of different isotopologues of succinate (M to M + 4) calculated from the relative areas of *m/z* 247–251 ions. The *upper* panel shows the molar ratios of succinate isotopologues (M to M + 4) derived from unlabelled parasites before and after correction for naturally occurring isotopes. The *lower* panel shows the molar ratios of isotopologues of succinate derived from $^{13C\text{-}U\text{-}glucose}$-fed parasites. Background correction is particularly important when labelling with the proffered isotope is low. The predominance of +3 and +4 labelled succinate in the ^{13}C-glucose-fed parasites reflects the conversion of phosphoenolpyruvate (PEP) to oxaloacetic acid by the glycosomal PEP carboxykinase and the operation of a complete TCA cycle, respectively

generated mass spectral libraries and the GC retention time of authentic standards (Fig. 2b). In CI mode, sample ionization is mediated by reagent gas ions, resulting in less fragmentation and improved detection of the molecular ion. GC-CI-MS is useful when information on the labelling of all carbons in a metabolite is needed.

The mass spectrometer can be operated in full scan and/or selected ion monitoring modes. Full scan is recommended for metabolite identification and initial validation of the metabolite labelling. SIM mode involves the collection of fewer ions with a concomitant increase in sensitivity and quantitative accuracy. This may be particularly important when there is a need to quantitate low-abundant, molecular ions or high mass fragments in EI mass spectrum.

3.4 NMR Analysis of the Culture Supernatants

1. An aliquot of the culture supernatant (540 μl) is gently transferred to a microfuge tube to which is added the DSS-d_6 (60 μl), ^{13}C-U-glycerol (5 μl), and imidazole (5 μl,) internal standards, all in D_2O. The spiked sample solution is then transferred to a clean NMR tube.
2. ^{13}C-spectra are collected at 200 MHz using an 800 MHz Bruker-Biospin Avance NMR fitted with a cryoprobe. Samples are maintained at 25 °C and spun at 20 Hz during sample collection. ^{13}C spectra were acquired using the Avance zgpg pulse program with power-gated ^{1}H decoupling. A pre-scan delay (DE) of 80.78 μs, a delay (D1) between pulses of 2.0 s, and an acquisition time (AQ) of 0.78 s were used. For each sample, 4 dummy scans were followed by 4,000 scans with a receiver gain set at 2050. The resulting ^{13}C free induction decays (FIDs) were processed with Bruker TopSpin™ version 2.0 (the exponential function with LB = 5.0 Hz was applied in the frequency domain prior to Fourier transform, baseline correction, and integration).

3.5 Data Analysis

GC-MS chromatograms are processed using ChemStation software (Agilent Technologies) and metabolite peaks detected in full scan mode and assigned based on mass spectra/diagnostic ions and retention times relative to authentic standards (*see* **Note 15**) (Fig. 2b). The fractional labelling of each metabolite is determined from selected fragments or molecular ions (Table 2). In each case, the peak areas for the monoisotopic (unlabelled) and associated isotopomer ions are determined using the ChemStation software (as extracted ion feature). Unlabelled metabolites contain a number of isotopomer peaks as a result of the presence of naturally occurring ^{13}C, ^{15}N, and $^{29/30}$Si (Fig. 2c, upper panel). These isotopes are present in both the metabolite and in the modifying TMS derivatization groups. Naturally labelled metabolite isotopologues can account for between 5 and 20 % of the monoisotopic species, being higher in carbon-rich metabolites such as fatty acids, and should be subtracted when calculating ^{13}C enrichment in the presence of ^{13}C-labelled carbon sources. Background labelling subtraction is achieved using the algorithm developed by Sauer and colleagues (*see* **Note 16**) [21]. An R-script that allows automated isotopologue corrections is available from the authors' laboratory upon request.

The operation of specific metabolic pathways can be inferred from careful analysis of the isotopomer distribution in individual metabolites. For example, analysis of the relative abundance of

Table 2
Fragment ions used to measure 13C enrichment in selected metabolites detected using GC-EI-MS

Metabolic pathway	Metabolite	Fragment ions used to measure enrichment[a]	Other fragment(s), m/z
Glycolysis	Glucose-6-phosphate Fructose-6-phosphate	357 357	471, 387, 315, 299, 217, 160 471, 387, 315, 299, 217, 160
Pentose phosphate pathway	Ribose-5-phosphate Ribulose-5-phosphate	357 357	459, 403, 315, 299, 217 387, 315, 299
Ate glycolysis	Phosphoenolpyruvate	369 (M–CH_3)	384 (M), 299, 225, 211, 133
Tricarboxylic acid/ glycosomal succinate fermentation	Citrate/isocitrate Malate Succinate Fumarate	465 (M–CH_3) 335 (M–CH_3) 247 (M–CH_3) 245 (M–CH_3)	480 (M), 375, 363, 347, 273 350 (M), 245, 233 262 (M)
Amino acids	Alanine (2TMS) Aspartate (3TMS) Glutamate (3TMS) Glycine (2TMS) Isoleucine Leucine Lysine (3TMS) Proline (2TMS) Threonine (2TMS) Valine (2TMS)	116 232 348 (M–CH_3) 204 (M–CH_3) 260 (M–CH_3) 260 (M–CH_3) 434 (M) 216 320 (M–CH_3) 218	233, 218, 190, 116 349 (M), 334, 218 363(M), 246, 230, 218 102 232, 218, 158 232, 218, 158 317, 230, 156, 128 244, 259 291, 218, 203, 117 144
Other	Myo-inositol	318	432, 305, 265, 217, 204, 191

M = molecular ion

[a]Fragment ions used to determine ^{13}C enrichment should contain most/all of the carbons in the parent metabolite and be abundant enough to allow robust quantitation

different citrate isotopologues in ^{13}C-glucose-fed *L. mexicana* promastigotes provided definitive evidence that a canonical oxidative TCA cycle operates in this stage, fuelled by acetyl-CoA and C4 dicarboxylic acids (malate) derived from glucose catabolism [19]. Relative metabolic fluxes can also be assessed using selectively labelled precursors. For example, labelling studies using a combination of ^{13}C-U-glucose and $^{13}C_1$-glucose or $^{13}C_{1,2}$-glucose can be used to measure the relative flux of hexose through glycolysis and the pentose phosphate pathway [22]. More complex computational methods have been developed to measure metabolic flux ratios from steady-state ^{13}C-isotopomer measurement data [23–25]. However, these frequently require a detailed model of the metabolism that is often missing for parasitic protozoa. An alternative approach, that is highly amenable to study of eukaryotic parasites, is kinetic flux profiling [26]. In this approach, metabolic fluxes are inferred from the rate of labelling of individual metabolites, with early intermediates being labelled with faster kinetics than later intermediates/end products. A major advantage of this approach is that specific metabolic fluxes can be determined using short labelling experiments (<1 h) without detailed prior knowledge of the entire metabolic network. This approach has now been used to measure metabolic fluxes from ^{13}C-labelling experiments in several eukaryotic systems [27, 28].

4 Notes

1. We have found that 4×10^7 cell equivalents are sufficient to detect major intermediates in *Leishmania* promastigote central carbon metabolism (sugars, sugar phosphates, organic acids, amino acids). However, it is important to the cellular equivalents needed for each new parasite species/developmental stages being measured.
2. Glucose-free RPMI is commercially available and can be used for preliminary labelling experiments. Replacement of iFBS with 0.5–1 % BSA in the labelling media reduces the level of ^{12}C-glucose in the medium. To further explore central carbon metabolism, labelled substrates other than glucose may be of interest. These can be added as a bolus over the concentrations usually found in RPMI or can be included in completely defined media (CDM) [29].
3. The supernatant can be removed using a 10 ml pipette and any residual liquid removed using a 200 μl pipette. It is important that this step is completed rapidly to reduce metabolic changes in densely packed parasite pellets.
4. Near maximal ^{13}C enrichment occurs in many *L. mexicana* central carbon intermediates within 3 h. However, true steady-state isotopic labelling can take >24 h in rapidly dividing stages (and much longer in slow/none growing stages) due to slow turnover of carbohydrate or lipid reserves. Time course experiments should be undertaken during the initial validation of the method to establish optimal labelling period(s).
5. It is critical that cell suspensions are not frozen during quenching, with resultant cell lysis and loss of intracellular metabolites. It is useful to monitor the temperature in the cell suspension, using an electronic or standard thermometer. To empirically determine the temperature at which the parasite suspension should be transferred from the dry ice-ethanol bath, perform trial experiments with aliquots of pre-warmed medium.
6. Extraction of parasite metabolites should be initiated immediately after cooling to 0 °C and harvesting by centrifugation to minimize continuing metabolic reactions and changes in labelling patterns.

7. ^{13}C-NMR analysis of the culture supernatants is used to calculate the uptake and secretion of metabolic end products. Samples can be stored at −70 °C prior to NMR analysis [19].
8. Parasite pellets are washed with 1 ml PBS without suspension of the cells and the first two supernatant washes removed using a fine-tipped plastic pipette. The third supernatant wash can be removed using a 200 μl pipette. This sequence is highly effective in minimizing carryover of media components.
9. The solvent-extracted pellet can be stored at −70 °C and subsequently used to quantify protein and/or DNA levels.
10. Incorporation of ^{13}C into lipid-linked fatty acids or sterols in the lipidic phase can be determined by GC-MS after methanolysis and TMS derivatization as previously described [19].
11. The metabolite mix typically contains the following metabolite standards at 1 nmol (scyllo-inositol, glucose, glucose 6-phosphate, fructose 6-phosphate, ribose 5-phosphate, ribulose 5-phosphate, sucrose), 10 nmol (alanine, arginine, aspartate, asparagine, glutamate, glutamine, glycine, histidine, isoleucine, leucine, methionine, proline, putrescine, serine, threonine, tryptophan, tyrosine, valine), and 5 nmol (citrate, malate, succinate, fumarate, pyruvate, phosphoenolpyruvate, lactic acid).
12. Automated methoximation and derivatization are possible if the GC is fitted with robotics, such as the GERSTEL MPS2 online derivatization module. Online derivatization decreases the amount of manual handling done by the experimenter and allows precise control over incubation times, temperatures, etc. While invaluable when large sample numbers (100 s) are routinely analyzed, it is not required for small-scale ^{13}C-labelling metabolomics experiments.
13. Polar metabolites can be converted to volatile derivatives using a number of different chemistries which is beyond the scope of this protocol report (but see technical summaries, such as Supelco (http://www.sigmaaldrich.com/etc/medialib/docs/Supelco/Application_Notes/4537.Par.0001.File.tmp/4537.pdf)). The N,O-bis(trimethylsilyl) trifluoroacetamide (BSTFA) reagent used in this protocol contains the catalyst trimethylchlorosilane (TMCS), and replaces the active hydrogen in polar organic compounds with a silyl group ($-Si(CH_3)_3$) (http://www.sigmaaldrich.com/etc/medialib/docs/Aldrich/General_Information/bstfa.Par.0001.File.tmp/bstfa.pdf).
14. For the experiment outlined here, one hexane wash every 5–6 samples is sufficient.
15. Care should be taken when using GC-MS mass spectral libraries as the best match may not be the top hit of the list. It is important to scroll through all the matches to find the best hit.

Greater weighting should be given to fragment ions derived from the metabolite rather than the derivatization groups (such as *m/z* 73 and 147 *m/z* derived from trimethylsilyl groups). The best mass spectra should be generated for each metabolite (i.e., by sampling left/right of the TIC peak apex) before searching the library. For a more targeted approach to metabolite identification, the "parametric retrieval" option allows the experimenter to obtain the mass spectra for any metabolite within the library. This information can then be used to search the TIC. Again, cross-referencing to authentic standards is critical. Accurate metabolite identification in the unlabelled samples is critical as a metabolite's mass spectrum will be different in a ^{13}C-enriched metabolite and therefore a library search may not correctly identify the metabolite.

16. Measurement correction for naturally occurring isotopes can be accomplished based on (1) empirical measurements of unlabelled samples or (2) published probabilities of isotopic occurrence. The first method can be used in nontargeted metabolic profiling where the precise chemical composition of some metabolites may not be known. The second method is amenable to automated signal processing and provides increased accuracy. Both methods require a construction of a correction matrix, left inverse of which is used to pre-multiply the measurement vector. In the first approach the correction matrix is constructed using unlabelled measurements (details can be found in [30]. In the second approach published natural mass isotopic distributions of atoms in the derivatization groups are used to construct the correction matrix [31], followed by the correction of the carbon skeleton itself [32]. Note that correction of the carbon skeleton is often not needed in ^{13}C-MFA, as carbon backbone natural isotopic abundance can be simulated by most modeling software.
17. All tubes, tweezers, thermometers, vials, inserts, etc. should be free of contaminants and stored under dust-free conditions to minimize spurious peaks in the GC-MS analyses. Tweezers and thermometers can be wiped down with 80 % ethanol using lint-free paper prior to use. Work areas should be kept clean and gloves should be worn when handling tubes, etc. We have found that Eppendorf-branded tips are resistant to organic solvents and generate minimal levels of plasticizers that are readily detected by GC-MS.

References

1. Stuart K, Brun R, Croft S, Fairlamb A, Gürtler RE, McKerrow J, Reed S, Tarleton R (2008) Kinetoplastids: related protozoan pathogens, different diseases. J Clin Invest 118:1301–1310
2. Creek DJ, Anderson J, McConville MJ, Barrett MP (2012) Metabolomic analysis of trypanosomatid protozoa. Mol Biochem Parasitol 181:73–84

3. Saunders EC, DE Souza DP, Naderer T, Sernee MF, Ralton JE, Doyle MA, Macrae JI, Chambers JL, Heng J et al (2010) Central carbon metabolism of Leishmania parasites. Parasitology 137:1303–1313
4. T'kindt R, Scheltema RA, Jankevics A, Brunker K, Rijal S, Dujardin J-C, Breitling R, Watson DG, Coombs GH, Decuypere S (2010) Metabolomics to unveil and understand phenotypic diversity between pathogen populations. PLoS Negl Trop Dis 4:e904
5. Murray HW, Berman JD, Davies CR, Saravia NG (2005) Advances in leishmaniasis. Lancet 366:1561–1577
6. Bern C, Maguire JH, Alvar J (2008) Complexities of assessing the disease burden attributable to leishmaniasis. PLoS Negl Trop Dis 2:e313
7. Croft SL, Sundar S, Fairlamb AH (2006) Drug resistance in leishmaniasis. Clin Microbiol Rev 19:111–126
8. Ivens AC, Peacock CS, Worthey EA, Murphy L, Aggarwal G, Berriman M, Sisk E, Rajandream MA, Adlem E et al (2005) The genome of the kinetoplastid parasite, *Leishmania major*. Science 309:436–442
9. Doyle MA, MacRae JI, De Souza DP, Saunders EC, McConville MJ, Likić VA (2009) LeishCyc: a biochemical pathways database for *Leishmania major*. BMC Syst Biol 3:57
10. Opperdoes FR, Szikora J-P (2006) *In silico* prediction of the glycosomal enzymes of Leishmania major and trypanosomes. Mol Biochem Parasitol 147:193–206
11. Chavali AK, Whittemore JD, Eddy JA, Williams KT, Papin JA (2008) Systems analysis of metabolism in the pathogenic trypanosomatid *Leishmania major*. Mol Syst Biol 4:177
12. Rogers MB, Hilley JD, Dickens NJ, Wilkes J, Bates PA, Depledge DP, Harris D, Her Y, Herzyk P et al (2011) Chromosome and gene copy number variation allow major structural change between species and strains of *Leishmania*. Genome Res 21:2129–2142
13. Cohen-Freue G, Holzer TR, Forney JD, McMaster WR (2007) Global gene expression in *Leishmania*. Int J Parasitol 37:1077–1086
14. Kramer S (2012) Developmental regulation of gene expression in the absence of transcriptional control: the case of kinetoplastids. Mol Biochem Parasitol 181:61–72
15. McConville MJ, Naderer T (2011) Metabolic pathways required for the intracellular survival of *Leishmania*. Annu Rev Microbiol 65: 543–561
16. Eisenreich W, Slaghuis J, Laupitz R, Bussemer J, Stritzker J, Schwarz C, Schwarz R, Dandekar T, Goebel W, Bacher A (2006) ^{13}C isotopologue perturbation studies of *Listeria monocytogenes* carbon metabolism and its modulation by the virulence regulator PrfA. Proc Natl Acad Sci U S A 103:2040–2045
17. Eisenreich W, Dandekar T, Heesemann J, Goebel W (2010) Carbon metabolism of intracellular bacterial pathogens and possible links to virulence. Nat Rev Microbiol 8:401–412
18. Eylert E, Schär J, Mertins S, Stoll R, Bacher A, Goebel W, Eisenreich W (2008) Carbon metabolism of *Listeria monocytogenes* growing inside macrophages. Mol Microbiol 69: 1008–1017
19. Saunders EC, Ng WW, Chamber JM, Ng M, Naderer T, Kroemer JO, Likic VA, McConville MJ (2011) Isoptopomer profiling of *Leishmania mexicana* promastigotes reveals important roles for succinate fermentation and aspartate uptake in TCA cycle anaplerosis, glutamate synthesis and growth. J Biol Chem 286:27706–27717
20. De Souza DP, Saunders EC, McConville MJ, Likić VA (2006) Progressive peak clustering in GC-MS metabolomic experiments applied to *Leishmania* parasites. Bioinformatics 22: 1391–1396
21. Zamboni N, Fendt SM, Ruhl M, Sauer U (2009) ^{13}C-based metabolic flux analysis. Nat Protoc 4:878–892
22. Bartek T, Blombach B, Lang S, Eikmanns BJ, Wiechert W, Oldiges M, Nöh K, Noack S (2011) Comparative ^{13}C metabolic flux analysis of pyruvate dehydrogenase complex-deficient, L-valine-producing *Corynebacterium glutamicum*. Appl Environ Microbiol 77:6644–6652
23. Tang YJ, Martin HG, Myers S, Rodriguez S, Baidoo EEK, Keasling JD (2009) Advances in analysis of microbial metabolic fluxes via ^{13}C isotopic labeling. Mass Spectrom Rev 28:362–375
24. Quek L-E, Wittmann C, Nielsen LK, Krömer JO (2009) OpenFLUX: efficient modelling software for ^{13}C-based metabolic flux analysis. Microb Cell Fact 8:25
25. Rantanen A, Rousu J, Jouhten P, Zamboni N, Maaheimo H, Ukkonen E (2008) An analytic and systematic framework for estimating metabolic flux ratios from ^{13}C tracer experiments. BMC Bioinformatics 9:266
26. Yuan J, Bennett BD, Rabinowitz JD (2008) Kinetic flux profiling for quantitation of cellular metabolic fluxes. Nat Protoc 3:1328–1340
27. Clasquin MF, Melamud E, Singer A, Gooding JR, Xu X, Dong A, Cui H, Campagna SR, Savchenko A et al (2011) Riboneogenesis in yeast. Cell 145:969–980

28. Lemons JMS, Feng X-J, Bennett BD, Legesse-Miller A, Johnson EL, Raitman I, Pollina EA, Rabitz HA, Rabinowitz JD, Coller HA (2010) Quiescent fibroblasts exhibit high metabolic activity. PLoS Biol 8:e1000514

29. Merlen T, Sereno D, Brajon N, Rostand F, Lemesre JL (1999) *Leishmania* spp: completely defined medium without serum and macromolecules (CDM/LP) for the continuous in vitro cultivation of infective promastigote forms. Am J Trop Med Hyg 60: 41–50

30. Jennings ME, Matthews DE (2005) Determination of complex isotopomer patterns in isotopically labeled compounds by mass spectrometry. Anal Chem 77:6435–6444

31. van Winden WA, Wittmann C, Heinzle E, Heijnen JJ (2002) Correcting mass isotopomer distributions for naturally occurring isotopes. Biotechnol Bioeng 80:477–479

32. Lee WN, Byerley LO, Bergner EA, Edmond J (1991) Mass isotopomer analysis: theoretical and practical considerations. Biol Mass Spectrom 20:451–458

Chapter 19

Molecular Genotyping of *Trypanosoma cruzi* for Lineage Assignment and Population Genetics

Louisa A. Messenger, Matthew Yeo, Michael D. Lewis, Martin S. Llewellyn, and Michael A. Miles

Abstract

Trypanosoma cruzi, the etiological agent of Chagas disease, remains a major public health problem in Latin America. Infection with *T. cruzi* is lifelong and can lead to a spectrum of pathological sequelae ranging from subclinical to lethal cardiac and/or gastrointestinal complications. Isolates of *T. cruzi* can be assigned to six genetic lineages or discrete typing units (DTUs), which are broadly associated with disparate ecologies, transmission cycles, and geographical distributions. This extensive genetic diversity is also believed to contribute to the clinical variation observed among chagasic patients. Unravelling the population structure of *T. cruzi* is fundamental to understanding Chagas disease epidemiology, developing control strategies, and resolving the relationship between parasite genotype and clinical prognosis.

To date, no single, widely validated, genetic target allows unequivocal resolution to DTU-level. In this chapter we present standardized methods for strain DTU assignment using PCR-restriction fragment length polymorphism analysis (PCR-RFLP) and nuclear multilocus sequence typing (MLST). PCR-RFLPs have the advantages of simplicity and reproducibility, requiring limited expertise and few laboratory consumables. MLST data are more laborious to generate but more informative; DNA sequences are readily transferable between research groups and amenable to recombination detection and intra-lineage analyses. We also recommend a mitochondrial (maxicircle) MLST scheme and a panel of 28 microsatellite loci for higher resolution population genetics studies.

Due to the scarcity of *T. cruzi* in blood and tissue, all of these genotyping techniques have limited sensitivity when applied directly to clinical or biological specimens, particularly when targets are single (MLST) or low copy number (PCR-RFLPs). We therefore describe essential protocols to isolate parasites, derive biological clones, and extract *T. cruzi* genomic DNA from field and clinical samples.

Key words *Trypanosoma cruzi*, PCR, Genotyping, Phylogenetics, Microsatellites, MLST, RFLP, Mitochondria, Sequencing

1 Introduction

Chagas disease is the most important parasitic infection in Latin America, where an estimated 10–12 million individuals are infected, with a further 80 million at risk [1]. The etiological agent, *Trypanosoma cruzi*, is a complex zoonosis, with a broad

Christopher Peacock (ed.), *Parasite Genomics Protocols*, Methods in Molecular Biology, vol. 1201,
DOI 10.1007/978-1-4939-1438-8_19, © Springer Science+Business Media New York 2015

endemic range that extends from the southern United States to Argentinean Patagonia. Disease transmission primarily occurs in areas where humans are exposed to the contaminated feces of domiciliated triatomine bug vectors. In the absence of chemotherapy, infection with *T. cruzi* is life-long and can lead to a spectrum of pathological sequelae ranging from subclinical to debilitation and death by irreversible cardiac and/or gastrointestinal syndromes [2]. Diagnosis and treatment options are further complicated by disproportionate distributions of disease pathologies; cardiomyopathies occur throughout South and Central America, whereas gastrointestinal complications are more common south of the Amazon. It has been suggested that this geographical heterogeneity is associated with genetic variation among *T. cruzi* strains [3–5]. However, the relationship between parasite genotype and clinical outcome remains controversial.

T. cruzi displays remarkable genetic diversity and a range of markers can be used to delineate this species. Typing of genetic polymorphisms in conserved housekeeping genes can define major genetic lineages [6–8], while analysis of hypervariable loci, such as microsatellites [9–11] or kDNA minicircle sequences [12–14], potentially allows identification of profiles specific to individual strains. Historically, the study of *T. cruzi* has been hindered by a lack of standardized molecular typing methods and the use of various alternative nomenclatures (recently reviewed in [15]). One useful conceptual development has been that of the discrete typing unit (DTU) which groups isolates using shared molecular characteristics but without explicitly defining their evolutionary relatedness [16]. For *T. cruzi* multilocus genotyping has consistently identified six DTUs, which are each correlated with distinct but not exclusive ecologies and geographical distributions [17]. Recently, DTU nomenclature has been revised by international consensus to reflect the current understanding of *T. cruzi* genetic diversity [18].

Molecular analyses suggest that *T. cruzi* has a predominantly clonal population structure, punctuated by infrequent genetic exchange events. DTUs TcI-TcIV form monophyletic clades and TcV and TcVI are known to be recent inter-lineage hybrids [19]. Multilocus sequence typing (MLST) data support these designations with TcI-TcIV characterized by substantial allelic homozygosity, likely resulting from recurrent, genome-wide and dispersed gene conversion, while TcV and TcVI display natural heterozygosity and minimal distinction, sharing intact alleles from their parental progenitors (TcII and TcIII) [20–22]. The origin(s) of these hybrid lineages is unresolved and it is presently contested whether they arose from two independent genetic exchange events [19, 23], or a single incidence of hybridization followed by clonal divergence [24] (recently reviewed in [25]).

The epidemiological relevance of the *T. cruzi* DTUs has also been the subject of considerable debate, with evidence emerging to support historical and contemporary associations of particular lineages with different transmission ecologies. In general, TcI, TcII, TcV, and TcVI are frequently isolated from domestic cycles and are responsible for the majority of human infections. The distribution of domestic TcI extends from the Amazon Basin northwards, where it is the primary cause of Chagas disease in Venezuela and Colombia [26, 27]. TcI is also ubiquitous among arboreal sylvatic transmission cycles throughout Latin America [28, 29], and commonly isolated from *Didelphis* species and the triatomine tribe *Rhodniini* [30]. By contrast, TcII, TcV, and TcVI appear restricted to domestic transmission in southern parts of South America. Strains from these three DTUs are rarely isolated from sylvatic reservoirs and their ecological niches are largely undefined [17]. TcIII has a dispersed terrestrial distribution that ranges from Amazonia to Argentina, where it is primarily transmitted by *Panstrongylus geniculatus* to *Dasypus novemcinctus* and other burrowing mammals [31–33]. TcIV is poorly understood, principally because several genotyping methods fail to distinguish this lineage from others, particularly from TcIII [6]. However, TcIV is known to circulate sympatrically with TcI in wild primates [34] and raccoons [29] in Amazonia and North America, respectively. It is also increasing in epidemiological importance and has been implicated in recent oral outbreaks in Amazonia [34, 35] and as a secondary agent of Chagas disease in Venezuela [3]. As yet, TcIII and TcIV only sporadically invade domestic transmission cycles, but this may reflect inadequate and/or inappropriate sampling and the insensitivity of conventional genotyping methods. Furthermore some of these ecological associations are complicated by overlapping sylvatic and domestic transmission cycles and frequent mixed infections in individual humans [36, 37], mammalian reservoirs [32, 38], and triatomine vectors [8, 39–41].

Elucidating the population structure and genetic diversity of *T. cruzi* is critical to furthering our understanding of the complex transmission dynamics, clinical variability and phylogeography underlying Chagas disease. Secondarily, detecting recombination among *T. cruzi* populations is also of profound epidemiological importance considering the expansion of the hybrid lineages within the domestic niche and the capacity for genetic exchange to drive the evolution of novel virulent recombinant strains. As yet, no single marker affords complete, unequivocal DTU resolution, and reliance on only one target is inadvisable given the potential confounding influence of hybridization [12, 21]. In this chapter we describe genotyping methods to assign *T. cruzi* isolates to DTU-level and those that can be used for higher resolution intra-lineage diversity studies.

For optimal genotyping results we strongly recommend the use of biologically cloned material, wherever possible. Multiclonality within individual *T. cruzi* strains can manifest as mixed infections of different DTUs [37–39, 41, 42] or multiple variants of the same genetic lineage [41, 43]. Infra-population genetic diversity is largely determined by levels of super-infection from discrete sources [44], inbreeding among closely related genotypes [45] and simultaneous transmission of mulitclonal populations between hosts [38]. We describe routine protocols to isolate *T. cruzi* parasites from infected patients/mammals and triatomine bugs. We then recommend methods to derive biological clones from *T. cruzi* strains, including plating on a solid medium [41], limiting dilution or micromanipulation of individual parasites [46] and also suggest techniques to extract genomic DNA from resulting axenic cultures as well as directly from clinical and field isolates.

To genotype *T. cruzi* isolates to DTU-level we recommend a standardized triple-assay comprising PCR product size polymorphism analysis of the 24Sα rRNA gene (LSU rDNA) and PCR-restriction fragment-length polymorphism analysis (PCR-RFLP) using heat shock protein 60 (*HSP60*) and glucose-6-phosphate isomerase (*GPI*) [47]. These PCR-based assays have the advantages of being easily reproducible and implemented with limited expertise, technical resources, and sample material. However, this methodology was developed using a panel of biologically cloned reference isolates and is reliant on the presence/absence of specific single-nucleotide polymorphisms (SNPs) and may be insensitive to mutations in as yet untested strains. In addition, both PCR-RFLPs are based on low copy targets and were evaluated using culture-extracted DNA and thus their sensitivity against field or clinical specimens and for resolving mixed infections may vary. The repertoire of PCR-based *T. cruzi* genotyping techniques is ever expanding and those recently described by D'Avila et al. [48], Burgos et al. [49], and Van der Auwera et al. [50] may be more appropriate for the aforementioned sample types.

Another technique that we advocate to unambiguously assign isolates to DTU-level is nuclear multilocus sequence typing (MLST). This is a sequence-based approach, which exploits conserved nucleotide diversity present in four single-copy housekeeping genes (3-hydroxy-3-methylglutaryl-CoA reductase (*HMCOAR*), glucose-6-phosphate isomerase (*GPI*), mitochondrial peroxidase (*TcMPX*), and rho-like GTP binding protein (*RHO1*)) [20, 51] and can be used as an adjunct to DTU allocation, in the rare cases when PCR-RFLPs fail to unequivocally genotype samples. MLST data offer minimal subjectivity in analysis and are transferable and electronically portable, allowing for interlaboratory comparisons without the exchange of reference isolates. Our research group, along with others [51], is presently expanding this panel of loci with the aim of formalizing an MLST scheme that can be used for high resolution genetic diversity studies [52].

We anticipate that with the rapid advancement of sequencing technology, current genotyping methods will imminently be superseded by comparative genomics of multiple representatives from each *T. cruzi* DTU [53]. However, in the interim, we recommend the use of a panel of 28 microsatellite loci (multilocus microsatellite typing, MLMT) and ten mitochondrial gene fragments (maxicircle MLST) to address intra-lineage population genetic hypotheses using appropriately assembled isolate cohorts. Microsatellites are short, neutrally evolving, codominant tandem repeats, with mutation rates several orders of magnitude higher than protein-coding genes [54]. These hypervariable markers provide a method of identifying and tracking individual strains as well as assessing the frequency of alleles in a given population. This MLMT scheme is highly discriminatory and has previously been used to describe intra-TcI and -TcIII population structuring on a continental scale [10, 33], to reveal genetic exchange within TcI domestic/peridomestic populations in Ecuador [11] and to expose the role of mammalian reservoirs in the diversification of *T. cruzi* genotypes [38]. Potential drawbacks associated with MLMT include limited transferability between laboratories and genotyping errors arising from homoplasy (when alleles are identical in sequence but not descent), allelic dropout, misprinting, artifact peaks, and stutter patterns [55]. Maxicircle MLST exploits inherent features of mitochondrial DNA, specifically uniparental inheritance and a faster mutation rate (compared to nuclear DNA), to detect directional gene flow among closely related isolates. Maxicircle MLST can be used in parallel with nuclear loci (MLMT and/or nuclear MLST) to identify phylogenetic incongruence, which is indicative natural recombination. This combined approach has uncovered novel mitochondrial introgression events occurring across geographically dispersed TcI populations [56] and revealed pervasive genetic exchange within Colombian TcI transmission cycles [44].

Herein, we describe the protocols used to (1) isolate *T. cruzi* samples from infected patients, mammalian hosts and triatomine bugs, (2) derive biological clones from *T. cruzi* strains by micromanipulation, plating on solid medium, or limiting dilution, (3) extract parasite DNA from cultured epimastigotes, human/mammalian hemocultures, or triatomine bug intestinal homogenates, (4) assign isolates to DTU-level using PCR-RFLP analysis, (5) amplify, sequence and analyze nuclear and maxicircle MLST targets, and (6) amplify, multiplex, and analyze microsatellite allele sizes.

2 Materials

Prepare all solutions using ultrapure water (purify deionized water to attain a resistivity of 18 MΩ cm at 25 °C) and analytical grade reagents.

All experimental work which involves handling live *T. cruzi* parasites should be conducted in a designated laboratory and in accordance with locally approved standard operating procedures (SOPs). All manipulation of live material should be carried out within a Class II microbiological safety cabinet. Accidental infection with *T. cruzi* can arise from inoculation of a single infectious metacylic trypomastigote or bloodstream-form trypomastigote and at least sixty-five cases of laboratory transmission have been recorded [57]. However, the risk of laboratory-acquired infection is minimal if appropriate guidelines are adhered to (*see* **Note 1**).

2.1 Isolation of T. cruzi

Here we present possible protocols for the isolation of *T. cruzi*, techniques for biologically cloning resulting parasites and methods of extracting *T. cruzi* genomic DNA. Choice of technique will depend upon the original source of the parasite and quality of DNA template required for downstream applications (*see* **Note 2**).

To maximize the likelihood of isolate recovery and minimize loss of clonal diversity, we strongly recommend processing all field and clinical samples by simultaneously (1) inoculating strains into axenic culture (proceed to (2) before the first re-passage), (2) biologically cloning strains, and (3) directly extracting genomic DNA (*see* Fig. 1).

2.1.1 Direct Hemoculture from Patients/Mammals

1. Blood agar base (Sigma-Aldrich, UK).
2. Agar (Sigma-Aldrich, UK).
3. Tryptone (Sigma-Aldrich, UK).

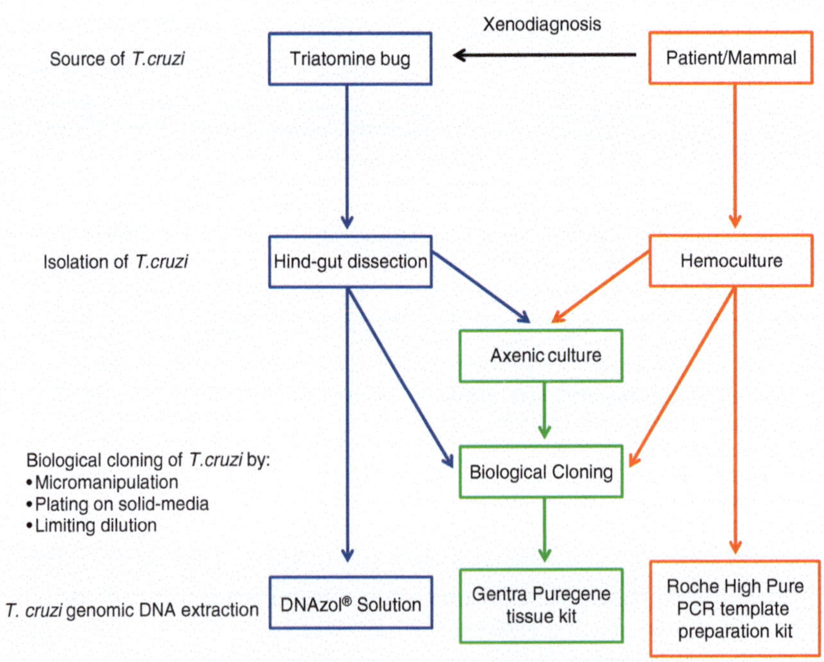

Fig. 1 Schematic of *T. cruzi* strain isolation, cloning, and DNA extraction protocols

4. Sodium chloride, NaCl (Sigma-Aldrich, UK).
5. Sterile defibrinated rabbit blood.
6. Gentamycin (Sigma-Aldrich, UK).
7. 5-Fluorocytosine (Sigma-Aldrich, UK).
8. Liver infusion broth (Difco™, Becton Dickinson, USA).
9. Glucose (Sigma-Aldrich, UK).
10. Potassium chloride, KCl (Sigma-Aldrich, UK).
11. Disodium hydrogen phosphate, Na_2HPO_4 (Sigma-Aldrich, UK).
12. Hemin (Sigma-Aldrich, UK).
13. Sodium hydroxide, NaOH (Sigma-Aldrich, UK).
14. Heat-inactivated fetal calf serum (Sigma-Aldrich, UK).
15. Ketamine hydrochloride (Sigma-Aldrich, UK).
16. Absolute ethanol (analytical reagent grade).
17. Iodine (Sigma-Aldrich, UK).
18. Guanidine hydrochloride (Sigma-Aldrich, UK).
19. Ethylenediaminetetraacetic acid (EDTA) disodium salt dihydrate (Sigma-Aldrich, UK).
20. Refrigerated centrifuge.
21. Sterile 15 ml centrifuge tubes (Greiner Bio-One, UK).
22. Sterile Nunclon™ Δ flat sided tubes (#734-2068, Nunc, UK).
23. Rubber caps from sodium heparin vacutainer tubes (#368480, Scientific Laboratory Supplies, UK).
24. Parafilm (VWR, UK).
25. Sterile 1, 2, 5, and 20 ml BD Plastipak™ syringes with needles (Becton Dickinson, USA).
26. BD Vacutainer® plus plastic K_2 EDTA tubes (Becton Dickinson, USA).
27. 28 °C humidified incubator.
28. Inverted microscope.
29. Sterile glycerol (VWR, UK).
30. Sterile cryovials (Nunc, Denmark).

2.1.2 Isolation from Triatomine Bugs (Xenodiagnosis)

1. Uninfected triatomine bug colony.
2. Mercuric chloride, $HgCl_2$ (Sigma-Aldrich, UK).
3. Hydrochloric acid sp.gr.1.18, HCl (VWR, UK).
4. Sodium chloride, NaCl (Sigma-Aldrich, UK).
5. Absolute ethanol (analytical reagent grade).
6. Gentamycin (Sigma-Aldrich, UK).
7. 5-Fluorocytosine (Sigma-Aldrich, UK).
8. Blood agar base (Sigma-Aldrich, UK).

9. Agar (Sigma-Aldrich, UK).
10. Tryptone (Sigma-Aldrich, UK).
11. Sterile defibrinated rabbit blood.
12. Sterile Nunclon™ Δ flat sided tubes (#734-2068, Nunc, UK).
13. Rubber caps from sodium heparin vacutainer tubes (#368480, Scientific Laboratory Supplies, UK).
14. Parafilm (VWR, UK).
15. Sterile broad forceps (Scientific Laboratory Supplies, UK).
16. Sterile Watchmakers' forceps (Scientific Laboratory Supplies, UK).
17. Perspex dissection screen.
18. Sterile microscope slides (VWR, UK).
19. Sterile 13 mm microscope cover glasses (VWR, UK).
20. Sterile broad microspatula (Scientific Laboratory Supplies, UK).
21. Sterile 1.5 ml graduated microcentrifuge tubes (Anachem, UK).
22. Sterile 1 ml plastic Pasteur pipettes (Scientific Laboratory Supplies, UK).
23. 28 °C humidified incubator.
24. Inverted microscope.

2.2 Biological Cloning of *T. cruzi*

2.2.1 Micromanipulation

1. Blood agar base (Sigma-Aldrich, UK).
2. Agar (Sigma-Aldrich, UK).
3. Tryptone (Sigma-Aldrich, UK).
4. Sodium chloride, NaCl (Sigma-Aldrich, UK).
5. Sterile defibrinated rabbit blood.
6. Gentamycin (Sigma-Aldrich, UK).
7. 5-Fluorocytosine (Sigma-Aldrich, UK).
8. Mercuric chloride, $HgCl_2$ (Sigma-Aldrich, UK).
9. Hydrochloric acid sp.gr.1.18, HCl (VWR, UK).
10. Absolute ethanol (analytical reagent grade).
11. Sterile microcapillary tubes (Sigma-Aldrich, UK).
12. Bunsen burner (Scientific Laboratory Supplies, UK).
13. Microscope slides (VWR, UK).
14. Sterile 13 mm microscope cover glasses (VWR, UK).
15. Sterile 7 ml Bijou tubes (Sterilin, UK).
16. Disposable hemocytometers (Immune Systems, UK).
17. Sterile Watchmakers' forceps (Scientific Laboratory Supplies, UK).
18. 28 °C humidified incubator.
19. Inverted microscope.

2.2.2 Plating on Solid Medium

1. RPMI-1640 liquid medium (Sigma-Aldrich, UK #R0883).
2. Tryptone (Sigma-Aldrich, UK).
3. HEPES sodium salt (Sigma-Aldrich, UK).
4. Hemin (Sigma-Aldrich, UK).
5. Sodium hydroxide, NaOH (Sigma-Aldrich, UK).
6. Heat-inactivated fetal calf serum (Sigma-Aldrich, UK).
7. Sodium glutamate (Sigma-Aldrich, UK).
8. Sodium pyruvate (Sigma-Aldrich, UK).
9. Streptomycin (Sigma-Aldrich, UK).
10. Penicillin (Sigma-Aldrich, UK).
11. Blood agar base (Sigma-Aldrich, UK).
12. Agar (Sigma-Aldrich, UK).
13. Sterile defibrinated rabbit blood.
14. Disposable hemocytometers (Immune Systems, UK).
15. Low melting point (LMP) agarose (Sigma-Aldrich, UK).
16. Sodium chloride, NaCl (Sigma-Aldrich, UK).
17. Gentamycin (Sigma-Aldrich, UK).
18. Parafilm (VWR, UK).
19. Sterile 90 mm petri dishes (Sterilin, UK).
20. Sterile 200 μl pipette tips (Star Laboratories, UK).
21. Sterile 48-well cell culture plates (Becton Dickinson, USA).
22. 28 °C humidified incubator.
23. Inverted microscope.

2.2.3 Limiting Dilution

1. RPMI-1640 liquid medium (Sigma-Aldrich, UK #R0883).
2. Tryptone (Sigma-Aldrich, UK).
3. HEPES sodium salt (Sigma-Aldrich, UK).
4. Hemin (Sigma-Aldrich, UK).
5. Sodium hydroxide, NaOH (Sigma-Aldrich, UK).
6. Heat-inactivated fetal calf serum (Sigma-Aldrich, UK).
7. Sodium glutamate (Sigma-Aldrich, UK).
8. Sodium pyruvate (Sigma-Aldrich, UK).
9. Streptomycin (Sigma-Aldrich, UK).
10. Penicillin (Sigma-Aldrich, UK).
11. Disposable hemocytometers (Immune Systems, UK).
12. Sterile 96-microwell culture plates (Nunc, UK).
13. 28 °C humidified incubator.
14. Inverted microscope.

2.3 Preparation of Parasite Genomic DNA

1. Gentra Puregene tissue kit (Qiagen, UK).
2. High Pure PCR template preparation kit (Roche, UK).
3. DNAzol® solution (Life Technologies, UK).
4. Centrifuge.
5. Microcentrifuge.
6. Vortex.
7. Water bath.
8. Phosphate-buffered saline (PBS) (Sigma-Aldrich, UK).
9. Absolute isopropanol (analytical reagent grade).
10. Absolute ethanol (analytical reagent grade).
11. Sodium hydroxide, NaOH (Sigma-Aldrich, UK).
12. Sterile 15 ml centrifuge tubes (Greiner Bio-One, UK).
13. Sterile 1.5 ml graduated microcentrifuge tubes (Anachem, UK).
14. Spectrophotometer.

2.4 PCR-RFLP Amplification

1. Oligonucleotides to amplify the D7 divergent domain of the 24Sα rRNA gene (LSU rDNA), heat shock protein 60 (*HSP60*), and glucose-6-phosphate isomerase (*GPI*) (*see* Table 1).
2. *T. cruzi* genomic DNA.
3. 10× NH_4 buffer (Bioline, UK).
4. 50 mM $MgCl_2$ solution (Bioline, UK).
5. Deoxynucleotide solution mix (10 mM stock of each dNTP) (New England Biolabs, UK).
6. BIOTAQ™ DNA polymerase (Bioline, UK).

Table 1
PCR-RFLP gene fragments and primer details

PCR-RFLP target	Primer name	Primer Sequence (5′→3′)
LSU rDNA[a]	D71	AAGGTGCGTCGACAGTGTGG (20)
	D72	TTTTCAGAATGGCCGAACAGT (21)
HSP60[b]	*HSP60*_for	GTGGTATGGGTGACATGTAC (20)
	*HSP60*_rev	CGAGCAGCAGAGCGAAACAT (20)
GPI[c]	*GPI*_for	GGCATGTGAAGCTTTGAGGCCTTTTTCAG (29)
	*GPI*_rev	TGTAAGGGCCCAGTGAGAGCGTTCGTTGAATAGC (34)

[a]Primer sequences according to Brisse et al. [73]
[b]Primer sequences according to Strurm et al. [74]
[c]Primer sequences according to Gaunt et al. [75]

7. Sterile 0.2 ml 96-well PCR reaction plates and adhesive plate seals (Fisher Scientific, UK) or 0.2 ml PCR tube strips and caps (VWR, UK).
8. PCR machine.
9. Microcentrifuge.
10. Sterile 1.5 ml graduated microcentrifuge tubes (Anachem, UK).

2.5 Nuclear MLST PCR Amplification

1. Oligonucleotides to amplify 3-hydroxy-3-methylglutaryl-CoA reductase (*HMCOAR*), glucose-6-phosphate isomerase (*GPI*), mitochondrial peroxidase (*TcMPX*), and rho-like GTP binding protein (*RHO1*) (*see* Table 2).
2. *T. cruzi* genomic DNA.
3. 5× colorless GoTaq® reaction buffer (Promega, UK).
4. Deoxynucleotide solution mix (10 mM stock of each dNTP) (New England Biolabs, UK).
5. GoTaq® DNA polymerase (Promega, UK).
6. Sterile 0.2 ml 96-well PCR reaction plates and adhesive plate seals (Fisher Scientific, UK) or 0.2 ml PCR tube strips and caps (VWR, UK).
7. PCR machine.
8. Microcentrifuge.
9. Sterile 1.5 ml graduated microcentrifuge tubes (Anachem, UK).

2.6 Maxicircle MLST PCR Amplification

1. Oligonucleotides to amplify ten maxicircle gene fragments (*see* Table 3 and Fig. 2).
2. *T. cruzi* genomic DNA.
3. 10× NH_4 buffer (Bioline, UK).
4. 50 mM $MgCl_2$ solution (Bioline, UK).
5. Deoxynucleotide solution mix (10 mM stock of each dNTP) (New England Biolabs, UK).
6. BIOTAQ™ DNA polymerase (Bioline, UK).
7. Sterile 0.2 ml 96-well PCR reaction plates and adhesive plate seals (Fisher Scientific, UK) or 0.2 ml PCR tube strips and caps (VWR, UK).
8. PCR machine.
9. Microcentrifuge.
10. Sterile 1.5 ml graduated microcentrifuge tubes (Anachem, UK).

2.7 MLMT PCR Amplification

1. Oligonucleotides to amplify 28 microsatellite loci. Five fluorescent dyes with different emission spectra are used to label the forward primers: 6-FAM and TET (Proligo, Germany) and NED, PET, and VIC (Applied Biosystems, UK) (*see* Table 4 and Fig. 3).

Table 2
Nuclear MLST gene fragments and primer details

Gene fragment	Primer name	Primer sequence (5′ → 3′)	Annealing temperature (°C)	Amplicon size (bp)	Sequence start 5′	Sequence start 3′	Sequenced fragment (bp)
HMCOAR	*HMCOAR* Fwd	AGGAGGCTTTTGAGTCCACA (20)	55	564	TGAGTCCA	TCCAACAA	554
	HMCOAR Rvs	TCCAACAACACCAACCTCAA (20)					
GPI	*GPI* Fwd	CGCCATGTTGTGAATATTGG (20)	55	424	TGAATATT	CAATGAGT	405
	GPI Rvs	GGCGGACCACAATGAGTATC (20)					
TcMPX	*TcMPX* Fwd	ATGTTTCGTCGTATGGCC (18)	55	678	TACATGGA	CGCACCGT	505
	TcMPX Rvs	TGCGTTTTTCTCAAAATATTC (21)					
RHO1	*RHO1* Fwd	AGTTGCTGCTTCCCATCAAT (20)	55	463	CTTCCCAT	TCTGCACA	455
	RHO1 Rvs	CTGCACAGTGTATGCCTGCT (20)					

Table 3
Maxicircle MLST gene fragments and primer details

Gene fragment	Genome position[a]	Primer name	Primer sequence (5′ → 3′)	Annealing temperature (°C)	Amplicon Size (bp)[b]	Sequence Start 5′	Sequence Start 3′	Sequenced Fragment (bp)[c]
12S rRNA	639-901	*12S* Fwd *12S* Rvs	GTTTATTAAATGCGTTTGTCTAAGAA (26) GCCCCAATCAAACATACAA (19)	50	299	GTCTAAGA	TACGTATT	263
9S rRNA	1077-1309	*9S* Fwd *9S* Rvs	TGCAATTCGTTAGTTGGGTTA (21) TCCACACCCATTAAATAGCACT (22)	50	302	TAAAATCG	TATTATTA	233
CYT b	4126-4733	*Sp18* Fwd *Sp18* Rvs	GACAGGATTGAGAAGCGAGAGAG (23) CAAACCTATCACAAAAAGCATCTG (24)	50	717	TTTGTYTT	TAATAYCA	608
*Murf1*a	6011-6393	*Murf1*a Fwd *Murf1*a Rvs	AAGGCRATGGGRATAGWRCCTATAC (25) TGGAACAATTRTATATCAGATTRGGA (26)	50	482	ACTAAGYA	ACTTTYTA	383
*Murf1*b	6528-6900	*Murf1*b Fwd *Murf1*b Rvs	ACMCCCATCCATTCTTCR (18) CCTTTGATYTATTGTGATTAACRKT (25)	50	423	CAAAAATT	GGATTTAT	373
ND1	7643-8011	*ND1* Fwd *ND1* Rvs	GCACTTTCTGAAATAATCGAAAA (23) TTAATCTTATCAGGATTTGTTAGCC (25)	50	400	TCGAAAAA	TTGTTAGC	369
COII	8194-8610	*COII* Fwd *COII* Rvs	GTTATTATCTTTTGTTTGTTTTGTGTG (27) AACAATTGGCATAAATCCATGT (22)	50	560	CTTTCTAC	ACCTRCCY	417
ND4	12153-12392	*ND4* Fwd *ND4* Rvs	TTTTTGAAAGTCTATTTTTCCCA (23) CTTCAACATGCATTTCCGGTT (21)	50	302	AATTTTAA	CGGTYRTC	240
*ND5*a	13829-14250	*ND5*a Fwd *ND5*a Rvs	TATGRYTAACYTTTTCATGYTCRG (24) GTCCTTCCATYGCATCYGG (19)	50	503	GTACATAY	TYTTYGTA	422
*ND5*b	14274-14640	*ND5*b Fwd *ND5*b Rvs	ARAGTACACAGTTTGGRYTRCAYA (24) CTTGCYAARATACAACCACAA (21)	50	444	TGATTRCC	GYARACCA	367

[a]Genome position according to the TcI Sylvio X10/1 reference maxicircle genome [76]
[b]Amplicon size according to TcI Sylvio X10/1. Indels in other strains may cause size variation
[c]Sequence length according to TcI Sylvio X10/1. Indels in other strains may cause length variation

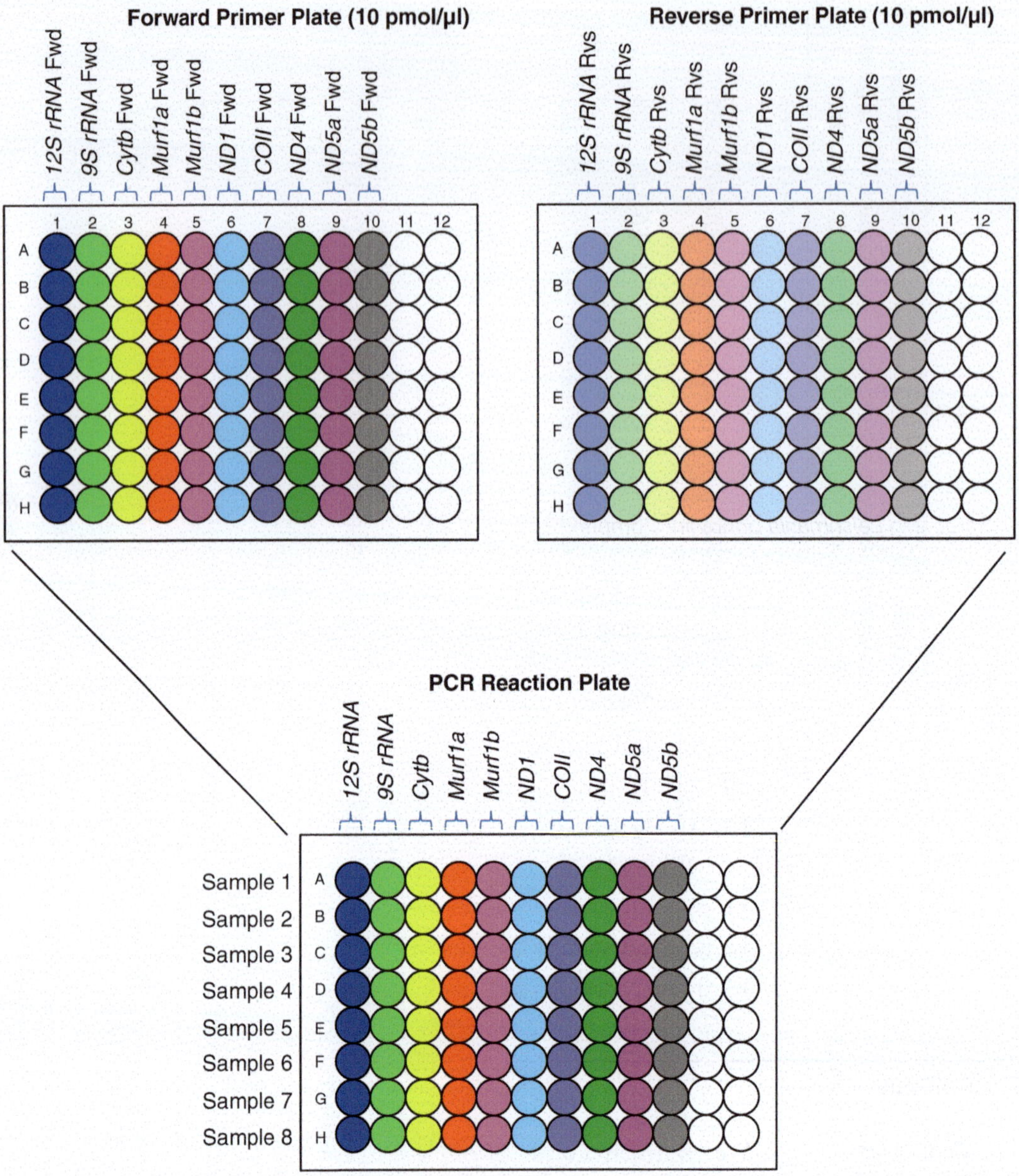

Fig. 2 Maxicircle primer positions in 96-well plate

2. 0.5× TE buffer (10 mM Tris–HCl and 1 mM EDTA (pH 8)) (both Sigma-Aldrich, UK).
3. *T. cruzi* genomic DNA.
4. 10× ThermoPol reaction buffer (New England Biolabs, UK).
5. 50 mM $MgCl_2$ solution (Bioline, UK).
6. Deoxynucleotide solution mix (10 mM stock of each dNTP) (New England Biolabs, UK).
7. *Taq* DNA polymerase (New England Biolabs, UK).

Table 4
***T. cruzi* microsatellite loci and primer sequences**

Chromosome[a]	Primer code	Repeat type	Forward/reverse primer (5′–3′)
6	6529(CA)a	$(CA)_n$	TGTGAAATGATTTGACCCGA
			AGAGTCACGCCGCAAAGTAT
6	6529(TA)b	$(TA)_n$	TGAAGGAGATTCTCTGCGGT
			CTCTCATCTTTTGTTGTGTCCG
6	mclf10	$(CA)_nA(CA)_n$	GCGTAGCGATTCATTTCC
			ATCCGCTACCACTATCCAC
10	6855(TA)(GA)	$(TA)_n(GA)_n$	TGTGATCAACGCGCATAAAT
			TTCCATTGCCTCGTTTTAGA
15	11863(CA)	$(CA)_n$	AGTTGACATCCCCAAGCAAG
			CCCTGATGCTGCAGACTCTT
19	TcUn3	Unknown	CTTAAAGAGATACAAGAGGGAAGG
			CTGTTATTTCAATAACACGGGG
19	10101(TA)	$(TA)_n$	AACCCGCGCAGATACATTAG
			TTCATTTGCAGCAACACACA
24	8741(TA)	$(TA)_n$	TGTAACGGTAGGTCTCAATTCG
			TTGCACTTGTGTATCTCGCC
27	10101(TC)	$(TC)_n$	CGTACGACGTGGACACAAAC
			ACAAGTGGGTGAGCCAAAAG
27	10101(CA)c	$(CA)_n$	GTGTCGTTGCTCCCAAACTC
			AAACTTGCCAAATGTGAGGG
27	10101(CA)a	$(CA)_n$	GTCGCCATCATGTACAAACG
			CTGTTGGCGAATGGTCATAA
34	6559(TC)	$(TC)_n$	CGCTCTCAAAGGCACCTTAC
			ATATGGACGCGTAGGAGTGC
37	10187(TTA)	$(TTA)_n$	GAGAGAGATTCGGAAACTAATAGC
			CATGTCCCTTCCTCCGTAAA
37	10187(CA)(TA)	$(CA)_n(TA)_n$	CATGTCATTAAGTGGCCACG
			GCACATGTTGGTTGTTGGAA
37	10187(TA)	$(TA)_n$	AGAAAAAGGTTTACAACGAGCG
			CGATGGAGAACGTGAAACAA
37	10187(GA)	$(GA)_n$	GTCACACCACTAGCGATGACA
			ACTGCACAATACCCCCTTTG

(continued)

Table 4 (continued)

Chromosome[a]	Primer code	Repeat type	Forward/reverse primer (5′–3′)
37	TcUn2	Unknown	AACAAAATCTAGCGTCTACCATCC
			GGTGTTGGCGTGTATGATTG
39	6925(TG)b	$(TG)_n$	GAAACGCACTCACCCACAC
			GGTAGCAACGCCAAACTTTC
39	7093(TC)	$(TC)_n$	CCAACATTCAACAAGGGAAA
			GCATGAATATTGCCGGATCT
39	6925(CT)	$(CT)_n$	CATCAAGGAAAAACGGAGGA
			CGGTACCACCTCAAGGAAAG
39	7093(TA)c	$(TA)_n$	CGTGTGCACAGGAGAGAAAA
			CGTTTGGAGGAGGATTGAGA
39	6925(TG)a	$(TG)_n$	TCGTTCTCTTTACGCTTGCA
			TAGCAGCACCAAACAAAACG
39	7093(TCC)	$(TCC)_n$	AGACGTTCATATTCGCAGCC
			AGCCACATCCACATTTCCTC
40	11283(TCG)	$(TCG)_n$	ACCACCAGGAGGACATGAAG
			TGTACACGGAACAGCGAAG
40	11283(TA)b	$(TA)_n$	AACATCCTCCACCTCACAGG
			TTTGAATGCGAGGTGGTACA

[a]Chromosomal assignment based on Weatherly et al. [79]

8. Sterile 0.2 ml 96-well PCR reaction plates (Fisher Scientific, UK) and adhesive plate seals or 0.2 ml PCR tube strips and caps (VWR, UK).
9. PCR machine.
10. Microcentrifuge.
11. Sterile 1.5 ml graduated microcentrifuge tubes (Anachem, UK).

2.8 Agarose Gel Electrophoresis

1. Molecular grade agarose (Bioline, UK).
2. NuSieve™ GTG™ agarose (Lonza, UK).
3. 1× TAE buffer (40 mM Tris–HCl, 20 mM acetic acid, and 1 mM EDTA (pH 8)) (Sigma-Aldrich, UK).
4. 10 mg/ml ethidium bromide (Sigma-Aldrich, UK) (*see* **Note 3**).
5. Molecular weight ladders: Hyperladder ™ IV and V (Bioline, UK).
6. 5× DNA loading buffer blue (Bioline, UK).

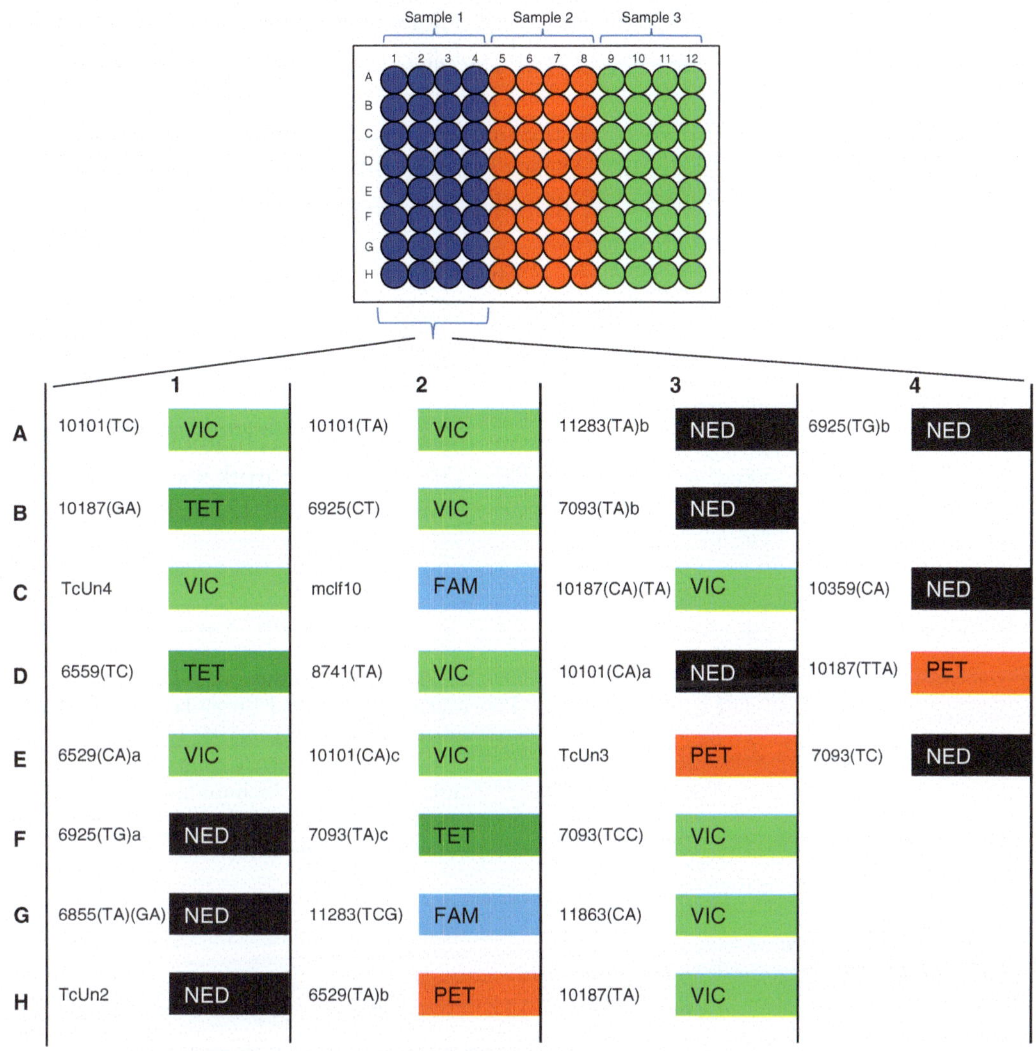

Fig. 3 Microsatellite primer positions in 96-well plate

7. Gel electrophoresis equipment (e.g., Jencons midi-horizontal gel electrophoresis system with 16-well combs and 13 × 15 cm casting trays) and power pack.
8. Microwave.
9. UV transilluminator.

2.9 PCR Purification

1. QIAquick PCR purification kit (Qiagen, UK).
2. Absolute ethanol (analytical reagent grade).
3. Absolute isopropanol (analytical reagent grade).
4. 0.5× TE buffer (10 mM Tris–HCl and 1 mM EDTA (pH 8)) (both Sigma-Aldrich, UK).
5. Microcentrifuge.

6. Sterile 1.5 ml graduated microcentrifuge tubes (Anachem, UK).
7. Sterile 0.5 ml graduated microcentrifuge tubes (Anachem, UK).

2.10 Restriction Enzyme Digestions

1. *Eco*RV restriction endonuclease, corresponding 10× buffer and 100× bovine serum albumin (BSA) (New England Biolabs, UK).
2. *Hha*I restriction endonuclease, corresponding 10× buffer and 100× BSA (New England Biolabs, UK).
3. Microcentrifuge.
4. Sterile 1.5 ml graduated microcentrifuge tubes (Anachem, UK).
5. 37 °C incubator.

2.11 Dye Terminator DNA Sequencing

1. BigDye™ Terminator v3.1 Cycle Sequencing Kit (Applied Biosystems, UK).
2. PCR machine.
3. Absolute ethanol (analytical reagent grade).
4. 96-well optical reaction plates with barcodes (Applied Biosystems, UK).
5. Hi-Di™ deionized formamide (Applied Biosystems, UK).
6. Refrigerated centrifuge.
7. Vortex.
8. 16-Capillary 3730 DNA Analyzer (Applied Biosystems, UK) (*see* **Note 4**).

2.12 MLMT PCR Product Multiplexing and Allele Size Determination

1. 96-well optical reaction plates with barcodes (Applied Biosystems, UK).
2. GeneScan™-500 LIZ™ size standard (Applied Biosystems, UK).
3. Hi-Di™ deionized formamide (Applied Biosystems, UK).
4. 16-Capillary 3730 DNA Analyzer (Applied Biosystems, UK).

3 Methods

3.1 Isolation of *T. cruzi*

3.1.1 Direct Hemoculture from Patients/Mammals

1. Prepare biphasic 4 N (USMARU) culture medium by adding 4 % (w/v) blood agar base, 0.6 % (w/v) agar, 0.6 % (w/v) NaCl, and 0.5 % (w/v) tryptone (all Sigma-Aldrich, UK) to H_2O and dissolve by autoclaving (121 °C for 15 min). Cool the medium to 50 °C and aspectically add 10 % (v/v) sterile defibrinated rabbit blood, 150 μg/ml gentamycin, and 150 μg/ml 5-fluorocytosine (both Sigma-Aldrich, UK) (*see* **Note 5**).
2. Aliquot 2 ml of biphasic 4 N culture medium into the bottom of a sterile Nunclon™ Δ flat sided tube (Nunc, UK) and allow to set at an angle, forming a slope.

3. Once set, overlay each culture with 500 μl of 0.9 % sterile NaCl, containing 150 μg/ml gentamycin and 150 μg/ml 5-fluorocytosine.
4. Prepare liver infusion tryptose (LIT) medium by dissolving 25 g liver infusion broth (Difco™, Becton Dickinson, USA), 5 g tryptone, 4 g NaCl, 2 g glucose (Sigma-Aldrich, UK), 0.4 g KCl (Sigma-Aldrich, UK), and 3.15 g Na_2HPO_4 (Sigma-Aldrich, UK) in 900 ml H_2O and adjust the pH to 7.4. Autoclave (121 °C for 15 min) and cool the medium to 50 °C. Add 25 g hemin, dissolved in 1 ml 1 N NaOH and 100 ml heat-inactivated fetal calf serum (both Sigma-Aldrich, UK).
5. For adult human samples, extract 15 ml venous blood using a sterile 20 ml BD Plastipak™ syringe with needle (Becton Dickinson, USA) and transfer to a BD Vacutainer® plus plastic K_2 EDTA tube (Becton Dickinson, USA) to prevent coagulation.
6. If isolating from mammals, take 1–2 ml blood by cardiac puncture, using a sterile 5 ml BD Plastipak™ syringe with needle (or 1 ml/2 ml syringes for smaller animals), after anesthetising the mammal by intramuscular administration of ketamine hydrochloride (100 mg/kg body weight) (Sigma-Aldrich, UK) and sterilizing the thorax first with iodised 70 % (v/v) ethanol (5 g iodine/l) and then non-iodized 70 % (v/v) ethanol.
7. For patient samples, transfer blood into a sterile 15 ml centrifuge tube (Greiner Bio-One, UK) and centrifuge for 10 min at 1,200 × *g* and 4 °C.
   (a) Discard all but 0.5 ml plasma and packed red cells.
   (b) Add 8 ml LIT medium to packed red cells.
   (c) Centrifuge for 10 min at 1,200 × *g* and 4 °C.
   (d) Carefully discard the supernatant.
   (e) Resuspend in 6 ml LIT medium.
   (f) Aliquot 2 ml of packed red cells to three separate 4 N culture tubes.
   (g) Seal each tube with a rubber cap from a 10 ml vacutainer tube (Scientific Laboratory Supplies, UK) and secure with Parafilm (VWR, UK). Rubber caps must be autoclaved prior to use.
   (h) Incubate cultures at 28 °C for 3–6 months, depending on strain growth rate. Once logarithmic phase cells become microscopically visible, parasites can be seeded into supplemented RPMI-1640 axenic culture medium (as described in Subheading 3.2.2).
   (i) For long-term cryopreservation of parasites, supplement late logarithmic phase cultures with sterile 10 % glycerol (v/v) (VWR, UK) and prepare aliquots in sterile cryovials (Nunc, Denmark). Store cryovials at –70 °C for 24 hours, before transfer to liquid nitrogen.

8. For mammal samples, inoculate the blood directly into several 4 N culture tubes.
9. If biologically cloning directly from blood (as described in Subheading 3.2), leave the whole blood to settle for 1 h in the BD Vacutainer® EDTA tube or centrifuge at a low speed (40 × *g* for 5 min) and then incubate at 37 °C for 45 min to ensure motile trypomastigotes have dispersed throughout the plasma, prior to cloning.
10. If directly extracting parasite genomic DNA from blood (as described in Subheading 3.3.2), dilute packed red cells in guanidine-EDTA (6 M guanidine, 0.2 M EDTA) (Sigma-Aldrich, UK) at a 1:1 ratio and store at 4 °C.

3.1.2 Isolation from Triatomine Bugs (Xenodiagnosis)

Xenodiagnosis can be undertaken by feeding up to 10–20 uninfected colony-reared triatomine bugs (third or fourth nymphal instars) on each suspected patient/mammal before isolating parasites after ~3 weeks as described below:

1. Prepare biphasic 4 N culture medium in sterile Nunclon™ Δ flat sided tubes as described in Subheading 3.1.1.
2. Prepare White's solution consisting of 0.025 g $HgCl_2$ (*see* **Note 6**), 0.65 g NaCl (both Sigma-Aldrich, UK), 0.125 ml conc. HCl (sp. gr. 1.18) (VWR, UK), 25 ml absolute ethanol, and 75 ml H_2O.
3. Immerse the bugs in White's solution for 10 min, rinse in 0.9 % sterile NaCl containing 300 μg/ml gentamycin and 300 μg/ml 5-fluorocytosine and dry (all Sigma-Aldrich, UK).
4. Aseptically dissect the intestinal contents of each bug into sterile saline (containing 300 μg/ml gentamycin and 300 μg/ml 5-fluorocytosine) on a sterile microscope slide (VWR, UK), behind a protective screen in a Class II microbiological safety cabinet. Dissection can be performed by holding the bug upside down in a pair of broad forceps, then using a pair of watchmakers' forceps (both Scientific Laboratory Supplies, UK) to pull the last abdominal segment away, extruding the gut onto a microscope slide.
5. Homogenize the intestinal contents using a sterile broad microspatula (Scientific Laboratory Supplies, UK) and discard the abdomen apex.
6. Remove the majority of intestinal homogenate from the dissection slide to a sterile 1.5 ml graduated microcentrifuge tube (Anachem, UK), using a sterile 1 ml plastic Pasteur pipette (Scientific Laboratory Supplies, UK) and place a sterile microscope cover glass over the remainder.
7. Examine slide microscopically and if parasites are observed, transfer 20 μl of inoculum to a 4 N culture tube.

8. Incubate cultures at 28 °C for 3–6 months, depending on strain growth rate. Once logarithmic phase cells become microscopically visible, parasites can be seeded into supplemented RPMI-1640 axenic culture medium (as described in Subheading 3.2.2).
9. If biologically cloning directly from triatomine intestinal contents (as described in Subheading 3.2), there is a high risk of contamination; ensure that 150 μg/ml gentamycin and 150 μg/ml 5-fluorocytosine are added to the relevant cloning medium.

3.2 Biological Cloning of *T. cruzi*

3.2.1 Biological Cloning of Parasites by Micromanipulation

1. Prepare biphasic 4 N culture medium as described in Subheading 3.1.1 but without gentamycin and 5-fluorocytosine.
2. Aliquot 2 ml of biphasic 4 N culture medium into the bottom of sterile 7 ml Bijou tubes (Sterilin, UK) and leave to set. Once set, overlay each culture with 750 μl of 0.9 % sterile NaCl, containing 100 μg/ml gentamycin (Sigma-Aldrich, UK) and 100 μg/ml 5-fluorocytosine (Sigma-Aldrich, UK).
3. Empirically prepare a dilute solution of logarithmic-phase *T. cruzi* epimastigotes (from axenic culture, patient blood or infected triatomine bug intestinal contents) such that microdrops delivered from microcapillaries contain a single parasite or no parasites (*see* **Note 7**).
4. Prepare fine microcapillaries by rotating a microcapillary tube (Sigma-Aldrich, UK) in a Bunsen flame, removing, and pulling apart the two ends to form a fine intervening microcapillary (each original microcapillary tube yields two microcapillaries).
5. On a microscope slide, place a sterile 13 mm microscope cover glass onto a small drop of sterile H_2O (for adhesion); dispense a microdrop of diluted culture onto the cover glass from a microcapillary tube and cover the drop with a second cover glass. Drops which occupy no more than one microscopic field at 400× magnification are ideal.
6. Microscopically examine the drop through multiple planes of vision, for the presence of parasites.
7. Transfer cover glass pairs with drops containing no organisms (control cultures) or a single parasite to 4 N cultures using sterile watchmakers' forceps. Discard all microdrops which contain more than one parasite.
8. Incubate all cultures at 28 °C for 3–6 months, depending on strain growth rate. Discard the entire series if any of the control cultures become positive. Once logarithmic phase cells become microscopically visible, parasites can be seeded into supplemented RPMI-1640 axenic culture medium (as described in Subheading 3.2.2).

3.2.2 Biological Cloning of Parasites on Solid Medium

Variations of this protocol, including different under- and over-lay media are published in full in [41]. We describe below a protocol which favors growth of *T. cruzi* strains from all DTUs:

1. Prepare sterile stock solutions (100×) of tryptone (0.175 g/ml, autoclaved), HEPES (1 M, pH 7.2, filter-sterilized), and hemin (2.5 mg/ml in 0.01 M NaOH, autoclaved) (all Sigma-Aldrich, UK).
2. Supplement RPMI-1640 medium (Sigma-Aldrich, UK #R0883) with 0.5 % (w/v) tryptone, 20 mM HEPES buffer (pH 7.2), 30 mM hemin, 10 % (v/v) heat-inactivated fetal calf serum, 2 mM sodium glutamate, 2 mM sodium pyruvate, 250 μg/ml streptomycin, and 250 U/ml penicillin (all Sigma-Aldrich, UK). Filter-sterilize the glutamine/pyruvate/penicillin solution before use.
3. Prepare blood agar plates by adding 10.8 ml biphasic 4 N culture medium (with 100 μg/ml gentamycin and 100 μg/ml 5-fluorocytosine) as described in Subheading 3.1.1 to sterile 90 mm petri dishes (Sterilin, UK).
4. Measure parasite density using a disposable hemocytometer (Immune Systems, UK).
5. Mix 10^2–10^3 logarithmic phase cells with 2.4 ml (w/v) supplemented RPMI-1640 medium and 0.6 ml molten 3 % (w/v) LMP agarose containing 0.9 % NaCl (w/v) (all Sigma-Aldrich, UK).
6. Pour this overlay onto a blood agar plate and allow to set.
7. Seal plates with Parafilm (VWR, UK) to minimize evaporation and incubate at 28 °C in a humidified atmosphere of 5 % CO_2.
8. Once colonies become visible (after 3–6 months, depending on strain growth rate), examine microscopically and remove clones using sterile 200 μl pipette tips. Inoculate each colony into 1 ml supplemented RPMI-1640 medium in a 48-well cell culture plate (Becton Dickinson, USA).

3.2.3 Biological Cloning of Parasites by Limiting Dilution

1. Serially dilute logarithmic phase cells to achieve a final concentration of 0.5 parasites/ml in a total volume of 20 ml supplemented RPMI-1640 medium (as described in Subheading 3.2.2).
2. Aliquot 200 μl of dilute culture into each well of a sterile 96-microwell culture plate (Nunc, UK).
3. Examine each well microscopically and mark those containing single organisms.
4. Seal each plate with Parafilm (VWR, UK) and incubate at 28 °C in an atmosphere of 5 % CO_2.
5. After 4–8 weeks, expand marked wells with sufficient numbers (~10^6/ml) of dividing cells into larger axenic culture volumes.

3.3 Preparation of Parasite Genomic DNA

3.3.1 Parasite Genomic DNA Extraction from Epimastigote Culture

Extraction of genomic DNA from 10 ml epimastigote cultures can be achieved using a Gentra Puregene tissue kit (Qiagen, UK), according to a modified version of the manufacturer's protocol (*see* **Note 8**). Additional necessary reagents are PBS (Sigma-Aldrich, UK), absolute isopropanol, and absolute ethanol. Cell lysis buffer, protein precipitation solution, and DNA hydration solution are all stored at room temperature. Proteinase K and RNase A are both stored at 4 °C. The modified manufacturer's protocol is as follows:

1. Centrifuge 10 ml of late log phase culture (~10^7–10^8 trypanosomes) in a sterile 15 ml centrifuge tube (Greiner Bio-One, UK) at 800 × *g* for 10 min.
2. Discard the supernatant by inverting tubes onto absorbent paper and resuspend fully in PBS, then centrifuge again as previously.
3. Resuspend in 3 ml cell lysis buffer (incubate at 37 °C and/or vortex to remove clumps, if necessary).
4. Cell suspensions are now stable and can be stored at −20 °C for 1–2 weeks.
5. Add 15 μl proteinase K solution (100 μg/ml) and incubate at 55 °C for 1 h, inverting periodically.
6. Leave to cool to room temperature.
7. Add 15 μl RNase A solution (20 μg/ml), invert 25 times and incubate at 37 °C for 15–60 min.
8. Cool on ice for 3 min and then add 1 ml protein precipitation solution (room temperature).
9. Vortex tubes vigorously for 20 s and then centrifuge at 2,000 × *g* for 10 min (ensure a tight pellet forms).
10. Remove the supernatant and transfer to a new sterile 15 ml centrifuge tube.
11. Precipitate DNA by the addition of 3 ml absolute isopropanol (room temperature) and invert 50 times.
12. Centrifuge at 2,000 × *g* for 3 min and discard the supernatant by inverting tubes onto absorbent paper.
13. Wash the DNA pellet in 3 ml 70 % (v/v) ethanol (room temperature), invert 10 times and centrifuge at 2,000 × *g* for 1 min.
14. Carefully remove the supernatant by inverting tubes and draining onto absorbent paper.
15. Air-dry the DNA pellet with tubes inverted at an angle for a maximum of 15 min.
16. Resuspend the DNA pellet in 250 μl DNA hydration solution, incubate at 65 °C for 1–2 h and then at room temperature overnight.
17. Estimate the DNA yield by spectrophotometry. Successful DNA extractions will yield 100 ng/μl or more and an A260/280 of 1.8–2.0.
18. Store extracted genomic DNA at −20 °C.

3.3.2 Parasite Genomic DNA Extraction from Patient Hemoculture

Extraction of genomic DNA from clinical hemocultures can be achieved using a High Pure PCR template preparation kit (Roche, UK), according to the manufacturer's protocol. Additional necessary reagents are PBS (Sigma-Aldrich, UK), absolute isopropanol, and absolute ethanol. Tissue lysis buffer, binding buffer, inhibitor removal buffer, wash buffer, and elution buffer are all stored at room temperature. Add absolute ethanol to the inhibitor removal buffer and the wash buffer, as instructed. Proteinase K is stored at 4 °C. Before beginning the DNA extraction, preheat the elution buffer to 70 °C. The manufacturer's protocol is as follows:

1. To a sterile 1.5 ml graduated microcentrifuge tube (Anachem, UK) mix 200 μl sample material (1:1 blood/guanidine-EDTA) with 600 μl binding buffer and 100 μl Proteinase K and incubate at 70 °C for 10 min.
2. Add 200 μl absolute isopropanol and mix well by vortexing.
3. Apply 550 μl to a High Pure filter tube and centrifuge at 8,000 × *g* for 1 min.
4. Discard the flow-through.
5. Repeat **steps 3** and **4** using the same High Pure filter tube.
6. Add 500 μl inhibitor removal buffer and centrifuge at 8,000 × *g* for 1 min.
7. Discard the flow-through.
8. Add 500 μl wash buffer and centrifuge at 8,000 × *g* for 1 min.
9. Discard the flow-through.
10. Repeat **steps** 7 and **8**.
11. Centrifuge at 13,000 × *g* for 10 s.
12. Place the High Pure filter tube in a clean 1.5 ml graduated microcentrifuge tube.
13. Add 200 μl pre-warmed elution buffer and centrifuge at 8,000 × *g* for 1 min.
14. Estimate the DNA yield by spectrophotometry. Successful DNA extractions will yield 3 ng/μl or more and an A260/280 of 1.8–2.0.
15. Store extracted genomic DNA at −20 °C.

3.3.3 Parasite Genomic DNA Extraction from Triatomine Bug Feces

DNAzol® solution (Life Technologies, UK) can be used to extract *T. cruzi* genomic DNA from triatomine bug feces, following hindgut dissection. Store DNAzol® solution, absolute ethanol, and NaOH at room temperature.

1. Lyse 50–100 μl of triatomine bug intestinal homogenate in a sterile 1.5 ml graduated microcentrifuge tube (Anachem, UK) by the addition of 1 ml DNAzol® solution.
2. Invert twice and incubate at room temperature for 3 min.

3. Precipitate DNA by the addition of 0.5 ml absolute ethanol (room temperature).
4. Pellet DNA by centrifuging at 13,000 × *g* for 4 min.
5. Discard the supernatant and wash twice with 1 ml 70 % (v/v) ethanol ensuring not to disturb the pellet.
6. Resuspend the DNA pellet in 50 μl 8 mM NaOH (Sigma-Aldrich, UK).
7. Estimate the DNA yield by spectrophotometry. Successful DNA extractions will yield 100 ng/μl or more and an A260/280 of 1.8–2.0.
8. Store extracted genomic DNA at −20 °C.

3.4 PCR-RFLP

3.4.1 PCR Amplification

1. Amplify the 24Sα rRNA (LSU rDNA) in a standard reaction containing: 1× NH_4 reaction buffer, 1.5 mM $MgCl_2$ (Bioline, UK), 0.2 mM dNTPs (New England Biolabs, UK), 1 pmol/μl of D71 and D72 primers (*see* Table 1), 1 U BIOTAQ™ DNA polymerase (Bioline, UK), and 10–100 ng of *T. cruzi* genomic DNA, made up to a total volume of 25 μl.
2. Reaction conditions for the 24Sα rRNA (LSU rDNA) are an initial denaturation step of 94 °C for 3 min and then 27 amplification cycles (94 °C for 1 min, 60 °C for 1 min, 72 °C for 1 min), followed by a final elongation step at 72 °C for 5 min.
3. Amplify both *HSP60* and *GPI* in a standard reaction containing: 1× NH_4 reaction buffer, 2 mM $MgCl_2$ (Bioline, UK), 0.2 mM dNTPs (New England Biolabs, UK), 1 pmol/μl of HSP60_for and HSP60_rev primers (for *HSP60*) or GPI_for and GPI_rev (for *GPI*) (*see* Table 1), 1 U BIOTAQ™ DNA polymerase (Bioline, UK), and 10–100 ng of *T. cruzi* genomic DNA, made up to a total volume of 25 μl.
4. Reaction conditions for both *HSP60* and *GPI* use a touchdown PCR strategy comprising an initial denaturation step of 3 min at 94 °C, followed by four cycles (94 °C for 30 s, 64 °C for 30 s, 72 °C for 1 min), followed by 28 cycles (94 °C for 30 s, 60 °C for 30 s, 72 °C for 1 min), and then a final elongation step at 72 °C for 10 min.

3.4.2 Agarose Gel Eletrophoresis

1. Visualize 10 μl of each 24Sα rRNA PCR product by gel electrophoresis using 3.5 % NuSieve™ GTG™ agarose gels (Lonza, UK) containing 0.5 μg/ml ethidium bromide (Sigma-Aldrich, UK) (*see* **Note 9**).
2. Visualize 5 μl of each *HSP60* and *GPI* PCR product by gel electrophoresis using 1.5 % agarose gels (Bioline, UK) containing 0.5 μg/ml ethidium bromide.
3. Load samples into gel wells with 1 μl of 5× DNA loading buffer (Bioline, UK) and run 5 μl of Hyperladder™ V (for 24Sα

rRNA) or IV (for *HSP60* and *GPI*) (Bioline, UK) as a molecular weight marker.

4. Run all gels at 100 V for 1–2 h in 1× TAE buffer and visualize under UV illumination, ensuring that the user is protected from the light source behind a UV shield.
5. If necessary, prior to restriction digestion, purify *HSP60* and *GPI* PCR products using a QIAquick PCR purification kit (Qiagen, UK) to remove nonspecific products, as described in Subheading 3.4.3.

3.4.3 PCR Purification

Purification of all PCR products can be achieved using a QIAquick PCR purification kit (Qiagen, UK) with spin columns to remove contaminating primers, nucleotides, DNA polymerases etc. (*see* **Note 10**) All of the necessary reagents are included within the kit (add ethanol to buffer PE as instructed) and are stored at room temperature. The manufacturer's protocol is as follows:

1. Add 5 volumes of buffer PB to 1 volume of the PCR reaction and mix.
2. Apply the sample to a QIAquick spin column placed in a 2 ml collection tube and centrifuge at >13,000 × *g* for 30–60 s.
3. Discard the flow-through.
4. Add 0.75 ml of buffer PE (with ethanol added) to the QIAquick column and centrifuge at >13,000 × *g* for 30–60 s.
5. Discard the flow-through and recentrifuge for 1 min at maximum speed.
6. Place the QIAquick column in a clean 1.5 ml graduated microcentrifuge tube (Anachem, UK).
7. To elute the DNA, add between 30 and 50 μl of buffer EB (10 mM Tris-Cl, 1 mM EDTA (pH 8)) to the center of the QIAquick membrane, incubate for 1–5 min and then centrifuge at >13,000 × *g* for 1 min (*see* **Note 11**).
8. Purified PCR products can be stored at −20 °C until required.

3.4.4 Restriction Enzyme Digestion

1. Digest 10 μl of *HSP60* or *GPI* PCR products (typically ~1 μg) in a reaction containing 0.25 U/μl of *Eco*RV or *Hha*I restriction endonucleases (New England Biolabs, UK), 100 ng/μl BSA and 1× quantity of the manufacturer's recommended reaction buffer in a total volume of 20 μl.
2. Incubate reactions at 37 °C for 4 h.

3.4.5 Restriction Fragment Length Polymorphism Analysis

1. Visualize 10 μl of each reaction using either 1.5 % (*GPI*/*Hha*I) or 3 % agarose gels (*HSP60*/*Eco*RV) (Bioline, UK) containing 0.5 μg/ml ethidium bromide (Sigma-Aldrich, UK) (*see* **Note 12**).
2. Load samples into agarose wells with 1 μl of 5× DNA loading buffer (Bioline, UK) and run 5 μl of Hyperladder™ V (for

24Sα rRNA), IV (*HSP60*), or I (*GPI*) (Bioline, UK) as a molecular weight ladder.

3. Run all gels at 100 V for 1–2 h in 1× TAE buffer and visualize under UV illumination, ensuring that the user is shielded from the light source (*see* **Note 13**).
4. The genotype assignment system based on the number and size of the restriction fragment bands is shown in Table 5, Figs. 4 and 5. For additional details please refer to Lewis et al. [47].

3.5 Nuclear MLST

3.5.1 PCR Amplification

1. Amplify each MLST target (*HMCOAR*, *GPI*, *TcMPX* and *RHO1*) in a standard reaction containing: 10 μl 5× colorless GoTaq® reaction buffer (Promega, UK), 0.2 mM dNTPs (New England Biolabs, UK), 0.2 μM of respective forward and reverse primers (*see* Table 2), 1 U GoTaq® DNA polymerase (Promega, UK), and 10–100 ng of *T. cruzi* genomic DNA, made up to a total volume of 50 μl.
2. Reaction conditions for all targets are an initial denaturation step of 94 °C for 5 min and then 35 amplification cycles (94 °C for 1 min, 55 °C for 1 min, 72 °C for 1 min), followed by a final elongation step at 72 °C for 5 min.

3.5.2 Agarose Gel Eletrophoresis

1. Visualize 5 μl of each PCR product by gel electrophoresis using 1.5 % agarose gels (Bioline, UK), as described in Subheading 3.4.2.

3.5.3 PCR Purification

1. Purify all PCR products, as described in Subheading 3.4.3.

3.5.4 Dye Terminator DNA Sequencing

Bidirectional sequencing can be performed using a BigDye™ Terminator v3.1 Cycle Sequencing kit (Applied Biosystems, UK). All of the necessary reagents are included within the kit. Big Dye Sequencing RR-100 is stored at −20 °C and sequencing buffer is stored at 4 °C (*see* **Note 14**). The modified version of the manufacturer's protocol is as follows:

1. Use 0.5–2 μl of PCR reaction template (~5–20 ng) in a standard reaction containing 0.5 μl Big Dye sequencing RR-100, 1.7 μl sequencing buffer, and 3.2 pmol of forward or reverse PCR primer (*see* **Note 15**), made up to a total volume of 10 μl.
2. Reaction conditions are as follows: 25 cycles of rapid thermal ramp to 96 °C (1 °C/s), 96 °C for 30 s, rapid thermal ramp to 55 °C (1 °C/s), 55 °C for 20 s, rapid thermal ramp to 60 °C (1 °C/s), and 60 °C for 4 min.
3. Purify samples in sterile 96-well optical reaction plates with barcodes (Applied Biosystems, UK).
4. Precipitate DNA by the addition of 8 μl of H_2O followed by 32 μl ice-cold 95 % (v/v) ethanol.

Table 5
***T. cruzi* genotype assignment of PCR amplification product sizes (bp)**

Target/enzyme	Expected PCR product (digestion product) band size (bp)					
	TcI	TcII	TcIII	TcIV	TcV	TcVI
LSU rDNA	110	125	110	117[a] or 120 or 125[b] or 130[c]	110 or 110 + 125[d]	125
*HSP60/Eco*RV	432–462 (432–462)	432–462 (432–462)	432–462 (314 + 148–118)	432–462 (432–462)	432–462 (432–462 + 314 + 148–118)	432–462 (432–462 + 314 + 148–118)
*GPI/Hha*I	1,264 (817 + 447)	1,264 (490 + 447 + 253)	1,264 (817 + 447)	1,264 (490 + 447 + 253)	1, 264 (817 + 490 + 447 + 253)	1,264 (817 + 490 + 447 + 253)

[a]According to Kawashita and others [77]
[b]According to Brisse et al. [73]
[c]For strains of North American origin, according to Brisse et al. [73]
[d]Double band pattern observed for most isolates; 125 bp band exhibits variable intensity

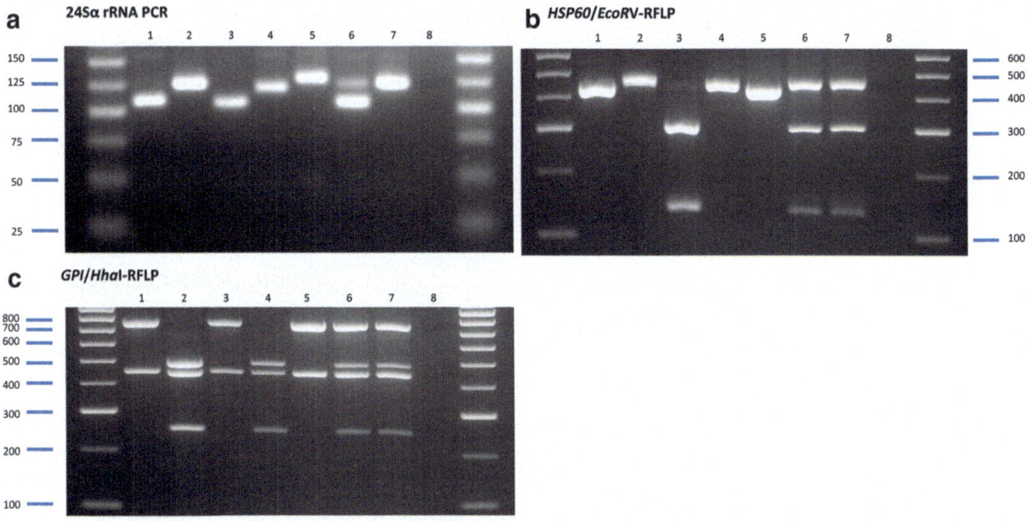

Fig. 4 Examples of PCR-RFLP genotyping profiles. (**a**) LSU rDNA. (**b**) *HSP60/Eco*RV digestion products are shown. (**c**) *GPI/Hha*I digestion products are shown. For all gels, Lanes: (*1*) Sylvio X10/1 (TcI), (*2*) Esm cl3 (TcII), (*3*) M5631 (TcIII), (*4*) CanIII cl1 (TcIV), (*5*) 92122102R (TcIV NA), (*6*) Sc43 cl1(TcV), (*7*) CL Brener (TcVI), (*8*) negative control

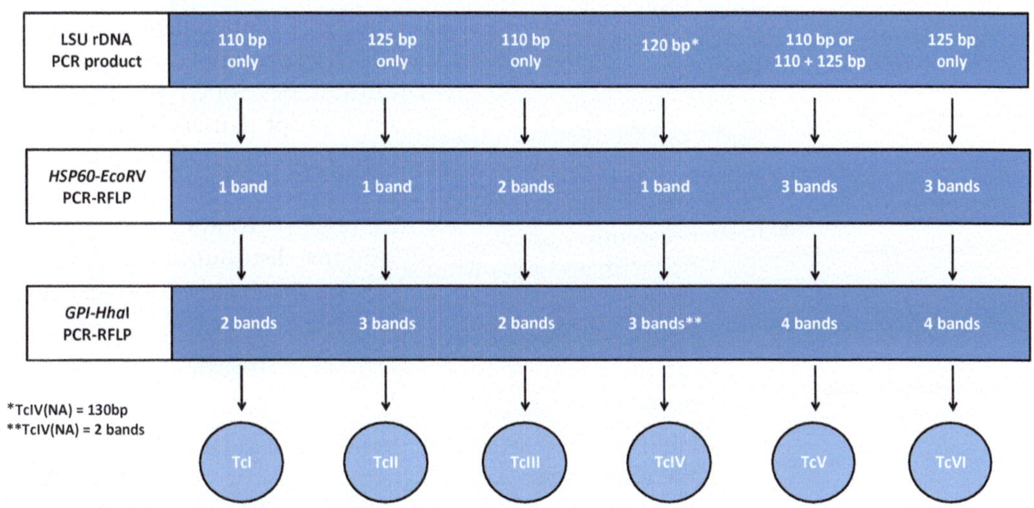

Fig. 5 Recommended triple-assay for discriminating *T. cruzi* DTUs

5. Incubate samples at 4 °C for 15 min and then centrifuge for 45 min at 3,000 × *g* and 4 °C.
6. Remove the supernatant by inverting plates onto absorbent paper and centrifuging at 20 × *g* for 10 s.
7. Wash DNA pellets by the addition of 50 μl ice-cold 70 % (v/v) ethanol and briefly vortex.
8. Spin plates for 30 min at 3,000 × *g* and 4 °C.

9. Discard supernatants as previously.
10. Dry pellets at room temperature until no visible ethanol remains (*see* **Note 16**).
11. Resuspend DNA pellets in 10 μl Hi-Di™ deionized formamide (Applied Biosystems, UK) (*see* **Note 14**).
12. DNA pellets can be stored at −20 °C until required.
13. Analyze DNA sequences using an automated 16-capillary 3730 DNA Analyzer (Applied Biosystems, UK) (*see* **Note 4**).

3.5.5 Analysis of Nuclear MLST Data

Nucleotide data can be assembled manually in BioEdit v7.0.9.0 sequence alignment editor software (Ibis Biosciences, USA) [58] and ambiguous peripheral regions of aligned sequences discarded to produce unambiguous consensus sequences for each isolate. Heterozygous positions are identified by the presence of two coincident peaks at the same locus ("split peaks"), verified in forward and reverse sequences and scored according to the one-letter nomenclature for nucleotides from the International Union of Pure and Applied Chemistry (IUPAC). If data for multiple gene targets have been generated, sequences can be concatenated for each isolate (*see* **Note 17**). Distance-based phylogenies can be constructed using individual or concatenated heterozygous diploid sequence data in SplitsTree4 (select the average states parameter to handle ambiguous sites) [59]. To aid DTU assignment, a reference panel of sequences from Yeo et al. [20] is electronically available to download from GenBank. In the absence of a formalized nuclear MLST scheme for population genetic studies, three additional targets (*LAP*, *RB19*, and *SODB*), described in [20, 51, 52] can be used for higher resolution genetic diversity studies. Additional analyses are described with accompanying software by Tomasini et al. [80].

3.6 Maxicircle MLST

3.6.1 PCR Amplification

1. Prepare a 96-well PCR reaction plate (Fisher Scientific, UK) containing maxicircle primer stocks at 10 pmol/μl (*see* Table 3) arranged according to Fig. 2.
2. Amplify all ten maxicircle genes in standard PCR reactions each containing: 1× NH_4 reaction buffer, 1.5 mM $MgCl_2$ (Bioline, UK), 0.2 mM dNTPs (New England Biolabs, UK), and 1 U BIOTAQ™ DNA polymerase (Bioline, UK), made to a final volume of 17 μl.
3. Prepare a PCR mastermix for 90 samples without DNA template and aliquot 145 μl per well across the first plate row of a sterile 96-well PCR reaction plate (A01-A10).
4. Use a 10–100 μl twelve-channel pipette to transfer 17 μl mastermix per well down the 96-well PCR reaction plate (A01-H01, A02-H02, etc.).
5. Add 1 μl of DNA template (10–100 ng of *T. cruzi* genomic DNA) for each isolate across the plate (sample 1 in A01-A10, sample 2 in B01-B10, etc.).

6. Use a 0.5–10 μl twelve-channel pipette to transfer 1 μl of each forward and 1 μl of each reverse primer per well from the respective primer plates to the corresponding row on the PCR reaction plate (A01-A10, B01-B10, etc.) (*see* **Notes 18** and **19**).
7. PCR reactions are performed with an initial denaturation step of 3 min at 94 °C, followed by 30 amplification cycles (94 °C for 30 s, 50 °C for 30 s, 72 °C for 30 s) and a final elongation step at 72 °C for 10 min.

3.6.2 Agarose Gel Eletrophoresis

1. Visualize 5 μl of each PCR product by gel electrophoresis using 1.5 % agarose gels (Bioline, UK), as described in Subheading 3.4.2.

3.6.3 PCR Purification

1. Purify all PCR products, as described in Subheading 3.4.3.

3.6.4 Dye Terminator DNA Sequencing

1. Use a 10–100 μl eight-channel pipette to transfer PCR products to a 96-well optical reaction plates with barcodes (Applied Biosystems, UK) for purification (*see* **Note 19**).
2. Sequence all PCR products, as described in Subheading 3.5.4.

3.6.5 Analysis of Maxicircle MLST Data

Assemble sequence data as described for nuclear loci (*see* Subheading 3.5.5). For each isolate maxicircle sequences can be concatenated according to their structural arrangement (*12S rRNA*, *9S rRNA*, *CYT b*, *MURF1*, *ND1*, *COII*, *ND4*, and *ND5*) and in the correct coding direction (*see* **Note 17**). The best-fit model of nucleotide substitution can be inferred in jMODELTEST 1.0 [60]. Phylogenies of increasing computational complexity can be constructed using MEGA 5 [61] (distance-based phylogenies), PhyML [62] (Maximum-Likelihood topologies) or MrBAYES v3.1 [63] (Bayesian topologies). A reference panel of maxicircle sequences is electronically available to download from GenBank under the accession numbers JQ581059-JQ581370 and JQ581403-JQ581480. For additional analyses please refer to Messenger et al. [56].

3.7 MLMT

3.7.1 PCR Amplification

1. Prepare a 96-well PCR reaction plate (Fisher Scientific, UK) with microsatellite primers diluted to 1 pmol/μl in 0.5× TE buffer (*see* Table 4) and arranged according to Fig. 3.
2. Amplify all microsatellite loci in a standard reaction containing: 1× ThermoPol Reaction Buffer (New England Biolabs, UK), 4 mM $MgCl_2$, 34 μM dNTPs, 1 U *Taq* polymerase (New England Biolabs, UK), and 1 ng of genomic DNA, made up to a final volume of 7 μl.
3. Prepare one PCR mastermix (for 32 loci) per DNA isolate and aliquot 74 μl per well across A01-A04 of a sterile 96-well PCR reaction plate.

4. Each PCR plate can be used to amplify microsatellite loci for three DNA samples; distribute the mastermixes for isolates 2 and 3 across A05-A08 and A09-A12, respectively.
5. Use a 0.5–10 μl twelve-channel pipette to transfer 8.5 μl mastermix per well from A01-A12 down the PCR reaction plate (A01-H01, A02-H02, etc.)
6. Use a 0.5–10 μl twelve-channel pipette to transfer 1.5 μl of each premixed primer pair from the primer plate to the corresponding row on the PCR reaction plate (A01-A04, B01-B04, etc.).
7. Repeat **step 6**, instead transferring primers to columns 5–8 and 9–12.
8. PCR reactions for all loci are performed with an initial denaturation step of 4 min at 95 °C, then 30 amplification cycles (95 °C for 20 s, 57 °C for 20 s, 72 °C for 20 s) and a final elongation step at 72 °C for 20 min.

3.7.2 MLMT Multiplexing and Allele Size Determination

1. Use a 10–100 μl eight-channel pipette to combine columns 2, 3, and 4 into column 1, columns 6, 7, and 8 into column 5, and columns 10, 11, and 12 into column 9.
2. Transfer the contents of column 1, 5, and 9 into columns 1, 2, and 3 of a new sterile 96-well PCR reaction plate (*see* Fig. 6) to form a stock plate.
3. Each 96-well stock plate can hold multiplexed microsatellite PCR products from 12 DNA samples.
4. Mix 25 μl GeneScan™-500 LIZ™ fluorescent size standard (Applied Biosystems, UK) with 950 μl Hi-Di™ deionized formamide (Applied Biosystems, UK) and aliquot 82 μl per well across A01-A12 of a sterile 96-well optical reaction plate with barcode (Applied Biosystems, UK) (*see* **Note 20**).
5. Use a 0.5–10 μl twelve-channel pipette to distribute 9.75 μl of GeneScan™/Hi-Di™ solution into each well from A01-A12 down the 96-well optical reaction plate.
6. Use a 0.5–10 μl twelve-channel pipette to transfer 0.5 μl of sample PCR product from the stock plate into each corresponding row of the optical reaction plate (A01-A12, B01-B12, etc.).
7. Determine allele sizes using an automated 16-capillary sequencer (AB3730, Applied Biosystems, UK), with a standard injection time of 10 s.

3.7.3 Analysis of MLMT Data

Allele sizes can be assembled in GeneMapper® v 4.0 (Applied Biosystems, UK) and isolates should be typed “blind” to control for user bias and checked manually for errors. A set of allele sizes for reference strains and bin sizes for each microsatellite locus are available online at: http://www.ki.se/chagasepinet/mlmt.html

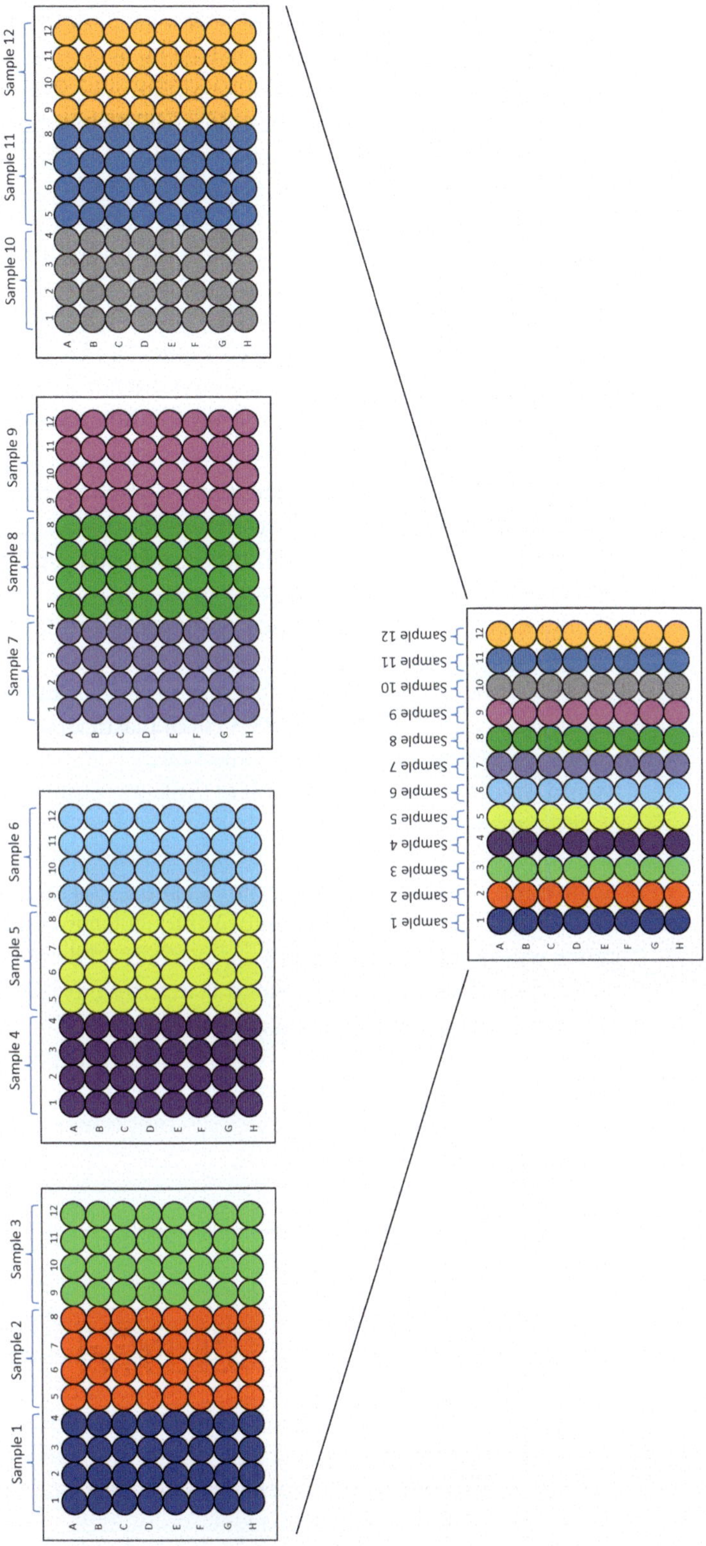

Fig. 6 Schematic of multiplexing microsatellite PCR products

Microsatellite data are highly amenable to quantitative analysis. Population structures between different geographical areas and transmission cycles can be inferred using pair-wise distance-based measurements, such as D_{AS} (infinite alleles model of (IAM)) or $\delta\mu^2$ (stepwise mutation model (SMM)) which can be calculated in MICROSAT v1.5d [64]. D_{AS} values can be assembled into a distance matrix and used to construct Neighbor-Joining trees in PHYLIP v3.67 [65]. Support for nodes in the Neighbor-Joining tree can be generated in PHYLIP v3.67 using 1000 bootstrap replicates of the data generated in MICROSAT v1.5d. The mean number of alleles per locus (MNA) and the sum number of occurrences of specific alleles for each locus can be calculated using the Microsatellite Toolkit add-in [66] for MS Excel. The software FSTAT 2.9.3.2 [67] can be used to estimate sample-size corrected allelic richness (A_r) and the inbreeding coefficient $F_{IS.}$ Heterozygosity indices, including deviations from Hardy–Weinberg equilibrium, and the extent of population differentiation (F_{ST}) can be calculated in ARLEQUIN v3.0 [68]. Multilocus linkage disequilibrium, estimated by the Index of Association (I_A), can be calculated in MULTILOCUS v1.3b [69]. Mantel's test to compare pair-wise geographical and genetic distances can be executed in GENALEX 6 [70].

We strongly discourage the use of model-based population assignment software (e.g., STRUCTURE and BAPS) as these programs use algorithms which assume Hardy–Weinberg expectations within populations and complete linkage equilibrium between genetic markers, two criteria that are largely violated by clonal reproduction in *T. cruzi*. Instead population subdivisions can also be inferred using a nonparametric (without Hardy–Weinberg constraints) *K*-means clustering algorithm [71], implemented in adegenet within the R 2.13 software package [78]. The number of "true" populations can be defined using the Bayesian Information Criterion (BIC) and the relationship between clusters can be evaluated using a Discriminant Analysis of Principal Components (DAPC), which first transforms allele frequencies at individual loci into uncorrelated variables (principal components), via a Principal Component Analysis (PCA) [72].

4 Notes

1. Infection with *T. cruzi* can only occur via direct inoculation or contamination of broken skin/intact mucosal membranes (conjunctiva, nose and mouth). Transmission via inhalation is highly unlikely as organisms do not readily aerosolize. In addition, parasites do not survive desiccation and are not free-living. Furthermore the predominant, but not exclusive form in exponentially growing axenic cultures is the non-infective epimastigote stage. To minimize risk of infection:

(a) Wear appropriate Personal Protective Equipment (PPE) at all times, including a Howie laboratory coat, eye-protection, and close-fitting disposable gloves.

(b) Conduct all manipulations of live material in a Class II microbiological safety cabinet, which should be fumigated regularly to prevent bacterial and fungal contamination.

(c) Do not touch the face or any exposed area while wearing contaminated gloves or handling live material.

(d) Routinely decontaminate work surfaces/cabinets with 70 % ethanol after use.

(e) Dispose of all contaminated material by immersing in 70 % ethanol or 10 % chloros (sodium hypochlorite) overnight.

(f) Restrict the use of sharps and glassware to avoid the risk of direct inoculation and dispose of all contaminated sharps in an appropriate sharpsafe bin.

(g) Avoid any procedures, e.g., centrifugation in open tubes or grinding of infected tissues, which may generate droplet suspensions.

(h) If necessary, wear a face visor or use a protective screen when directly handling infectious material, e.g., dissecting infected triatomine bugs.

(i) Establish full written risk assessments and emergency accident procedures before commencing work with live *T. cruzi*.

2. *T. cruzi* genomic DNA can be extracted from cultured epimastigotes, human hemocultures, or triatomine bug feces. The Gentra Puregene tissue kit (Qiagen, UK) and High Pure PCR template preparation kit (Roche, UK) both produce high quality template but with some loss of DNA yield and are most appropriate to extract DNA from cultured parasites and human clinical samples, respectively. DNA extracted using DNAzol® is typically of a higher yield but of lesser quality and is thus more suitable for extracting DNA from samples with low parasite density, including those derived from bug feces homogenate.
3. Ethidium bromide is mutagenic and toxic, so PPE must be worn at all times when handling this reagent.
4. We assume that the researcher has access to an automated fluorescent sequencer either through affiliations with an academic institution or by outsourcing to a commercial sequencing company.
5. Ideally, sterile test 5 % of each 4 N culture batch, by incubating at 37 °C for 3 days and checking for contamination.

6. Mercuric chloride is highly toxic and must be handled while wearing PPE (Howie laboratory coat, disposable gloves, and eye-protection) and with extreme care.
7. Some *T. cruzi* strains have a predilection to grow in clumps, therefore cultures should be checked microscopically and if clumpy, parasites can be separated by low-speed centrifugation (~200 × *g*) prior to cloning.
8. Parasite genomic DNA can also be extracted from smaller culture volumes using the Gentra Puregene tissue kit. This protocol can be modified to extract DNA from 1 ml of *T. cruzi* culture in 1.5 ml graduated microcentrifuge tubes, by decreasing reagent volumes tenfold and performing all centrifugation steps in a microcentrifuge at >13,000 × *g*.
9. Prepare NuSieve™ GTG™ low melting temperature agarose (Lonza, UK) by first soaking the agarose in chilled 1× TAE buffer for 15 min; this prevents the agarose from foaming during heating. Heat the agarose and buffer in a microwave on medium power for 2 min. Gently swirl the solution to resuspend any settled powder/gel pieces and reheat on high power until the solution begins to boil. Hold at boiling point for 1 min or until all of the agarose particles are dissolved. Allow the solution to cool to 50–60 °C prior to the addition of 0.5 μg/ml ethidium bromide and casting.
10. If consumable costs are restricted, PCR products can also be purified using absolute isopropanol. Add an equal volume of absolute isopropanol to PCR product in a sterile 0.5 ml graduated microtubes (Anachem, UK). Incubate at room temperature for 15 min. Spin tubes at >13,000 × *g* in a microcentrifuge for 20 min and discard the supernatant. Wash the pellet in 70 % (v/v) ethanol by spinning for 10 min. Discard the supernatant and air-dry the pellet. Resuspend the pellet in H_2O or 0.5× TE buffer.
11. Heating buffer EB to 55 °C before applying to the column and incubating for 1–5 min, prior to elution, can increase the yield from QIAquick columns.
12. If PCR-RFLP genotyping will be routinely performed it may be useful to prepare a stock of digested DNA size standards from *T. cruzi* reference isolates for each DTU. These can be stored at −20 °C and run as positive controls alongside unknown samples where necessary.
13. Ideally, *GPI*-RFLP gels should be run for long as possible in order to clearly separate bands at 490 and 447 bp (TcV and TcVI genotypes). In addition, the smallest *HSP60* band (118–148 bp; TcIII, TcV, and TcVI genotypes) can be difficult to

visualize, in which case it may be necessary to run a larger volume of digest reaction.

14. Aliquot both the Big Dye Sequencing RR-100 (e.g., 20 μl/aliquot) and Hi-Di™ deionized formamide (e.g., 1 ml/aliquot) and store at −20 °C in order to minimize the number of freeze–thaw and exposure cycles for each tube. An appropriate volume aliquot will receive less than five freeze–thaw cycles and contain sufficient quantity for 1 week's worth of reactions.
15. This modified protocol is for a reaction that is half the manufacturer's recommended volume. In addition, the reagent mix (Big Dye Sequencing RR-100) has been reduced by one-eighth of the recommended amount to save considerable costs.
16. It is important to ensure that no ethanol remains in the sequencing reaction plate but equal care must be taken not to overdry the DNA pellets as this may inhibit their resuspension in Hi-Di™ deionized formamide.
17. Ensure that in the nucleotide alignment, isolate sequences are placed in the same order for each gene, otherwise it is very easy to mistakenly concatenate sequences from different isolates across multiple loci.
18. To speed manipulations, we strongly recommend the use of eight- and twelve-channel multichannel pipettes. Although it is possible to perform all pipetting individually, the multichannel renders the process much less laborious and more robust. Our current choice of pipette is the ErgoOne® range (Star Labs, UK) and we use 0.5–10 μl twelve-channel (S7112-0510), 10–100 μl twelve-channel (S7112-1100), and 10–100 μl eight-channel pipettes (S7108-1100) for maxicircle MLST and MLMT PCR amplifications.
19. When transferring between plates ensure that the plates are first lined up in the same orientation as each other (A01 to A01 and H12 to H12) as it is remarkably easy to accidentally reverse a plate.
20. It may be useful to create a set of allele size standards prepared from reference strains to run alongside samples as internal controls.

Acknowledgments

Research detailed in this chapter was funded by support from the Wellcome Trust, the BBSRC and the European Union Seventh Framework Programme grant 223034 ("ChagasEpiNet").

References

1. Rassi A Jr, Rassi A, Marin-Neto JA (2010) Chagas disease. Lancet 375:1388–1402
2. Prata A (2001) Clinical and epidemiological aspects of Chagas disease. Lancet Infect Dis 1:92–100
3. Miles MA, Cedillos RA, Póvoa MM et al (1981) Do radically dissimilar *Trypanosoma cruzi* strains (zymodemes) cause Venezuelan and Brazilian forms of Chagas disease? Lancet 1:1338–1340
4. Campbell DA, Westenberger SJ, Sturm NR (2004) The determinants of Chagas disease: connecting parasite and host genetics. Curr Mol Med 4:549–562
5. Macedo AM, Machado CR, Oliveira RP et al (2004) *Trypanosoma cruzi*: genetic structure of populations and relevance of genetic variability to the pathogenesis of Chagas disease. Mem Inst Oswaldo Cruz 99:1–12
6. Fernandes O, Souto R, Castro J et al (1998) Brazilian isolates of *Trypanosoma cruzi* from humans and triatomines classified into two lineages using mini-exon and ribosomal RNA sequences. Am J Trop Med Hyg 58:807–811
7. Souto RP, Fernandes O, Macedo AM et al (1996) DNA markers define two major phylogenetic lineages of *Trypanosoma cruzi*. Mol Biochem Parasitol 83:141–152
8. Hamilton PB, Lewis MD, Cruickshank C et al (2011) Identification and lineage genotyping of South American trypanosomes using fluorescent fragment length barcoding. Infect Genet Evol 11:44–51
9. Oliveira RP, Broude NE, Macedo AM et al (1998) Probing the genetic population structure of *Trypanosoma cruzi* with polymorphic microsatellites. Proc Natl Acad Sci U S A 95: 3776–3780
10. Llewellyn MS, Miles MA, Carrasco HJ et al (2009) Genome-scale multilocus microsatellite typing of *Trypanosoma cruzi* discrete typing unit I reveals phylogeographic structure and specific genotypes linked to human infection. PLoS Pathog 5:e1000410
11. Ocaña-Mayorga S, Llewellyn MS, Costales JA et al (2010) Sex, subdivision, and domestic dispersal of *Trypanosoma cruzi* lineage I in Southern Ecuador. PLoS Negl Trop Dis 4:e915
12. Burgos JM, Altcheh J, Bisio M et al (2007) Direct molecular profiling of minicircle signatures and lineages of *Trypanosoma cruzi* bloodstream populations causing congenital Chagas disease. Int J Parasitol 37:1319–1327
13. Telleria J, Lafay B, Virreira M et al (2006) *Trypanosoma cruzi*: sequence analysis of the variable region of kinetoplast minicircles. Exp Parasitol 114:279–288
14. Lages-Silva E, Ramírez LE, Pedrosa AL et al (2006) Variability of kinetoplast DNA gene signatures of *Trypanosoma cruzi* II strains from patients with different clinical forms of Chagas disease in Brazil. J Clin Microbiol 44: 2167–2171
15. Zingales B, Miles MA, Campbell DA et al (2012) The revised *Trypanosoma cruzi* subspecific nomenclature: rationale, epidemiological relevance and research applications. Infect Gent Evol 12:240–253
16. Tibayrenc M (1998) Genetic epidemiology of parasitic protozoa and other infectious agents: the need for an integrated approach. Int J Parasitol 28:85–104
17. Miles MA, Llewellyn MS, Lewis MD et al (2009) The molecular epidemiology and phylogeography of *Trypanosoma cruzi* and parallel research on *Leishmania*: looking back and to the future. Parasitology 136:1509–1528
18. Zingales B, Andrade SG, Briones MR et al (2009) A new consensus for *Trypanosoma cruzi* intraspecific nomenclature: second revision meeting recommends TcI to TcVI. Mem Inst Oswaldo Cruz 104:1051–1054
19. Lewis MD, Llewellyn MS, Yeo M et al (2011) Recent, independent and anthropogenic origins of *Trypanosoma cruzi* hybrids. PLoS Negl Trop Dis 4:e1363
20. Yeo M, Mauricio IL, Messenger LA et al (2011) Multilocus sequence typing (MLST) for lineage assignment and high resolution diversity studies in *Trypanosoma cruzi*. PLoS Negl Trop Dis 5:e1049
21. Brisse S, Henriksson J, Barnabé C (2003) Evidence for genetic exchange and hybridization in *Trypanosoma cruzi* based on nucleotide sequences and molecular karyotype. Infect Genet Evol 2:173–183
22. Machado CA, Ayala FJ (2001) Nucleotide sequences provide evidence of genetic exchange among distantly related lineages of *Trypanosoma cruzi*. Proc Natl Acad Sci U S A 98:7396–7401
23. De Freitas JM, Augusto-Pinto L, Pimenta JR et al (2006) Ancestral genomes, sex, and the population structure of *Trypanosoma cruzi*. PLoS Pathog 2:e24
24. Westenberger SJ, Barnabé C, Campbell DA et al (2005) Two hybridization events define the population structure of *Trypanosoma cruzi*. Genetics 171:527–543
25. Sturm NR, Campbell DA (2010) Alternative lifestyles: the population structure of *Trypanosoma cruzi*. Acta Trop 115:35–43

26. Añez N, Crisante G, Da Silva FM et al (2004) Predominance of lineage I among *Trypanosoma cruzi* isolates from Venezuelan patients with different clinical profiles of acute Chagas disease. Trop Med Int Health 9:1319–1326
27. Ramirez JD, Guhl F, Rendón LM et al (2010) Chagas cardiomyopathy manifestations and *Trypanosoma cruzi* genotypes circulating in chronic Chagasic patients. PLoS Negl Trop Dis 4:e899
28. Barnabé C, Brisse S, Tibayrenc M (2000) Population structure and genetic typing of *Trypanosoma cruzi*, the agent of Chagas disease: a multilocus enzyme electrophoresis approach. Parasitology 120:513–526
29. Roellig DM, Brown EL, Barnabé C et al (2008) Molecular typing of *Trypanosoma cruzi* isolates, United States. Emerg Infect Dis 14: 1123–1125
30. Gaunt M, Miles M (2000) The ecotopes and evolution of triatomine bugs (Triatominae) and their associated trypanosomes. Mem Inst Oswaldo Cruz 95:557–565
31. Marcili A, Lima L, Valente VC et al (2009) Comparative phylogeography of *Trypanosoma cruzi* TcIIc: new hosts, association with terrestrial ecotopes and spatial clustering. Infect Genet Evol 9:1265–1274
32. Yeo M, Acosta N, Llewellyn M et al (2005) Origins of Chagas disease: *Didelphis* species are natural hosts of *Trypanosoma cruzi* I and armadillos hosts of *Trypanosoma cruzi* II, including hybrids. Int J Parasitol 35:225–233
33. Llewellyn MS, Lewis MD, Acosta N (2009) *Trypanosoma cruzi* IIc: phylogenetic and phylogeographic insights from sequence and microsatellite analysis and potential impact on emergent Chagas disease. PLoS Negl Trop Dis 3:e510
34. Marcili A, Valente VC, Valente SA et al (2009) *Trypansoma cruzi* in Brazilian Amazonia: lineages TCI and TCIIa in wild primates, *Rhodnius* spp. and in humans with Chagas disease associated with oral transmission. Int J Parasitol 39:615–623
35. Valente SA, Valente VC, Neves Pinto AY et al (2009) Analysis of an acute Chagas disease outbreak in the Brazilian Amazon: human cases, triatomines, reservoir mammals and parasites. Trans R Soc Trop Med Hyg 103:291–297
36. Vagos AR, Andrade LO, Leite AA et al (2000) Genetic characterization of *Trypanosoma cruzi* directly from tissues of patients with chronic Chagas disease: differential distribution of genetic types into diverse organs. Am J Pathol 156:1805–1809
37. Burgos JM, Begher S, Silva HM et al (2008) Molecular identification of *Trypanosoma cruzi* I tropism for central nervous system in Chagas reactivation due to AIDS. Am J Trop Med Hyg 78:294–297
38. Llewellyn MS, Rivett-Carnac JB, Fitzpatrick S et al (2011) Extraordinary *Trypanosoma cruzi* diversity within single mammalian reservoir hosts implies and mechanism of diversifying selection. Int J Parasitol 41:609–614
39. Bosseno MF, Telleria J, Vargas F et al (1996) *Trypanosoma cruzi*: study of the distribution of two widespread clonal genotypes in Bolivian *Triatoma infestans* vectors shows a high frequency of mixed infections. Exp Parasitol 83:275–282
40. Cardinal MV, Lauricella MA, Ceballos LA et al (2008) Molecular epidemiology of domestic and sylvatic *Trypanosoma cruzi* infection in rural northwestern Argentina. Int J Parasitol 38:1533–1543
41. Yeo M, Lewis MD, Carrasco HJ et al (2007) Resolution of multiclonal infections of *Trypansoma cruzi* from naturally infected triatomine bugs and from experimentally infected mice by direct plating on a sensitive solid medium. Int J Parasitol 37:111–120
42. Herrera L, D'Andrea PS, Xavier SC et al (2005) *Trypanosoma cruzi* infection in wild mammals of the National Park "Serra da Capivara" and its surroundings (Piaui, Brazil), an area endemic for Chagas disease. Trans R Soc Trop Med Hyg 99:379–388
43. Macedo AM, Pimenta JR, Aguiar RS et al (2001) Usefulness of microsatellite typing in population genetic studies of *Trypanosoma cruzi*. Mem Inst Oswaldo Cruz 96:407–413
44. Ramirez JD, Guhl F, Messenger LA et al (2012) Contemporary cryptic sexuality in *Trypanosoma cruzi*. Mol Ecol 21:4216–4226
45. Rougeron V, De Meeûs T, Hide M et al (2009) Extreme inbreeding in *Leishmania braziliensis*. Proc Natl Acad Sci U S A 106:10224–10229
46. Miles MA (1993) Culturing and biological cloning of *Trypanosoma cruzi*. In: Hyde JE (ed) Protocols in molecular parasitology, vol 21. Springer, London, pp 15–28
47. Lewis MD, Ma J, Yeo M et al (2009) Genotyping of *Trypanosoma cruzi*: systematic selection of assays allowing rapid and accurate discrimination of all known lineages. Am J Trop Med Hyg 81:1041–1049
48. D'Avila DA, Macedo AM, Valadares HM et al (2009) Probing population dynamics of *Trypanosoma cruzi* during progression of the

chronic phase in chagasic patients. J Clin Microbiol 47:1718–1725

49. Burgos JM, Diez M, Vigliano C (2010) Molecular identification of *Trypanosoma cruzi* discrete typing units in end-stage chronic Chagas heart disease and reactivation after heart transplantation. Clin Infect Dis 51:485–495
50. Van der Auwera G., Maes I., Lewis M.D. et al. (2012) Standardized method for direct determination of *Trypanosoma cruzi* discrete typing units. Trans R Soc Trop Med Hyg submitted
51. Lauthier JL, Tomasini N, Barnabé C et al (2012) Candidate targets for Multilocus Sequence Typing of *Trypanosoma cruzi*: validation using parasite stocks from the Chaco Region and a set of reference strains. Infect Genet Evol 12:350–358
52. Diosque P, Tomasini N, Lauthier JJ et al (2014) Optimized multilocus sequence typing scheme (MLST) for *Trypanosoma cruzi*. PLoS Negl Trop Dis (in press.
53. Andersson B (2011) The *Trypanosoma cruzi* genome; conserved core genes and extremely variable surface molecule families. Res Microbiol 162:619–625
54. Ellegren H (2000) Microsatellite mutations in the germline: implications for evolutionary inference. Trends Genet 16:551–558
55. Hoffman J, Amos W (2005) Microsatellite genotyping errors: detection approaches, common sources and consequences for paternal exclusion. Mol Ecol 14:599–612
56. Messenger LA, Llewellyn MS, Bhattacharyya T et al (2012) Multiple mitochondrial introgression events and heteroplasmy in *Trypanosoma cruzi* revealed by maxicircle MLST and next generation sequencing. PLoS Negl Trop Dis 6:e1584
57. Herwaldt BL (2001) Laboratory-acquired parasitic infections from accidental exposures. Clin Microbiol Rev 14:659–688
58. Hall TA (1999) Bioedit: a user-friendly biological sequence alignment edit and analysis program for Windows 95/98/NT. Nucleic Acids Symp Ser 41:95–98
59. Huson DH, Bryant D (2006) Application of phylogenetic networks in evolutionary studies. Mol Biol Evol 23:254–267
60. Posada D (2008) jModelTest: phylogenetic model averaging. Mol Biol Evol 25:1253–1256
61. Tamura K, Peterson D, Peterson N et al (2011) MEGA5: molecular evolutionary genetics analysis using maximum likelihood, evolutionary distance and maximum parsimony methods. Mol Biol Evol 28:2731–2739
62. Guindon S, Dufayard JF, Lefort V et al (2010) New algorithms and methods to estimate maximum-likelihood phylogenies: assessing the performance of PhyML 3.0. Syst Biol 59: 307–321
63. Ronquist F, Huelsenbeck JP (2003) MrBayes 3: Bayesian phylogenetic inference under mixed models. Bioinformatics 19:1572–1574
64. Minch E, Ruiz-Linares A, Goldstein D et al (1997) MICROSAT v1.5d: a computer programme for calculating various statistics on microsatellite allele data. Department of Genetics, Stanford University, Stanford, CA
65. Felsenstein J (1989) PHYLIP – phylogeny inference package (version 3.2). Cladistics 5:164–166
66. Park SDE (2001) Trypanotolerance in West African cattle and the population genetic effects of selection. Ph.D. thesis, University of Dublin
67. Goudet J (1995) FSTAT (version 1.2): a computer program to calculate F-statistics. J Hered 86:485–486
68. Excoffier L, Schneider S (2005) Arlequin ver. 3.0: an integrated software package for population genetics data analysis. Evol Bioinform Online 1:47–50
69. Agapow PM, Burt A (2001) Indices of multilocus linkage disequilibrium. Mol Ecol Notes 1: 101–102
70. Peakall R, Smouse P (2006) GENALEX 6: genetic analysis in Excel. Population genetic software for teaching and research. Mol Ecol Notes 6:288–295
71. Liu N, Zhao H (2006) A non-parametric approach to population structure inference using multilocus genotypes. Hum Genomics 2:353–364
72. Jombart T, Devillard S, Balloux F (2010) Discriminant analysis of principal components: a new method for the analysis of genetically structured populations. BMC Genet 11:94
73. Brisse S, Verhoef J, Tibayrenc M (2001) Characterisation of large small subunit rRNA and mini-exon genes further supports the distinction of six *Trypanosoma cruzi* lineages. Int J Parasitol 31:1218–1226
74. Sturm NR, Vargas NS, Westenberger SJ et al (2003) Evidence for multiple hybrid groups in *Trypanosoma cruzi*. Int J Parasitol 33:269–279
75. Gaunt MW, Yeo M, Frame IA et al (2003) Mechanism of genetic exchange in American trypanosomes. Nature 421:936–939
76. Franzén O, Ochaya S, Sherwood E et al (2011) Shotgun sequencing analysis of *Trypanosoma cruzi* I Sylvio X10/1 and comparison with *T. cruzi* VI CL Brener. PLoS Negl Trop Dis 5:e984
77. Kawashita SY, Sanson GFO, Fernandes O et al (2001) Maximum-likelihood divergence date estimates based on rRNA gene sequences sug-

gests two scenarios of *Trypanosoma cruzi* intraspecific evolution. Mol Biol Evol 18: 2250–2259

78. Jombart T (2008) adegenet: a R package for the multivariate analysis of genetic markers. Bioinformatics 24:1403–1405
79. Weathery DB, Boehlke C, Tarleton RL (2009) Chromosome level assembly of the hybrid *Trypanosoma cruzi* genome. BMC Genomics 10:255
80. Tomasini N, Lauthier JJ, Llewellyn MS et al. (2013) MLSTest: novel software for multi-locus sequence data analysis in eukaryotic organisms. Infect Genet Evol 20:188–196

Chapter 20

Screening *Leishmania donovani* Complex-Specific Genes Required for Visceral Disease

Wen-Wei Zhang and Greg Matlashewski

Abstract

Leishmania protozoan parasites are the causing agent of leishmaniasis. Depending on the infecting species, *Leishmania* infection can causes a wide variety of diseases such as self-healing cutaneous lesions by *L. major* and fatal visceral leishmaniasis by *L. donovani* and *L. infantum*. Comparison of the visceral disease causing *L. infantum* genome with cutaneous disease causing *L. major* and *L. braziliensis* genomes has identified 25 *L. infantum* (*L. donovani* complex) species-specific genes that are absent or pseudogenes in *L. major* and *L. braziliensis*. To investigate whether these *L. donovani* complex species-specific genes are involved in visceral infection, we cloned these genes from *L. donovani* and introduced them into *L. major* and then determined whether the transgenic *L. major* had an increased ability to survive in liver and spleen of BALB/c mice. Several of these *L. donovani* complex specific genes were found to significantly increase *L. major* survival in visceral organs in BALB/c mice including the A2 and Ld2834 genes, while down regulation of these genes in *L. donovani* by either antisense RNA or gene knockout dramatically reduced *L. donovani* virulence in BALB/c mice. This demonstrated that *L. donovani* complex species-specific genes play important roles in visceral infection. In this chapter, we describe procedures to screen *L. donovani* complex specific genes required for visceral infection by cross species transgenic expression, gene deletion targeting and measuring infection levels in mice.

Key words Leishmania, Visceral leishmaniasis, Cutaneous leishmaniasis, Tissue tropism, Functional genomics, Comparative genomics, Species-specific genes, Crossing species expression, Transgenic expression, Gene targeting, Virulence genes

1 Introduction

Leishmania protozoan infection can cause a broad range of diseases depending on the infecting *Leishmania* species. *L. donovani*, *L. infantum*, and *L. chagasi* from the *L. donovani* complex cause fatal visceral leishmaniasis involving the liver and spleen. *L. major* and *L. tropica* infections result in cutaneous lesions at the site of the sand fly bite. *L.* (*Viannia*) *braziliensis* causes cutaneous lesions and in some individuals also cause highly destructive mucocutaneous leishmaniasis in the nasopharyngeal tissue [1, 2]. The host health status including HIV infection and the genetic background

Christopher Peacock (ed.), *Parasite Genomics Protocols*, Methods in Molecular Biology, vol. 1201,
DOI 10.1007/978-1-4939-1438-8_20,

can influence the outcome of infection [3–5]; however, the major factor that determines the tropism and pathology of infection is the species of *Leishmania* [1, 2]. Although *Leishmania* genome sequencing projects have identified genetic differences among different *Leishmania* species [6–10], it has not been established which genetic differences are responsible for the distinct tissue tropisms and pathologies.

A2 represents the prototype *L. donovani* species-specific gene initially isolated because its expression is upregulated in the amastigote stage [11]. A2 was subsequently shown to be a pseudogene in *L. major* and *L. tropica* and was required for *L. donovani* visceral infection in mice [12–14]. Transfection of the *L. donovani* A2 gene into *L. major* rendered *L. major* more virulent in visceral infections but less virulent at the cutaneous site [14, 15]. This demonstrated that species-specific genes could influence virulence and pathology of *Leishmania* infection.

Comparison of the *L. infantum* genome with *L. major* and *L. braziliensis* genomes has identified about 200 differentially distributed genes between these three species. Among these differentially distributed genes, 5 genes are *L. major* specific, 25 genes *L. infantum* specific (*L. donovani* complex species specific), and 49 genes *L. braziliensis* specific [7]. These 25 *L. donovani* complex specific genes, including A2, are absent or present as pseudogenes in *L. major* and *L. braziliensis*. The function of some of these *L. donovani* complex specific genes can be predicted by sequence similarity searches; however, the majority of these genes encode hypothetical proteins with no known function.

To investigate whether these *L. donovani* complex species-specific genes are involved in tissue tropism of *Leishmania* infection, we cloned orthologs of these genes from *L. donovani* and introduced them into *L. major* and determined whether the transgenic *L. major* were phenotypically better able to survive in the liver and spleen of BALB/c mice [16]. Several of these *L. donovani* genes that are pseudogenes in *L. major* were found to significantly increase *L. major* survival in the visceral organs. Moreover, gene deletion of one of these genes in *L. donovani* dramatically reduced its virulence in visceral infection in BALB/c mice, further confirming the importance of *L. donovani* complex specific genes in infection tissue tropism and pathology.

With more *Leishmania* genomes of various species (and clinical isolates) being sequenced, it is important to compare these *Leishmania* genomes and determine what roles species-specific genes play; whether they are essential, important or redundant for a particular *Leishmania* species, or whether they are involved in infection tissue tropism and virulence? To answer these questions, one relatively simple approach would be to introduce species-specific genes into a closely related species (or isolate), which does not contain these genes to determine whether crossing species transgenic expression would alter the receipt species growth and virulence phenotype. In this chapter, we describe experimental procedures to screen *L. donovani* complex species-specific genes important for visceral infection, including crossing species expression in *L. major*, gene deletion targeting in *L. donovani* and assessing virulence phenotype of these *Leishmania* mutants in the animal model, BALB/c mice.

2 Materials

2.1 Common Lab Equipment and Lab Ware

1. Microcentrifuge (Eppendorf Centrifuge 5415C).
2. Low and high speed centrifuges.
3. Microcentrifuge and centrifuge tubes (1.5 ml, 15 ml, and 50 ml such as Falcon conical centrifuge tubes).
4. 4 °C refrigerator, −20 °C and −80 °C freezers.
5. Fume hood.
6. Class II Biological Safety Cabinet.
7. Protein and DNA gel electrophoresis and blotting apparatus.
8. 27 °C and 37 °C incubators supplemented with 5 % CO_2.
9. 37 °C incubators with or without shaker.
10. Water bath.
11. Microscope with 10×, 40×, and 100× objective lens.
12. Hemocytometer (Thoma cell).
13. Lab counters with two or more counting units.
14. Spectrophotometer.
15. Sterile glass bottles.
16. Pipetting aid and sterile pipets.
17. 25-cm^2 and 75-cm^2 sterile culture flasks.
18. 96-Well and 24-well microplates.
19. Micropipet, sterile tips.
20. Distilled water.
21. Phosphate-buffered saline (PBS), pH 7.2.
22. Ice.

2.2 Leishmania Strains and Culture Mediums

1. We have *Leishmania major* Friedlin V9 and *L. donovani* 1S/Cl2D strains in the lab.
2. *L. donovani* promastigotes culture medium: 1× M199 medium, 10 % heat-inactivated fetal bovine serum (FBS), 100-U/ml penicillin, 100-μg/ml streptomycin, 0.1 mM adenosine, 2 mM glutamine, 10 μg/ml folic acid, and 25 mM HEPES, pH 7.4.

3. Two *L. donovani* axenic amastigotes culture mediums are used in the lab:
   (a) 1× RPMI-1640 medium, 20 % heat-inactivated FBS, 1× RPMI-1640 Vitamin Mix, 1× RPMI-1640 Amino Acid Mix, 15 mM KCl, 114.6 mM KH_2PO_4, 10.38 mM K_2HPO_4, 0.5 mM $MgSO_4$, 24 mM $NaHCO_3$, 2 mM glutamine, 22 mM d-glucose, 20-U/ml penicillin, 20 μg/ml streptomycin, 0.25 mM adenosine, 5.36 g/l MES, and 10μg/ml folic acid, pH 5.5.
   (b) 1× M199 medium, 25 % non-heat-inactivated FBS, 100-U/ml penicillin, 100 μg/ml streptomycin, 2 mM $MgCl_2$, and 10 mM succinic acid and pH 5.5.
4. *L. major* promastigotes culture medium: 1× M199 medium, 10 % heat- inactivated FBS, 100 U/ml penicillin, 100 μg/ml streptomycin, 0.1 mM adenine, 5 mg/l haemin, 1 mg/l Biotin, 1 mg/l biopterin, and 40 mM HEPES, pH 7.4.
5. Selection antibiotics: stock solutions are 50 mg/ml for hygromycin B, puromycin dihydrochloride, and geneticin disulfate(G418), and 20 mg/ml for phleomycin. Aliquot and store at −20 °C.

2.3 Leishmania DNA Isolation

1. *Leishmania* cell lysis buffer (TELT buffer): 50 mM Tris–HCl pH 8.0, 62.5 mM EDTA pH 9.0, 2.5 M LiCl, and 4 % (v/v) Triton X-100.
2. Water-equilibrated phenol–chloroform mixture (1:1 v/v).
3. Ethanol, 100 and 70 %.

2.4 Polymerase Chain Reaction

1. 10× PCR buffer minus Mg.
2. 50 mM $MgCl_2$.
3. 10 mM dNTP mixture.
4. Taq DNA polymerase (5 U/μl, Invitrogen).
5. Primers (20 μm each).
6. 200 μl thin wall PCR tubes.
7. Thermocycler (Biometra *TGRADIENT*).

2.5 DNA Gel Electrophoresis (Analytical and Preparative)

1. Agarose.
2. 1× TAE buffer (stock solution 10×: 400 mM Tris-acetate, 10 mM EDTA, pH 8.4).
3. Ethidium bromide, 0.1 mg/ml (stock solution 10 mg/ml).
4. DNA Ladder.
5. Electrophoresis Power supply (200 V/2A, Bio Rad).
6. Agarose gel electrophoresis apparatus (e.g., Bio-Rad).

7. Gel documentation equipment.
8. UV box for examining the agarose gel and cutting out the DNA bands from the gel.

2.6 Restriction Enzymes and Ligase

We use restriction enzymes from Invitrogen, MBI Fermentas and New England Biolabs.

1. Restriction enzymes such as *Hind* III, *Kpn* I, *Xba* I, *Bam* HI, and *Bgl* II (10 U/μl) and corresponding 10× buffers.
2. T4 DNA ligase (1 U/μl) and 5× DNA Ligase buffer (Invitrogen).

2.7 Bacterial Culture and Cells

1. Luria Bertani (LB) medium: 10 g/l Bacto-Tryptone, 5 g/l yeast extract, 10 g/l NaCl.
2. SOC medium: 20 g/l Bacto-Tryptone, 5 g/l yeast extract, 0.5 g/l NaCl.
3. LB-agar: LB + 15 g/l agar.
4. Ampicillin (100 mg/ml stock solution).
5. DH5α-competent *Escherichia coli* cells (purchased or prepared in the lab).

2.8 DNA Minipreparation (Minipreps) and Maxipreparation (Maxipreps) Kits (Qiagen or GE Healthcare Products)

1. Solution I: 50 mM glucose, 25 mM Tris–HCl, pH 8.0, 10 mM EDTA, pH 8.0, 50 μg/ml RNase A. Store at +4 °C.
2. Solution II: 0.2 N NaOH freshly diluted from a 10 N stock, 1 % sodium dodecyl sulfate (SDS); prepare immediately before use.
3. Solution III: 5 M potassium acetate (60 ml), glacial acetic acid (11.5 ml), distilled H_2O (28.5 ml). Store at +4 °C.

2.9 Sequencing

We send our DNA samples (PCR products or plasmid constructs) to University Genome Center for sequencing.

2.10 Leishmania Transfection

1. Cytomix electroporation buffer: 120 mM KCl, 0.15 mM $CaCl_2$, 10 mM K_2HPO_4, 25 mM HEPES, 2 mM EDTA, and $MgCl_2$; pH 7.6. Autoclave-sterilize and keep at +4 °C.
2. Electroporator (Bio-Rad Gene pulser ™).
3. Electroporation cuvets (4 mm gap).
4. Plasmid DNA or DNA fragment resuspended in 10–30 μl sterile H_2O at a concentration of 0.5–2 μg/μl.

2.11 Analysis of Transfectants

2.11.1 Western Blotting

1. 1 % SDS-PAGE sample loading buffer.
2. Protein gel electrophoresis and blotting apparatus.
3. Blotting membrane such as Hybond™ ECL (nitrocellulose) from GE Healthcare.
4. Orbital shaker.

5. Forceps with rounded, non-serrated tips.
6. Nonfat dried milk.
7. Diluent and wash buffer PBS-T (0.1 % Tween 20 in PBS).
8. Primary antibodies, such as anti-A2 Tag antibody.
9. Anti-mouse IgG, Horseradish Peroxidase-Linked whole antibody (from sheep).
10. ECL Western blotting detection reagents.
11. X-ray film cassettes.
12. Timer.
13. X-ray film developer.
14. Or Gel documentation equipment able to directly detect the light from Chemiluminescence reaction on the Western blot membrane.

2.11.2 Southern Blotting

1. DNA gel electrophoresis and blotting apparatus.
2. 1× TE buffer: 10 mM Tris–HCl, pH 8.0, 1 mM EDTA.
3. Denaturing solution: 1.5 M NaCl, 0.5 M NaOH.
4. Neutralizing solution: 1.5 M NaCl, 0.5 M Tris–HCl pH 7.2, 1 mM EDTA.
5. Nylon-membranes: Hybond N+(Amersham).
6. Chromatography paper: Whatman 3MM.
7. Oligolabelling kit (Amersham).
8. Radiolabeled 32 P-dCTP.
9. DNA polymerase 1 Klenow fragment.
10. 37 °C water bath.
11. 20× SSC: 3 M NaCl, 0.3 M Na_3citrate
12. 20× SSPE: 3.6 M NaCl, 200 mM NaH_2PO_4, 20 mM EDTA, pH 7.7.
13. SDS.
14. 65 °C incubator with shaker.
15. Hyperfilm (Amersham).
16. Autoradiography cassettes and amplifying screens.

2.12 Characterization of Leishmania Transfectants in Mice

1. 16–20 g female BALB/c mice.
2. Syringes with needle size: 25G or smaller (We use 0.5 ml insulin syringes).
3. 70 % ethanol.
4. Facial tissues.
5. A table light with an incandescent light bulb.

6. A lab designed mouse restrainer (a 50 ml plastic tube with breathing holes on the wall and a hole in the middle of the cap).
7. A caliper with readability of 0.05 mm.
8. Surgical Forceps and Scissors.
9. 35 × 10 mm culture dishes.
10. An electronic toploading balance with weighing range 0–32 g and readability of 0.002 g.
11. Glass microslides with frosted writing area at one end.
12. Giemsa stain kit (Diff-Quick).
13. Pyrex* Brand Tenbroeck Tissue Grinders (7 ml, 18 × 130 mm).

3 Methods

3.1 Culture for L. donovani and L. major Promastigotes

We routinely culture *Leishmania* promastigotes in 25 cm^2 flasks which contain 4–12 ml medium at 27 °C incubator without 5 % CO_2. The flask is stood up when it contains 5 ml or less medium and laid down when has more than 10 ml culture. The *Leishmania* promastigotes culture is passaged once a week in 1:40 to 1:10 dilutions in fresh culture medium or when the culture reaches late stationary phase.

3.2 Preparation of Leishmania Genomic DNA

We use the mini-prep method [17] to isolate *Leishmania* genomic DNA. All the purification procedures are carried out at room temperature.

1. Harvest stationary *Leishmania* promastigotes culture (1.5 ml; 2×10^7 to 8×10^7 cells/ml) in a microcentrifuge tube by centrifugation in a bench top centrifuge at maximum speed for 1 min.
2. Suspend the pelleted cells in 150 μl of lysis buffer (50 mM Tris–HCl, pH 8.0/62.5 mM EDTA, pH 9.0/2.5 M LiCl/4 % (v/v) Triton X-100; termed TELT buffer) by inverting the tube several times to completely resuspend the cells, incubate for 5 min.
3. Add 150 μl of a water-equilibrated phenol–chloroform mixture (1:1 v/v, Invitrogen) into the tube and inverting the tube several times, and incubate for 5 min.
4. Separate the phases by centrifugation at 13,000 × *g* for 5 min.
5. Carefully transfer the upper phase (about 140–150 μl) to a new tube. Precipitate the nucleic acids from the upper phase by adding 300 μl of absolute ethanol. Mix by inverting the tube several times and incubate for 5 min (*see* **Note 1**).

6. Collect the nucleic acids by centrifugation at 13,000 × *g* for 10 min. Wash the pellet once with 1 ml of absolute ethanol at 13,000 × *g* for 5 min.
7. Dry the DNA pellet with the cap open for 10 min.
8. Dissolve the nucleic acids in 50–100 μl of TE-8 (10 mM Tris–HCl, pH 8.0/11 mM EDTA, pH 8.0) or H_2O.

The isolated *Leishmania* genomic DNA can be readily used as template DNA for PCR or digested by restriction enzymes. Recovery of DNA ranges from 20 to 50 μg from 1×10^8 cells. The procedure can be conveniently scaled up for the isolation of DNA from large cultures. The TELT lysis buffer can be prepared in large volumes and kept at room temperature for storage (months) with no apparent deterioration.

3.3 Leishmania Expression Vectors

We use *Leishmania* expression vector pLpneo (*see* Fig. 1) to express *L. donovani* specific genes in *L. major* [18] (*see* **Note 2**). In the pLpneo vector, the multiple cloning sites are flanked by two 1 kb A2rel intergenic sequences which provide 5′-*trans*-splicing and 3′-polyadenylation sites for proper transcript processing of cloned gene in *Leishmania*. The Neomycin phosphotransferase gene (NEO) placed downstream of the second A2rel intergenic sequence is used as the selectable marker to confer resistance to neomycin. To facilitate detection of expression of *L. donovani* specific gene in transfected *L. major* cells, a sequence encoding the ten amino acids

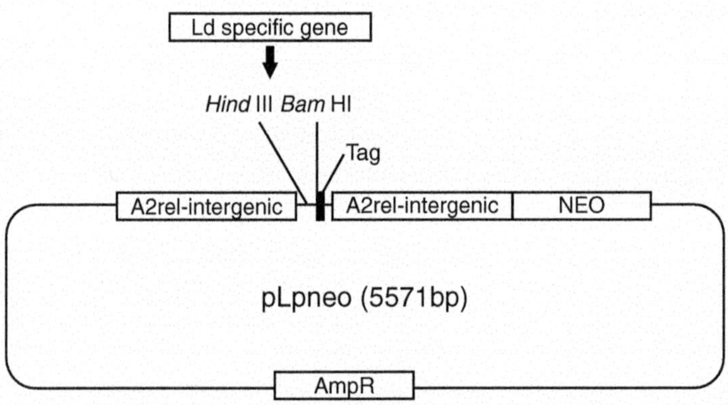

Fig. 1 Schematic representation of the *Leishmania* expression vector pLpneo. The multiple cloning sites are flanked by two 1 kb A2rel intergenic sequences (A2rel-intergenic), the drug selection marker Neomycin resistance gene (NEO) is placed downstream of the second A2rel intergenic sequence. To simplify the drawing, only the restriction enzymes (*Hind* III and *Bam* HI) used for cloning *Leishmania* genes are shown in the multiple cloning site. The complete list of restriction enzymes in the multiple cloning sites (according to the sequence order) are: *Bgl* II, *Xho* I, *Hind* III, *Sal* I, *Apa* I, *Sma* I, *Bam* HI and *Not* I. The ten amino acids A2 tag coding sequence is indicated as tag in the figure. See text for details

A2 epitope-tag is inserted to the multiple cloning sites so the expressed *L. donovani* specific gene product is fused at its C terminal with A2 tag and can be detected by Western blot analysis with anti-A2 antibody (*see* **Note 3**). In our lab, we also have pLGFPN and pLGFPC vectors, which can be used to express GFP fusion proteins at either N or C terminus [16, 19, 20]. In some cases, the *L. donovani* specific genes are fused with GFP at either its N or C terminus to help localize these proteins in *Leishmania* cells.

3.4 Comparison of Leishmania Genomes and Selection of L. donovani Complex Specific Genes for Transgenic Crossing Species Expression Study in L. major

About 200 differentially distributed genes have been identified by comparing the *L. infantum* genome with *L. major* and *L. braziliensis* genomes [7]. Among these differentially distributed genes, 25 genes are *L. infantum* specific, 5 genes *L. major* specific and 49 genes uniquely present in *L. braziliensis* (*see* **Note 4**). These *L. infantum* specific (*L. donovani* complex specific) genes are absent or present as pseudogenes in *L. major* and *L. braziliensis*. The majority of these genes encode hypothetical proteins with no known function [7]. The basic sequence analysis information for these *L. infantum* specific genes can be found in the TriTrypDB website (http://tritrypdb.org) and the GeneDB website (http://www.genedb.org). These websites provide information such as DNA sequence (coding and flanking genomic sequences), predicted protein sequences, protein features (signal peptide, transmembrane domains, Hydropathy Plot, Secondary Structure and Blastp hits), predicted molecular weight, isoelectric point, similarities to protein data bank chains, chromosome location, orthologs and paralogs within TriTryDB, the gene ontology, and the expression profiling in promastigotes and amastigotes. We design primers for these *L. infantum* specific genes based on the sequence information from these genome databases.

3.5 Cloning of L. donovani Specific Genes into Leishmania Expression Vector

3.5.1 Gene Specific Primers Design

Since *Leishmania* genes have no introns, *Leishmania* genes can be directly amplified by PCR from its genomic DNA and cloned into expression vectors described above. Usually, oligonucleotide primers with about 20 nucleotides specific for N terminus (forward primer) and C terminus (reverse primer) coding sequences are synthesized (*see* **Note 5**).

To better explain how the specific primer pairs are designed, the *L. donovani* complex specific gene LinJ.28.0340 [16] is used as an example. To facilitate cloning into the pLpneo vector, a restriction enzyme site is added to the 5′ end of forward and reverse primers (*see* **Note 6**). The forward primer for LinJ.28.0340 is designed as 5′ ccc***aagctt***acaATGGCCGATGTGCAGCTC. Here the first three nucleotides added to the left of *Hind* III restriction endonuclease site (*aagctt*) are random chosen nucleotides to increase restriction enzyme recognition and digestion efficiency for the PCR product (*see* **Note 7**). The three nucleotides (aca) immediately upstream of the translation initiation codon ATG is from the 5′ flanking sequence of the gene and could be part of the Kozak consensus sequence which is required for efficient translation in eukaryotes (*see* **Note 8**). ATGGCCGATGTGCAGCTC is the N terminus coding sequence of LinJ.28.0340 gene. We typically set up the primer melting temperature (Tm) at around 66 °C. The Tm is estimated by counting the number of nucleotides in the primer sequence, each G or C nucleotide accounts for 4 °C and A or T nucleotide for 2 °C. Therefore, the Tm for the forward primer of LinJ.28.0340 gene will be 66 °C (4 °C × 12 (G&C) = 48 °C; 2 °C × 9(A&T) = 18 °C; 48 °C + 18 °C = 66 °C). The restriction enzyme site and the other nucleotides added to the 5′ end of primer should not be included for calculating the Tm. The reverse primer for the LinJ.28.0340 gene is 5′cg*agatct*gtCATATCCAT-CAAGATTTC GTTGAT. Again, the cg nucleotides added to the left of *Bgl* II site are to improve enzyme digestion efficiency; the gt nucleotides added to the right of *Bgl* II site are to ensure (in this case) that LinJ.28.0340 protein expressed in *L. major* would be in the same frame as the A2 epitope-tag in the vector (i.e., correct LinJ.28.0340 A2 tag fusion protein); the Tm for the reverse primer is estimated to be at 66 °C.

3.5.2 PCR Amplification and Restriction Enzyme Digestion

We use *Taq* DNA polymerase to set up PCR amplification (*see* **Note 9**).

1. The PCR is set up in a total reaction volume of 50 µl in a 200 µl PCR tube including 1× PCR buffer, 2.5 mM $MgCl_2$, 200 µM dNTPs, 1–5 µl (80 ng) of genomic DNA, 50 pmol of each primer, and 1.25 U Taq DNA polymerase. The PCR tubes are incubated in a thermal cycler (Biometra $T_{GRADIENT}$) at 95 °C for 3 min to completely denature the genomic DNA, followed by 35 cycles of PCR amplification as follows: Denature 95 °C for 30 s; Anneal 55 °C for 1 min; Extend 72 °C for 2–3 min (depending on the gene size). Incubate for an additional 10 min at 72 °C and maintain the reaction at 4 °C. The PCR samples can be stored at −20 °C until use.
2. 15–25 µl PCR products are loaded into a 1–1.5 % agarose gel for electrophoresis and visualized by ethidium bromide (or SYBR Green) staining with appropriate molecular weight standards. The amplified gene bands with expected size are excised from the agarose gel and purified with Sephaglas™ BandPrep Kit (Amersham Pharmacia Biotech) which is designed for the rapid extraction of DNA from agarose gels (*see* **Note 10**).
3. The entire 20 µl of purified PCR product (eluted from purification) is subjected to restriction enzyme double digestion. A typical digestion reaction is set up as follows: in a 1.5 ml microcentrifuge tub, add 20 µl purified PCR product, 3 µl 10× REACT buffer 2, 0.5 µl (5 U) *Hind* III, 0.5 µl (5 U) *Bgl* II, and 6 µl H_2O to make the final reaction volume of 30 µl and incubate at 37 °C over night (*see* **Note 11**).

4. Simultaneously, the *Leishmania* expression vector pLpneo is subjected to *Hind* III and *Bam* HI sequential digest. As *Bam* HI requires higher salt concentration, the reaction is set up as follows: in a 1.5 ml tube, add 2 μl (1 μg) pLpneo plasmid, 2 μl REACT buffer 2, 15.5 μl H_2O, 0.5 μl (5 U) *Hind* III and mix well, incubate at 37 °C for 2–4 h, then add 5.5 μl H_2O, 1 μl REACT buffer 2, 3 μl 500 mM NaCl, and 0.5 μl (5 U) *Bam* HI to the tube to make the final reaction volume of 30 μl and incubate at 37 °C for additional 2 h (or overnight for convenience) (*see* **Note 12**).
5. After overnight digestion, the reactions are loaded into a 1 % agarose gel for electrophoresis to separate the gene and vector bands from the digested short nucleotides fragments. The double digested gene band and the vector band are excised and purified with the DNA purification kit.

3.5.3 Ligation Reaction, Transformation, and Positive Clone Selection

Set Up Ligation Reaction

We set up ligation reactions as follows: in 1.5 ml tube, add 1 μl (3–30 fmol) *Hind* III and *Bam* HI double digested pLpneo vector, 3 μl (9–90 fmol) *Hind* III and *Bgl* II (or *Bam* HI) digested *L. donovani* specific gene PCR product, 2 μl 5× T4 DNA ligase buffer, 3 μl H_2O and 1 μl T4 DNA ligase (1 U) to make final volume of 10 μl, mix gently and incubate at room temperature for 2–3 h (*see* **Note 13**).

Transform Chemically Competent *E. coli* Cells

1. Thaw on ice 1 vial of DH5α competent *E. coli* cells for each transformation.
2. Add the whole ligation reaction (10 μl) to the *E. coli* vial and mix gently.
3. Incubate on ice for 5–30 min.
4. Heat-shock the cells for 30 s at 42 °C without shaking.
5. Add 250 μl of room temperature SOC medium to the cells.
6. Cap the tubes and shake at 37 °C for 1 h.
7. Spread the whole transformation on pre-warmed LB plates containing 100 μg/ml ampicillin.
8. Incubate plates overnight at 37 °C.
9. An efficient ligation reaction should produce 50–100 colonies. Pick about 10 colonies for analysis.

Analyzing Positive Clones

1. Culture 10 colonies overnight in 3 ml LB medium containing 100 μg/ml ampicillin.
2. Isolate plasmid DNA with a MiniPrep Kit of your choice.
3. Analyze the plasmid by restriction analysis and by sequencing to confirm the correct sequence.

3.5.4 Plasmid Mini and Maxi Preparation

We use plasmidPrep Mini spin Kit from GE Healthcare for plasmid mini preparation and QIAGEN kit for plasmid maxi preparation.

3.6 Transfecting L. donovani Specific Gene Expression Vector into L. major

L. major promastigotes growing in later-log phase are counted with a hemocytometer, harvested at 1,300 × *g* for 5 min and washed once with ice cold cytomix electroporation buffer [21]. The promastigotes are resuspended in ice-cold cytomix buffer to a concentration of 2×10^8 cells/ml. 500 μl of cells are aliquoted into a 4-mm gap electroporation cuvette on ice, then 5–20 μg of expression vector DNA is added into cuvette and mixed. The cells are electroporated twice at 25 μF, 1500 V (3.75 kV/cm), pausing 10 s between pulses in an electroporation apparatus (Bio-Rad Gene pulser™). Following electroporation, cells are allowed to sit on ice for 10 min, transferred to a flask containing 10 ml *Leishmania* culture media and incubated at 26 °C overnight. Next day morning, 10 μl of G418 at 50 mg/ml is added into each flask to make final G418 concentration at 50 μg/ml. 4 days after adding G418, the culture medium is replaced with fresh medium containing 100 μg/ml G418. The growth of transfected *Leishmania* cells are monitored under microscope. It typically takes 2–3 weeks to establish a transfected *Leishmania* cell culture. To avoid selecting clones with unexpected mutation, we use pooled transfectants for subsequent expression and virulence studies. Gene expression can often be improved due to increased episomal plasmid copy number with increasing concentration of G418 (up to150μg/ml) in the medium and passage times.

3.6.1 Determine Expression of Transfected Genes

Western Blot Analysis

Since a 10 amino acid A2 epitope-tag coding sequence is added into the 3′ end position of each *L. donovani* specific gene in the expression vector, it is possible to use the anti-A2 monoclonal antibody to examine whether the *L. donovani* specific gene is properly expressed in *L. major* because *L. major* does not express endogenous A2 genes.

1. Harvest 1×10^{8} transfected *L. major* cells by centrifuge at 1300 g for 5 min and wash once with PBS.
2. Lyse cells in 40 μl of 1 % SDS sample loading buffer, boil in water bath for 3 min, and centrifuge at high speed for 10 min.
3. Load 10–20 μl of the cell lysate (supernatant) into each well of a mini SDS-PAGE gel (10–12 %). Run SDS-Page gel electrophoresis and transfer into nitrocellulose membrane.
4. Block the membrane in 10 ml 10 % skim dry milk in PBS-T buffer for 1 h and wash with 10 ml PBS-T, 10 min each wash for total 3 times.
5. Incubate the membrane for 2 h in 10 ml anti-A2 monoclonal hybridoma culture media diluted in PBS-T containing 5 % skim dry milk, wash 3 times with PBS-T.
6. Incubate the membrane for 1 h in 10 ml HRP labelled anti-mouse IgG antibody diluted in PBS-T containing 5 % skim dry milk, wash 5 times with PBS-T (15 min for first wash and 5 min for each subsequent 4 washes).

7. Incubate the membrane in ECL (GE Healthcare) detection mix for 1 min.
8. Drain the reagent, cover the membrane with Saran Wrap, and immediately expose to film for 30 s to 10 min, develop the film.
9. Check if *L. donovani* specific genes are expressed in transfected *L. major*, the gene product size and expression level.

Fluorescence Microscope Analysis

If the *L. donovani* specific gene is cloned into a *Leishmania* expression vector such as pLGFPC or pLGFPN, the gene product will be fused with green fluorescent protein at either the N terminus (pLGFPC) or the C terminus (pLGFPN). The *L. donovani* specific gene product-GFP fusion protein can be viewed under a fluorescence microscope to determine its localization in living *Leishmania* cells [16, 19, 20].

3.7 Characterization of Recombinant L. major Cells Expressing L. donovani Specific Genes

After introducing *L. donovani* specific genes into *L. major*, it is important to determine whether *L. donovani* specific genes can alter *L. major* growth property in vitro and virulence in mice. Since the transfected *L. major* cells are selected and cultured in medium containing G418, it is important to have a *L. major* cell line transfected with empty pLpneo vector alone as control.

3.7.1 Growth Curves in Promastigotes and Amastigotes Culture Mediums

To determine whether expression of *L. donovani* specific genes would affect *L. major* growth in in vitro culture medium, we measure the growth curves of these recombinant *L. major* cells in 96 well culture plates (*see* **Note 14**).

1. Determine the concentration of transfected *L. major* promastigotes growing in stationary phase by counting the cells on a hemocytometer.
2. Inoculate *L. major* cells into wells of a 96 well plate containing 200 μl/well fresh *L. major* promastigote culture medium with 100 μg/ml G418 to a concentration of 2×10^6/ml (4×10^5/200 μl), triple wells for each cell line, incubate in a 27 °C incubator.
3. Inoculate *L. major* cells into wells of a 96 well plate containing 200 μl/well fresh *L. donovani* axenic amastigote culture medium with 100 μg/ml G418 (*see* **Note 15**) to a concentration of 2×10^6/ml (4×10^5/200 μl), triple wells for each cell line, incubate at 37 °C in a 5 % CO_2 incubator.
4. Measure the OD600 value directly from the plates in a spectrophotometer daily for total 7 days or until the cultures reach stationary phase.
5. Use Microsoft Excel to plot the growth curves and determine whether the growth of *L. major* cells in vitro is affected by the presence of the *L. donovani*-specific genes.

3.7.2 Infecting BALB/c Mice

The main objective of such a study is to investigate whether *L. donovani*-specific genes play a role in visceral *Leishmania* infection. Therefore, it is important to examine whether expression of *L. donovani* specific genes increase *L. major* survival/virulence in visceral organs in mice.

Visceral Infection in BALB/c Mice

1. 5 Female BALB/c mice (4–6 weeks old weighing 15–20 g) are used for each transgenic *L. major* cell line (*see* **Note 16**).
2. Harvest stationary phase *L. major* cells by centrifugation and resuspend cells in PBS at a concentration of 1×10^9/ml.
3. Infect mice by tail vein injection with 1×10^8 cells/100 μl/mouse (*see* **Note 17**). We restrain mouse in a 50 ml plastic tube with several breathing holes drilled on the wall, the mouse tail is let out through a hole drilled on the center of the cap, use some wrinkled paper to fill the space of tube tip so the mouse in the tube is fully restrained after the cap is closed and the entire tail is out of the tube. Sterilize the tail with 70 % ethanol, hold the tail near the base tightly with left hand fingers, warm the tail slightly with a light bulb so the lateral tail veins can be clearly seen (a regular table light works well which provides not only heat but also light for precise injection). Use a syringe with needle size: 25G or smaller (such as 0.5 ml insulin syringe) to inject transgenic *L. major* cells.
4. 4 and 6 weeks after infection, mice are sacrificed. When BALB/c mice are viscerally infected with *L. donovani* by tail vein injection, the liver parasite burden usually reaches a peak level at around 4 weeks post infection then slowly decreases, the spleen parasite burden however may continue to increase for 2 or 3 more weeks before it starts to decrease, the infection will be finally cleared without treatment 3–6 months post infection. After sacrifice, the liver and spleen are removed for measuring their weights and parasite burden. We determine the liver parasite burden by either counting the amastigotes number in liver imprints (liver) or limiting dilution culture (liver, footpad or spleen).
   (a) *Counting amastigotes number in liver imprints to calculate the L. donovani unit (LDU)*:
      - Stain mouse liver imprints on glass slide with a Giemsa stain kit (Diff-Quick).
      - Examine the liver imprints under microscope with 100× objective lens.
      - Count the amastigotes number over a thousand liver cell nuclei.
      - The liver parasite burden LDU is calculated by multiplying the amastigotes number per thousand liver cell nuclei to the liver weight (g).

(b) *Limiting dilution culture* (*see* **Note 18**):

- Cut liver or spleen into several pieces into a Pyrex* Brand Tenbroeck Tissue Grinder containing 5 ml *Leishmania* culture medium.
- Homogenize until no liver or spleen tissue is visible.
- Transfer the entire cell suspension into a 15 ml centrifuge tube and add about 4 ml medium to make final volume of 10 ml and mix well.
- Fill 96 well plates with *Leishmania* promastigotes culture medium (100 μl/well) (*see* **Note 19**).
- Add 100 μl cell suspension of **step 3** into the first well of 96 well plate, mix and transfer 100 μl mix to the second well for twofold dilution, mix then transfer the same volume to the third well for fourfold dilution, continue until the last column (#12) for 2,048-fold dilution. Triple rows for each liver or spleen (*see* **Note 20**).
- Seal the plates with parafilm and incubate at 27 °C for 2–3 weeks.
- Determine the highest diluted wells under a microscope where *Leishmania* promastigotes grow.
- Calculate liver or spleen parasite number. For example, if the last growing well is in column 8 (128-fold dilution from the first column well), the number of *Leishmania* cells in a liver or spleen can be calculated as: 100 (dilution factor in **step 3**) × 128 = 12,800.

Cutaneous Infection in BALB/c Mice

After introducing *L. donovani* specific genes into *L. major*, it is also interesting to determine whether these *L. donovani* specific genes can affect *L. major* virulence in cutaneous infection in BALB/c mice.

1. 5 Female BALB/c mice (4–6 weeks old weighing 15–20 g) for each recombinant transgenic *L. major* cell line.
2. Infect mice on one of its hind footpads by subcutaneous injection with 5×10^6 stationary *L. major* promastigotes/50 μl/footpad/mouse.
3. Monitor lesion development by weekly caliper measurement of footpad swelling.

Mice can also be sacrificed at chosen time to determine footpad parasite burden by limiting dilution culture.

3.8 Generating *L. donovani* Specific Gene Null Mutants by Gene Targeting

To investigate whether *L. donovani* specific genes are involved in visceral *Leishmania* infection, it is important to determine whether deletion (or down regulation of expression) of *L. donovani* specific genes would reduce *L. donovani* virulence.

1. Like in other organisms, we use homologous recombination gene targeting to delete the gene of interest in *Leishmania* (*see* **Note 21**). A gene-targeting construct usually contains a drug resistance gene flanked by sequences derived from 5′ and 3′ flanking regions of the gene to be targeted. Since *Leishmania* is a diploid organism, two rounds of gene targeting are required to generate a gene null mutant. We use two of the following backbone plasmids to make gene targeting constructs: pSPY-neo, pSPYhyg [22] and pSPY-Ble [14] (*see* Fig. 2).
2. Design primers for 5′ and 3′ flanking sequences of the gene to be targeted:
   (a) Download from *Leishmania* genome database (TriTrypDB) the segment of genomic DNA sequence which includes 2,000 bp upstream 5′ flanking sequence from the gene open reading frame and 2,000 bp downstream 3′ flanking sequence (geneStart-2000 to geneEnd + 2000).
   (b) Use the primer design program Primer3 (http://frodo.wi.mit.edu/primer3/) to choose primer pairs for 5′ and 3′ flanking sequences so that the PCR products will be around 1,000 bp in length (*see* **Note 22**).
   (c) Add a *Hind* III site (5′ ccc*aagctt*) to the 5′ end position of the forward primer for the 5′ flanking sequence; add a *Bam* HI (or *Bgl* II) site (*5′ cgggatcc*) to the 5′ end position of reverse primer for 5′ flanking sequence; add a *Bam* HI site (5′cg*ggatcc*) to the 5′ end position of the forward primer for the 3′ flanking sequence; add a *Bgl* II site (5′cg*agatct*) to the 5′ end position of the reverse primer for the 3′ flanking sequence (*see* Fig. 2a).
3. Obtain 5′ and 3′ flanking sequence PCR products as described (*see* Subheading 3.5.2).
4. Insert the 3′ flanking sequence into the backbone vector:
   (a) Digest the backbone plasmid with *Bgl* II and treat the digested plasmid with Calf Intestinal Alkaline Phosphatase (CIAP) by following manufactor's instruction; digest 3′ flanking sequence PCR product with *Bam* HI and *Bgl* II.
   (b) Purify the digested backbone vector and the 3′ flanking sequence PCR product.
   (c) Set up ligation reaction, perform transformation and confirm that the 3′ flanking sequence is cloned into the vector in correct orientation.
5. Insert the 5′ flanking sequence into the backbone vector which already contains the 3′ flanking sequence:
   (a) Digest backbone plasmid and 5′ flanking sequence PCR product with *Hind* III and *Bam* HI.

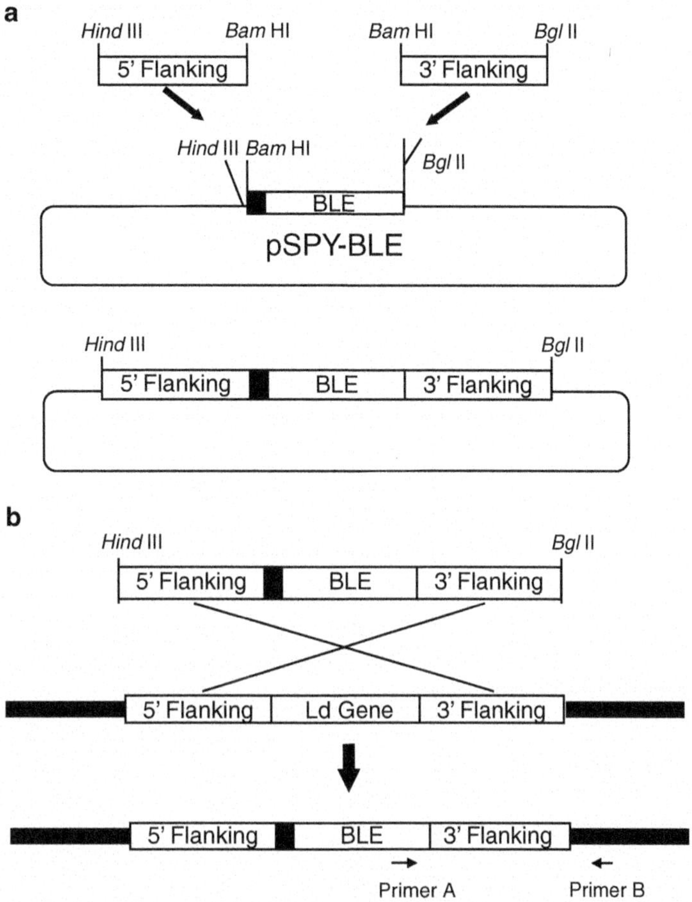

Fig. 2 Schematic representation of *Leishmania* gene targeting. (**a**) Gene targeting construct, the pSPY-BLE plasmid is used as the backbone vector to make the *Leishmania* gene targeting construct. BLE: Blemycin (Phleomycin) resistance gene; 5′ Flanking: *L. donovani* specific gene 5′ flanking sequence; 3′ Flanking: *L. donovani* specific gene 3′ flanking sequence; the filled black rectangle represents a stretch of 92 bp of pyrimidines and a splice acceptor site. Only the restriction enzymes used for cloning are shown. (**b**) PCR strategy to confirm correct gene targeting. The *Hind* III and *Bgl* II fragment containing BLE gene flanked by the *L. donovani* specific gene 5′ and 3′ flanking sequences is used to transfect *L. donovani* promastigotes. Once transfectants are obtained, a primer pair (as indicated, one specific for the BLE gene, the other from the downstream 3′ flanking sequence) is used to verify the correct homologous replacement event. Other selectable marker genes can also be used, conferring resistance to hygromycin B (HYG gene), the aminoglycoside G418 (NEO), Puromycin (PAC), or Streptothricin (SAT). These resistance genes can be readily interchanged by *Bam* HI/*Bgl* II double digestion

(b) Purify the digested plasmid and PCR fragment.

(c) Set up ligation reaction, perform transformation to create the complete gene-targeting construct.

6. Digest the plasmid with *Hind* III and *Bgl* II to release the gene targeting fragment and use 5–15 μg of gene targeting fragment to transfect *L. donovani* promastigotes as described (*see* Subheading 3.6.1).
7. Culture transfected cells in a flask containing the selection drug with a minimum concentration required to kill untransfected cells (*see* **Note 23**).
8. Perform limiting dilution cloning in 96 well plates (*see* Subheading 3.7.2.1, **step 4b**) as soon as live drug resistant cells can be seen in the flask about 7–10 days post transfection. Expand the culture in a 12 well plate with 2 ml medium per well.
9. Prepare genomic DNA and confirm correct first round gene targeting by PCR analysis with a drug resistance gene specific primer and a primer further down 3′ flanking sequence (50–100 bps beyond the targeting flanking sequence) (*see* Fig. 2b).
10. Second round gene targeting:

    (a) A simplest method is to rapidly increase drug concentration three- to fourfold to select high drug resistant clones for loss of heterologous. This method is simple, there is no need to make second round targeting construct. However, it may not work for all the genes.

    (b) Perform second round gene targeting with second drug selection marker to obtain the double drug resistance clones (*see* **Note 24**).

11. Confirmation of null mutant:

    (a) PCR analysis:

    - With primer pair specific for the gene of interest.
    - With one primer specific for the second drug resistance gene and other primer beyond the targeting flanking sequence (*see* **Note 25**).

    (b) Southern blot: if the PCR analysis indicates that the gene of interest has been deleted by gene targeting, the putative knock outs can be subjected to a Southern blot analysis to further confirm the correct targeting event with a gene specific probe or probe specific for the flanking sequence or the drug selection marker gene.

    - Digest genomic DNA (~5 μg/lane) with proper restriction enzymes. Ideally, the enzyme digestion (single or double enzymes) will generate a DNA fragment which contains the entire or partial gene open reading frame and ranges from 200 to 6,000 bps (a size easy to be separated in a regular agarose gel).

- Separate the digested genomic DNA in agarose gel, denature the DNA, transfer the DNA to a nylon membrane (HybondTM-N, Amersham) by setting up a capillary blot with 20× SSC blotting buffer.
- Hybridize the membrane with radioactive or nonradioactive labelled probe.
- Expose the membrane to a film for certain period (usually overnight, depending on the sensitivity of the probe).
- Develop film and analyze the Southern blot data. Ideally, the gene containing band is absent in the knock outs but present in wildtype cells.

(c) Western blot: If an antibody specific for the gene of interest is available, one can perform Western blot analysis.

- Freezing down cells: It is important to freeze down some cells in 10 % DMSO FBS solution to prevent loss of cell lines.

3.9 Characterize the Null Mutants

Characterize the null mutants as described (*see* Subheading 3.7), i.e., in vitro growth curves, visceral and cutaneous infections in BALB/c mice.

Specific assays: if the gene product may have a specific function, one may characterize the knock outs with specific assays such as for example enzyme or transport assays.

4 Notes

1. Cotton like DNA precipitate will form immediately after mixing.
2. Several other *Leishmania* expression vectors are available, such as pALT-neo, pXG-HYG, and pIR-SAT. pIR-SAT is a ribosomal DNA locus integration vector and reportedly allows high levels of gene expression through transcription by RNA polymerase I in the rDNA locus [21]. Since all these vectors have been successfully used in *Leishmania*, selection of these vectors would depend on the convenience of selection marker and cloning site. In the pLpneo vector, the neomycin resistance gene marker can be easily replaced with other drug selection marker such as hygromycin resistance gene.
3. Other tags such as HA tag, Flag tag, GFP tag can also be used to make fusion protein for detection.
4. Rogers et al. [9] have recently sequenced the *L. mexicana* genome and re-sequenced the reference *L. infantum*, *L. major*, and *L. braziliensis* genomes and corrected the species-specific

gene numbers present in each species. The current number of unique genes present in each *Leishmania* species are: 2 for *L. mexicana*, 14 for *L. major*, 19 for *L. infantum*, and 67 for *L. braziliensis* (for further details, *see* ref. 9).

5. If adding a tag (fusion protein) is not under consideration, the primers can be chosen from the 5′ and 3′ flanking regions of the coding sequence.
6. It is important to select the cloning enzymes which are present in the multiple cloning sites in the plasmid vector or produce compatible cohesive ends to the multiple cloning site that are absent in the gene sequence.
7. *New England* BioLabs has shown that most restriction endonucleases require one to three nucleotide base pairs flanking their recognition sequences for an efficient digestion.
8. The Kozak consensus sequence is a sequence which occurs on eukaryotic mRNA and has the consensus (gcc)gccRccAUGG, where R is a purine (adenine or guanine) three bases upstream of the start codon (AUG), which is followed by another "G." The Kozak consensus sequence plays a major role in the initiation of the translation process [23]. Therefore, we usually include the three base pairs of the 5′ flanking sequence immediately upstream of the start codon (ATG) for forward primers.
9. Since there are a variety of DNA polymerases available today, one should carefully follow manufacture's instruction as each DNA polymerase could have different optical reaction conditions. The protocol shown here serves as a guideline and a starting point for PCR amplification. Optimal reaction conditions (incubation times and temperatures, concentration of DNA polymerase, primers, $MgCl_2$, and template DNA) can vary for different PCR amplifications.
10. Depending on convenience and cost, one may use other commercial DNA extraction kits to purify the PCR products.
11. It is important to ensure that the PCR product and expression vector are completely digested for high-efficiency insertion into the vector. An alternative, more lengthy procedure is to clone the PCR product into a TA-cloning type vector and carry out a double digestion on the recombinant plasmid.
12. When digesting with two separate restriction enzymes, the most rigorous procedure is to perform the digests one at a time in the recommended React buffer and purify the DNA before the second digest. However, in an effort to save time, many researchers perform sequential digests or double digests.
13. Insert to vector ratio should be around 3:1 for an efficient ligation reaction. We use the intensities of DNA bands in the agarose gel to estimate the amount of vector and insert gene DNA to be added to the ligation reaction.

14. Alternatively, the *Leishmania* cell growth curve can be measured in a 25 cm^2 flask containing 10 ml culture medium by taking out 200 μl of culture daily to measure OD600 value, or by counting the cell number daily with a hemocytometer, although the later method is time consuming while dealing with large number of samples.
15. Unlike *L. donovani*, *L. major* promastigotes cannot differentiate into amastigote-like cells in in vitro culture conditions. However, it is interesting to examine whether expression of *L. donovani*-specific genes can allow *L. major* to grow better in in vitro amastigote culture media.
16. Maxim 5 adult mice are allowed per standard cage in most animal facilities, therefore, 4–5 mice per group is convenient.
17. Most university animal care facilities provide training on animal handling and injection techniques; however, some practice is required before one can perform mouse tail vein injection well.
18. If the infection level is expected to be low, limiting dilution culture is the preferred method to estimate the parasite burden.
19. We sometimes also include additional plates filled with G418 containing medium so the percentage of the recovered *L. major* cells retaining the *L. donovani* specific gene (transfected plasmid) can be determined.
20. If the parasite burden is expected to be high, the culture can be further diluted with second plate or the cell suspension in **step 3** can be further diluted before adding it into the first column well of a 96 well plate in **step 5**.
21. We have also successfully used antisense RNA to block A2 gene expression in *L. donovani* [13]. However, antisense RNA appears only to work for a few genes in *Leishmania*. The A2 highly repeated coding sequences could be easier targets for antisense RNA. RNAi appears to works in *L. brazilensis* but not in *L. donovani* and *L. major* [24].
22. The size of the homologous sequences can be longer but should not be less than 200 bp. Below this threshold (200 bp), the efficiency of integration into the *Leishmania* genome can be very low [25].
23. It is important to use the minimum drug concentration to select the initial transfectants since the drug resistance gene may not be expressed well in the targeted locus.
24. Third or even fourth rounds of gene targeting may be necessary if the gene of interest is located in chromosomes with 3 or 4 copies in chromosome aneuploidy cases such as for example chromosome 31 in *L donovani* [10]. Sometimes, the gene of

interest could be essential. In this case, one may provide the *Leishmania* cell with an extra chromosome copy of the gene before carrying out second round of targeting. If the second drug selection marker can be correctly targeted in the presence of an additional copy of the gene, it suggests the gene of interest could indeed be essential.

25. To help interpret the PCR data, it is important to use wildtype *L. donovani* genomic DNA as positive control for [1] and negative control for [2].

Acknowledgments

We would like to thank the Canadian Institutes of Health Research for supporting this research.

References

1. Herwaldt B (1999) Leishmaniasis. Lancet 354: 1191–1199
2. Murray H, Berman J, Davis C et al (2005) Advances in Leishmaniasis. Lancet 366: 1561–1577
3. Desjeux P (2001) The increase in risk factors for leishmaniasis worldwide. Trans R Soc Trop Med Hyg 95:239–243
4. Lipoldova M, Demant P (2006) Genetic susceptibility to infectious disease: lesions from mouse models of leishmaniasis. Nat Rev Genet 7:294–305
5. Chappuis F, Sundar S, Hailu A, Ghalib H, Rijal S, Peeling RW et al (2007) Visceral leishmaniasis: what are the needs for diagnosis, treatment and control? Nat Rev Microbiol 5:873–882
6. Ivens AC, Peacock CS, Worthey EA et al (2005) The genome of the kinetoplastid parasite, *Leishmania major*. Science 309:436–442
7. Peacock CS, Seeger K, Harris D et al (2007) Comparative genomic analysis of three Leishmania species that cause diverse human disease. Nat Genet 39:839–847
8. Smith DF, Peacock CS, Cruz AK (2007) Comparative genomics: from genotype to disease phenotype in the leishmaniasis. Int J Parasitol 37:1173–1186
9. Rogers MB, Hilley JD, Dickens NJ et al (2011) Chromosome and gene copy number variation allow major structural change between species and strains of *Leishmania*. Genome Res 21: 2129–2142
10. Downing T, Imamura H, Decuypere S et al (2011) Whole genome sequencing of multiple *Leishmania donovani* clinical isolates provides insights into population structure and mechanisms of drug resistance. Genome Res 21: 2143–2156
11. Charest H, Matlashewski G (1994) Developmental gene expression in *Leishmania donovani*: differential cloning and analysis of an amastigote-stage-specific gene. Mol Cell Biol 14:2975–2984
12. Ghedin E, Zhang WW, Charest H, Sundar S, Kenney RT, Matlashewski G (1997) Antibody response against a *Leishmania donovani* amastigote-stage-specific protein in patients with visceral leishmaniasis. Clin Diagn Lab Immunol 4:530–535
13. Zhang WW, Matlashewski G (1997) Loss of virulence in *Leishmania donovani* deficient in an amastigote-specific protein, A2. Proc Natl Acad Sci U S A 94:8807–8811
14. Zhang WW, Matlashewski G (2001) Characterization of the *A2–A2rel* gene cluster in *Leishmania donovani*: involvement of A2 in visceralization during infection. Mol Microbiol 39:935–948
15. Zhang WW, Mendez S, Ghosh A et al (2003) Comparison of the *A2* gene locus in *Leishmania donovani* and *L. major* and its control over cutaneous infection. J Biol Chem 278:35508–35515
16. Zhang WW, Matlashewski G (2010) Screening *Leishmania donovani*-specific genes required for visceral infection. Mol Microbiol 77(2): 505–517
17. Medina-Acosta E, Cross GAM (1993) Rapid isolation of DNA from trypanosomatid protozoa using a simple "mini-prep" procedure. Mol Biochem Parasitol 59:327–329

18. Zhang WW, Charest H, Matlashewski G (1995) The expression of biologically active human p53 in *Leishmania* cells: a novel eukaryotic system to produce recombinant proteins. Nucleic Acids Res 23:4073–4080
19. Zhang WW, Peacock CS, Matlashewski G (2008) A genomic-based approach combining in vivo selection in mice to identify a novel virulence gene in Leishmania. PloS Negl Trop Dis 2(6):e248
20. Zhang WW, Chan KF, Song Z, Matlashewski G (2011) Expression of a Leishmaniadonovani nucleotidesugartransporterinLeishmaniamajor enhances survival in visceral organs. Exp Parasitol 129:337–345
21. Robinson KA, Beverley SM (2003) Improvements in transfection efficiency and tests of RNA interference (RNAi) approaches in the protozoan parasite *Leishmania*. Mol Biochem Parasitol 128:217–228
22. Papadopoulou B, Roy G, Ouellette M (1994) Autonomous replication of bacterial DNA plasmid oligomers in *Leishmania*. Mol Biochem Parasitol 65:39–49
23. Sambrook J, Fritsch EF, Maniatis T (1989) Molecular cloning. A laboratory manual. Cold Spring Harbor Laboratory Press, Cold Spring Harbor, NY
24. Lye LF, Owens K, Shi H et al (2010) Retention and loss of RNA interference pathways in trypanosomatid protozoans. PLoS Pathog 6(10): e1001161
25. Papadopoulou B, Roy G, Dumas C (1997) Parameters controlling the rate of gene targeting frequency in the protozoan parasite Leishmania. Nucleic Acids Res 25:4278–4286

ERRATUM

From Sequence Mapping to Genome Assemblies

Thomas D. Otto

Christopher Peacock (ed.), *Parasite Genomics Protocols*, Methods in Molecular Biology, vol. 1201, DOI 10.1007/978-1-4939-1438-8_2, © Springer Science+Business Media New York 2015

DOI 10.1007/978-1-4939_1438-8_21

Figures 1, 2 and 3 of this chapter is incorrect. The correct figures are as shown below.

Also there are 7 additional references that were missed to be added in the final version of the book. They are as listed below.

CITATION 1

Assemblathon 2: evaluating de novo methods of genome assembly in three vertebrate species.

Bradnam KR, Fass JN, Alexandrov A, Baranay P, Bechner M, Birol I, Boisvert S, Chapman JA, Chapuis G, Chikhi R, Chitsaz H, Chou WC, Corbeil J, Del Fabbro C, Docking TR, Durbin R, Earl D, Emrich S, Fedotov P, Fonseca NA, Ganapathy G, Gibbs RA, Gnerre S, Godzaridis E, Goldstein S, Haimel M, Hall G, Haussler D, Hiatt JB, Ho IY, Howard J, Hunt M, Jackman SD, Jaffe DB, Jarvis ED, Jiang H, Kazakov S, Kersey PJ, Kitzman JO, Knight JR, Koren S, Lam TW, Lavenier D, Laviolette F, Li Y, Li Z, Liu B, Liu Y, Luo R, Maccallum I, Macmanes MD, Maillet N, Melnikov S, Naquin D, Ning Z, Otto TD, Paten B, Paulo OS, Phillippy AM, Pina-Martins F, Place M, Przybylski D, Qin X, Qu C, Ribeiro FJ, Richards S, Rokhsar DS, Ruby JG, Scalabrin S, Schatz MC, Schwartz DC, Sergushichev A, Sharpe T, Shaw TI, Shendure J, Shi Y, Simpson JT, Song H, Tsarev F, Vezzi F, Vicedomini R, Vieira BM, Wang J, Worley KC, Yin S, Yiu SM, Yuan J, Zhang G, Zhang H, Zhou S, KorfIF.Gigascience. 2013 Jul 22;2(1):10. doi: 10.1186/2047-217X-2-10. PMID:23870653[PubMed]

The online version of the original chapter can be found at http://dx.doi.org/10.1007/978-1-4939-1438-8_2

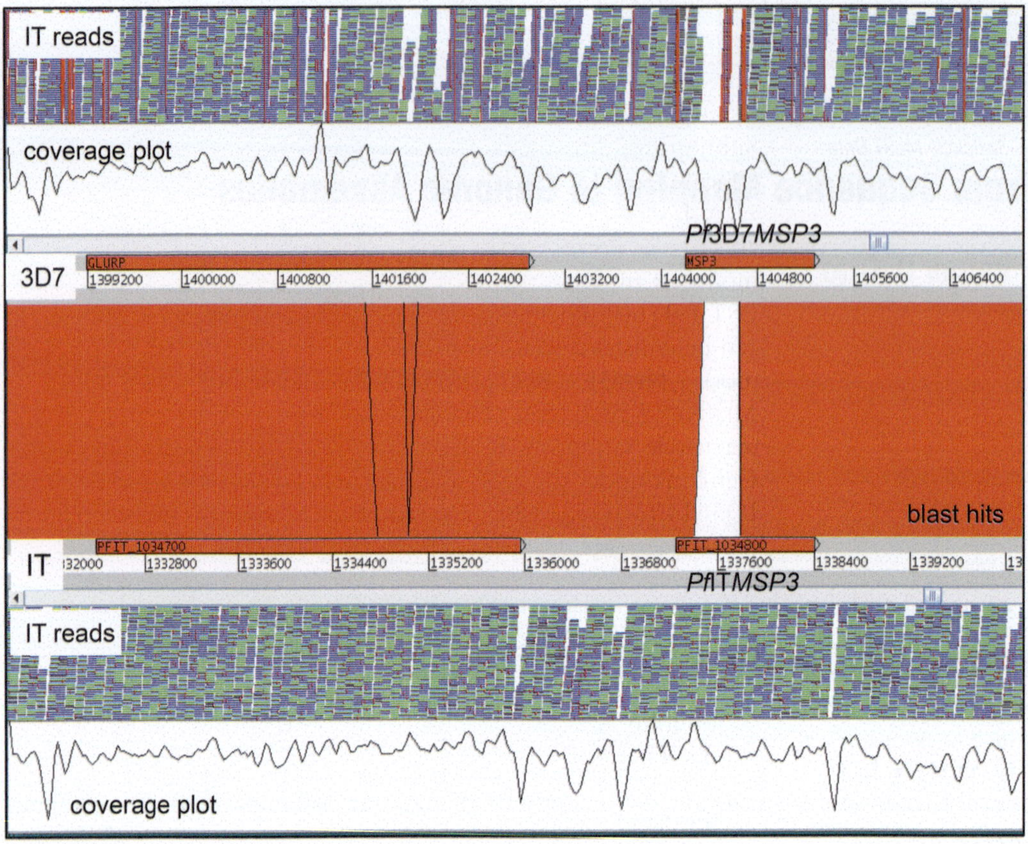

Fig. 1 Mapping versus assembly. Two genes of *P. falciparum* 3D7 (*red boxes*) can be seen at the top. The *horizontal green* and *blue lines* are mapped sequencing reads from the IT clone. *Red points* in the reads are differences between the IT reads and the 3D7 reference. The lower part shows the *de novo* assembly of IT. The *vertical bars* are blast hits. The graphs are the coverage plots. Some regions of *MSP3* in 3D7 are not covered by mapped IT reads. The de novo assembly has an insertion, indicated by the shape of the blast hit. Reads map even over this new assembled region

CITATION 2

A comprehensive evaluation of assembly scaffolding tools.Hunt M, Newbold C, Berriman M, Otto TD.Genome Biol. 2014 Mar 3;15(3):R42. [Epub ahead of print]PMID:24581555

CITATION 3

REAPR: a universal tool for genome assembly evaluation.Hunt M, Kikuchi T, Sanders M, Newbold C, Berriman M, Otto TD.Genome Biol. 2013 May 27;14(5):R47. doi: 10.1186/gb-2013-14-5-r47.PMID:23710727

CITATION 4

A tale of three next generation sequencing platforms: comparison of Ion Torrent, Pacific Biosciences and IlluminaMiSeq sequencers.

Quail MA, Smith M, Coupland P, Otto TD, Harris SR, Connor TR, Bertoni A, Swerdlow HP, Gu Y.

BMC Genomics. 2012 Jul 24;13:341. doi: 10.1186/1471-2164-13-341.PMID:22827831

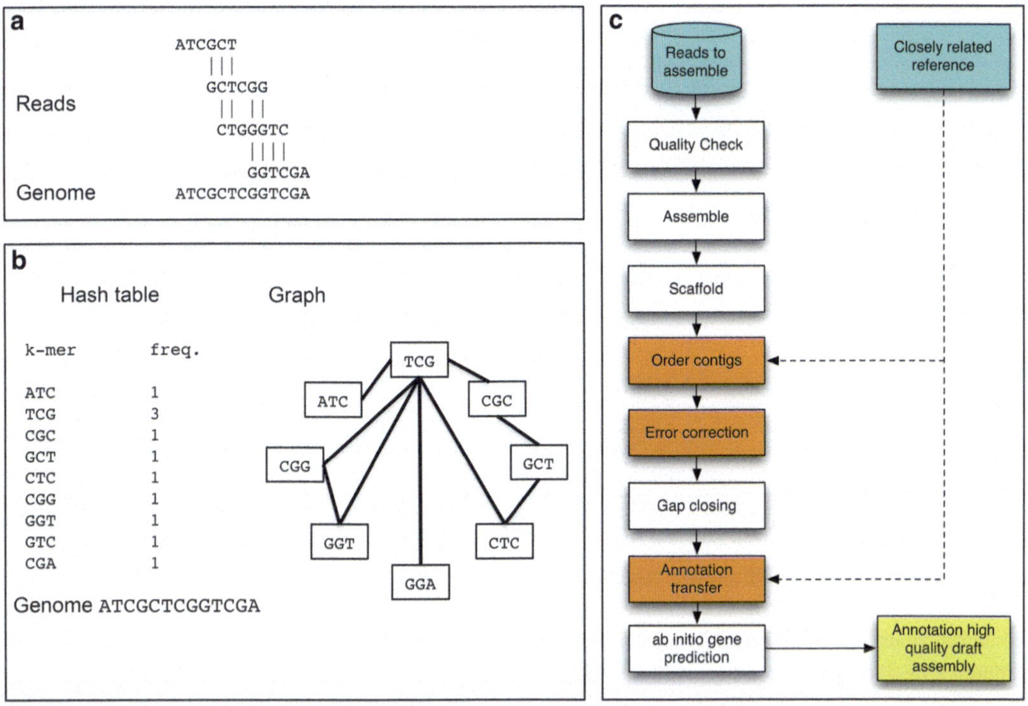

Fig. 2 (**a**) Assembly with longer reads: Nearly identical overlap between reads enable the generation of the consensus. (**b**) Assembly with short reads, using de Bruijn graph: First the reads are index and the k-mer are stored in a hash table, including the k-mer and the frequency. With a k-mer length of 3 the k-mer TCG is non unique. Due to this non unique k-mer, the graph quite complicated. (**c**) Overview of typical pipeline for de novo assembly and annotation

CITATION 5 and 6

- Iddo Friedberg Automated protein function prediction—the genomic challenge Brief Bioinform (2006) 7 (3): 225-242 first published online May 23, 2006 doi:10.1093/bib/bbl004

Nat Rev Genet. 2012 Apr 18;13(5):329-42. doi: 10.1038/nrg3174.

A beginner's guide to eukaryotic genome annotation.

Yandell M1, Ence D.

CITATON 7 (prokka)

Bioinformatics. 2014 Jul 15;30(14):2068-9. doi: 10.1093/bioinformatics/btu153. Epub 2014 Mar 18.

Prokka: rapid prokaryotic genome annotation.

Seemann T.PMID:

24642063

[PubMed - in process]

a

```
>Pf3D7_10_v3
taaaccctgaaccctaaaccctgaaccctaaaccactaaccctaaaccctgaaccctgaa
ccctgaaccctaaaccctaaacctaaaccctgaaccctaaaacctaaaccctgaaccctg
aaccctaaaccctaaaccctaaaccctaaaccctgaaccctaaaccctaaaccctgaacc
ctaaaccctgaaccctgaaccctaaaccctgaaccctgaaccctaaaccctgaaccacta
```

b

```
@IL39_6014:8:110:3699:4595#3/1
TATTTGAACTGACAATTTTATAAGATCCATATATATGAAGATCTCAAAAAAATATATGTTTTTTTGAAAATTTTCA
+
CCB@@*CCC<GCHHEHGGGGGEEGDDGGGGGEGGHEGCEGGFFCHGDBCB?BH?GGBH=BEBB@@E=B>=EBECGA
@IL39_6014:8:88:4857:8768#5/1
TATTTAAACTGACAATTTTATAAGATCCATATATATGAAGATCTCAAAAAAATATATGTTTTTTTGAAAATTTTCA
```

c

```
IL39_6014:8:110:3699:4595#3    83        PfIT_10_v2          1404405    60         76M
           =          1404336    -145
           TGAAAATTTTCAAAAAAACATATATTTTTTTGAGATCTTCATATATATGGATCTTATAAAATTGTCAGTTCAAATA
           AGCEBE=>B=E@@BBEB=HBGG?HB?BCBDGHCFFGGECGEHGGEGGGGGDDGEEGGGGGHEHHCG<CCC*@@BCC
           AS:i:73
```

Fig. 3 Examples of different file formats. (**a**) fasta: Each sequence starts with a ">" and a name. Then the sequence is followed. (**b**) fastq: Similar to fasta, but with the quality coded in ASCII. (**c**) SAM format: First column is the name of the read. Next column is the mapping flag that can be used for querying a BAM file. Third and fourth, seven and eight columns are mapped to the reads and its mate, respectively. Column nine is the fragment size. The information how well the reads map is in column five and six, mapping quality and cigar string, respectively. The sequence and the quality of the reads are stored in column ten and eleven. The last column can have many different information, like an alignment score, other possible position to map repetitively. This depends on the mapper

CITATION 8 trimmomatic

Bioinformatics. 2014 Aug 1;30(15):2114-20. doi: 10.1093/bioinformatics/btu170. Epub 2014 Apr 1.

Trimmomatic: a flexible trimmer for Illumina sequence data.

Bolger AM1, Lohse M2, Usadel B1.

PMID:

24695404

Index

Christopher Peacock (ed.), *Parasite Genomics Protocols*, Methods in Molecular Biology, vol. 1201, DOI 10.1007/978-1-4939-1438-8, © Springer Science+Business Media New York 2015

N

O

P

Q

R

V

W

X

The manufacturer's authorised representative in the EU is Springer Nature Customer Service Centre GmbH, Europaplatz 3, 69115 Heidelberg, Germany. If you have any concerns regarding our products, please contact ProductSafety@springernature.com

Printed and bound by CPI Group (UK) Ltd, Croydon, CR0 4YY
12/08/2026
02194518-0005